SUSTAINABLE FORESTRY
Emerging Challenges

SUSTAINABLE FORESTRY
Emerging Challenges

Editors

Prodyut Bhattacharya

Professor

University School of Environment Management

GGS Indraprastha University

New Delhi

A.K. Kandya

Professor (Retd.)

Department of Botany

Dr. H.S. Gour University

Sagar (M.P.)

I.K. International Publishing House Pvt. Ltd.

NEW DELHI

Published by

I.K. International Publishing House Pvt. Ltd.
S-25, Green Park Extension
Uphaar Cinema Market
New Delhi–110 016 (India)
E-mail:info@ikinternational.com
Website: www.ikbooks.com

ISBN 978-93-84588-99-1

Published by Krishan Makhijani for I.K. International Publishing House Pvt. Ltd., S-25, Green Park Extension, Uphaar Cinema Market, New Delhi–110 016 and Printed by Rekha Printers Pvt. Ltd., Okhla Industrial Area, Phase II, New Delhi–110 020.

Dedication

We are happy to dedicate this book to Dr. Ram Prasad, for his lifetime contribution towards sustainable forest management. Dr. Prasad was born on January, 1, 1944 at Basti, Uttar Pradesh, India. He was the topper of the 1968 batch of the Indian Forest Services (IFS). He obtained his basic training of Field Forestry during his Masters Degree (1968-70) and later in his PhD programme (1972-76). This resulted in his devotion and commitment to the development and conservation of forests in India during his placements in various positions, viz., Deputy Conservator of Forest to the highest designation in a state, i.e., Principal Chief Conservator of Forest in Madhya Pradesh. He also contributed as Director of Tropical Forestry Research Institute, Jabalpur, MP, State Forest Research Institute, Jabalpur; Chief Wildlife Warden of Madhya Pradesh and Director, Indian Institute of Forest Management, Bhopal. He has been the National Fellow and Project Advisor at the International Centre of Community Forestry at IIFM, Bhopal and Chief Technical Advisor to the Non-Timber Forest Produce Centre of Excellence, Government Forest Department, Agartala, Tripura. The benefit of his deep knowledge and vast experience in forest conservation has been harnessed by his 25 PhD students.

Dr. Prasad was associated with the Panel of Experts of the Department of Science and Technology (DST), Department of Biotechnology (DBT) and Ministry of Environment and Forest (MoEF), Government of India, as well as the International Union for the Conservation of Nature and Natural Resources (IUCN). His contribution as the Member of the Scientific Committee of Food and Agriculture Organisation (FAO, Rome) for the Outlook Study of Future of Forestry in Asia-Pacific, 2020 has been remarkable and outstanding. He also served as Principal National Consultant for International Tropical Timber Organisation (ITTO, Japan). His authored books, *Aerial Seeding, Wasteland Development in India, Forestry Seed Technology and Social Forestry in India* are widely used by teachers, students, researchers, academicians and policy makers.

Preface

Sustainable Forest Management (SFM) addresses forest degradation and deforestation while increasing direct benefits to people and the environment. It is an interdisciplinary concept covering ecological, social and economic dimensions of forest management. At the social level, sustainable forest management contributes to livelihoods, income generation and employment. At the environmental level, it contributes to important services such as carbon sequestration and water, soil and biodiversity conservation. Over the last three decades, growing awareness and concern about forest science and their management have been pushing nations into global dialogue focusing on management initiatives to promote sustainable practices. The linkage between management and sustainable development evolved during the Rio Earth Summit in 1992 (the United Nations Conference on Environment and Development - UNCED) and was reiterated at the World Summit on Sustainable Development in Johannesburg in 2002, and recently on Sustainable Development Goals in 2015. The instruments of change facilitating sustainable development at international, national and local levels needed to be identified. Multi-stakeholder agreements arising from UNCED, such as the UN Millennium Declaration, the Intergovernmental Panel on Forests (IPF), the Intergovernmental Forum on Forests, the World Bank's revised forest policy and strategy and the United Nations Forum on Forests, demonstrate the consensus in different countries.

The sustainable management of natural resources, especially forests, is of vital importance to global, regional, and national efforts to achieve sustainable development and should play a key role in efforts to mitigate and adapt to climate change and further low-carbon development. Managing forests sustainably means; increasing their benefits, including timber and food, to meet society's needs in a way that conserves and maintains forest ecosystems for the benefit of present and for the future generations. Throughout the world, forests and forest ecosystems provide timber, other raw materials, non-timber benefits, and protection against natural and human-induced threats. Forests are also an important energy source, providing fuelwood and energy biomass. In industrialized countries, fossil fuels have, however, replaced wood and become the dominant source of energy. Currently, coal, oil and natural gas provide cheap options for most human energy needs. Use of fossil fuels is increasing atmospheric concentrations of greenhouse gases (GHGs), especially carbon dioxide (CO_2), with the consequent warming of global climate and changes in precipitation. Global efforts are needed to mitigate the climate change and minimize the impacts of climate change. The future of forestry should lie in pluralism, not a single strategy. Every forest system is different in its biophysical, economic, social, and political attributes. Every situation needs a response tailored to its present needs and these needs will inevitably change over time. The capacity and skill of the forest manager is to be able to draw upon multidisciplinary approach, the linkages of forest science, with all stakeholders to develop the best management regime for today and for the future generation to come.

The global debate on instruments of sustainable forest management during the past three decades has generated concern about evolving sustainable development models, and eliciting people's participation has become a mantra for sustainability. The scope of the book is to refresh the meanings and perceptions of SFM against the background of the rapid changes in our natural and social environment. A crucial issue therefore is how forest management can contribute to forest conservation in light of changing climatic conditions. Moreover, policy changes justify the re-evaluation of the role of SFM in delivering ecological goods and services from our forests.

The book comprises 19 contributions covering the whole range of SFM core contributions of forests to sustainable development. The first chapter emphasizes on the intensive efforts needed for capacity building of the personnel and also the stakeholder groups including the government functionaries working with people to curb the forestry issues for addressing the present needs. The second chapter lays importance on the need of forest certification as a voluntary market tool for sustainable forestry and also to examine the role and contribution of forest certification, present status of forest certification in India and how the government, local communities and industries can get benefit out of it. The third chapter highlights the REDD+ initiatives in India, a new opportunity for the country to get some financial support to address deforestation and degradation, the chapter highlighted its thrust areas, and what the challenges faced by the country so far. The fourth chapter explains the need of integration of non-timber forest products in reference to collection, production, silviculture technique, transportation, storage, ownership, benefit sharing, monopoly, transit rules, tax, conservation need, etc., in forest management to achieve sustainable harvesting. The fifth chapter highlights various strategies to be adopted by the paper industry, to meet its ever-growing demand of wood on continuous and sustainable basis is to enlarge social and farm forestry plantation apart from raising plantations by private industry from their captive plantation. The sixth chapter discusses important gum yielding species (*Sterculiaurens*) of Madhya Pradesh which in recent decades have declined due to the unsustainable use and overexploitation for commercial trade, the paper analyses the role of forest department and traders in the marketing. The seventh chapter discusses the participatory initiative of JFM programme and its linkage with rural livelihoods across forest fringe communities in JFM areas of Madhya Pradesh, the paper highlights how forestry and livelihood development can go together. The eighth chapter embodies the innovative techniques developed for sustainable harvesting of commercially important NTFPs for the development of local NTFP resource development. The ninth chapter provides opportunity to grow potential NTFP species in agricultural fields of the farmers, how such scientific domestication process under agroforestry system can improve the status of NTFP tree production in the outside natural forest with ecological, social and cultural aspect, it will not only increase the number of trees, it will provide economic support to local inhabitants for the future sustainability. The tenth chapter brings the role of indigenous practices and contribution of wild food plants in Thakur tribes to cope up with extreme environmental conditions, describes the utilization of forest products during severe drought and famine. The eleventh chapter addresses the need of conserving medicinal plants by certification which requires standards and good management practices for sustainable harvesting of the economically important medicinal plants and particularly lichen. The twelfth chapter discusses the ecological status of forest species, endemicity of tree species against the land-use changes of the Achanakmar and Amarkantak forests in Central India. The thirteenth chapter provides an insight that how bamboo plantation helps to improve carbon absorption in vegetation and soil, reduces the emission of carbon and extends the storage time of carbon in soil thus helping in conserving and restoring the forest resources. The chapter fourteenth highlighted the importance of sustainable management, their vegetation status, as valuable forest resources that can only be ensured by the safeguard and maintenance of the valuable biodiversity, also their association present in a biodiversity hotspot. The fifteenth chapter argues how the ecosystem services (ES) depend on the effective functioning of ecosystem, human intervention influences the natural ecosystem, the problem of ES or socio-ecological processes cannot be understood in isolation, it also shares the future scope of the subject in forest management. The sixteenth chapter discusses about need of an effective intervention of afforestation, particularly where regeneration is poor, a holistic view for sustainable development of fragiledesert ecosystem is provided. The seventeenth chapter emphasises on the need to understand the non-market value provided by urban forests or urban greens which are responsible for various environmental, social and educational benefits to the human society, why green space planning is important for biodiversity improvement in cities. The eighteenth chapter provides an overview of the renewable and non-renewable energy resources in India and to conserve these energy resources what various measures have to be taken. The nineteenth chapter provides the conservation importance of sacred groves, it maps out the availability of sacred groves in the state of Madhya Pradesh and provides the inventory of various sacred groves, their total area, species composition and other details, which are important for future conservation based on cultural and religious value of a society.

The approach involves addressing site-specific ecological, socio-political, geographical and cultural requirements for the sustainable forestry; the topic is such important international fora should further facilitate regional and global partnerships to provide a synthesis of similar effort to promote sustainable forest management. The volume aims to make the wide body of collective knowledge and experience about sustainable forest management more accessible to forest managers and other stakeholders, thereby supporting SFM dissemination and implementation on the ground in India and other part of the globe.

Prodyut Bhattacharya
A.K. Kandya

Contents

1

Training for Sustainable Forestry in India: Emerging Areas for Holistic Capacity Building

Prodyut Bhattacharya
University School of Environment Management
GGS Indraprastha University, New Delhi - 110078
Email: prodyutbhattacharya@yahoo.com

ABSTRACT

The present forestry in India reached a longways, though the dependency of the economy on forest has changed, still it plays an integral part of environment and livelihoods. The changes, due to crucial paradigm shift in broad ecosystem all across and other social and economic sectors linked to the forestry. Such a shift needs intensive efforts for capacity building of the personnel and also the stakeholder groups including the government functionaries working with people. A number of studies in recent years have drawn attention to the declining state of professional forestry education, both in developing and developed countries. Forestry sector, like any other sector, requires new technologies, new approach and new management tools to keep pace with the current developments at the local, national, regional and global levels. With the shift towards sustainable development and participatory mode of forest management, the forestry has been facing new challenges which are fundamentally different from earlier ones. Capacity building, in addition to upgradation of their technical knowledge and expertise is required in the changing scenario. This paper deals with the social and managerial issues hindering professionalism of the foresters and suggested strategies. It will be imperative to pool forestry resources and establish linkages with various training and academic institutions not only from within the country but also with international organizations. It is suggested that the training curriculum may be simple but relevant to the present situation, address real issues of the sector and appreciate for the forest department and the curriculum should also be such that people can adopt in varied geographic situations.

INTRODUCTION

The forestry sector in India is passing through several changes, due to crucial paradigm shift in broad ecosystem all across and other social and economic sectors linked to forestry. Such a shift necessitates fundamental changes in the orientation and capabilities of the personnel. This includes attitudinal change towards the various stakeholders on core forestry issues, and acquiring social understanding. This is a challenging task and necessitates intensive efforts for capacity building of the personnel and also the stakeholder groups including the government functionaries working with people.

Role of forestry profession is multifarious due to tremendous changes that have taken place in this profession over the years. Everyone realizes the crucial role of forests in environmental amelioration as they reduce the effect of greenhouse gases to mitigate global warming. While the ecologists and environmentalists are searching for strategies for overall ecosystem conservation, the foresters are deliberating upon the conservation and regeneration of degraded forests and development of wastelands outside the traditional frontiers of forests.

Since the forests are an integral part of environment, it has been clearly mentioned in its mandate about environmental relevance of the forestry, such as climate change, biodiversity conservation and rehabilitation of degraded forests or mined out areas. Apart from protected areas, management of natural forests (which in many countries, especially in the tropics, has been predominantly exploited) or plantations has largely focused on producing industrial wood. Until recently, traditional resource use systems that cater to the needs of local communities largely remained outside the framework of the formal knowledge system. Although some broadening of curricula has taken place during the past two decades through the inclusion of community forestry and allied subjects, the core of forestry education in most countries still remains timber-focused.

A number of studies in recent years have drawn attention to the declining state of professional forestry education, both in developing and developed countries. Symptoms of this decline include a significant reduction in funding to educational institutions, low student enrolment (Van Lierop, 2003), inability to attract talented students, and most important, the declining demand for forestry graduates. Confronted with rapidly changing circumstances, many academic institutions are adopting survival strategies and primarily attempting to compete with other similar institutions (Nair, 2004).

In the present scenario, there is global divide of the rich and poor countries with respect to climate change and adaptation issues. Acute poverty, malnutrition, unemployment coupled with exponential growth of population, increasing livestock pressures, expanding urbanization, industrialization and globalization of markets are putting tremendous pressure on the tropical forests. It is therefore, essential to develop expertise in the field of forestry and wildlife and related sectors which are directly or indirectly connected with forest resources. In the modern era of expanding frontiers of knowledge and globalization of forestry issues, the horizon of Indian forestry profession cannot remain in oblivion and confine itself to classical conventional forestry practices. Many stakeholders of forestry sector have suggested that it is essential that the frontline foresters learn from the experience of other countries in order to strike a balance between conservation of these natural resources and meeting the developmental needs of different sectors of economy. Changes in policies on the perspective of livelihood promotion and natural resource sustainability are key drivers which have helped to pump good amount of public money through various projects from Rural Development Ministry and Ministry of Environment, Forest, and Climate Change.

In India, forestry has undergone a metamorphosis since the last few decades. The Environment Protection Act was enacted in 1986 for improving the environment of the country. It has shown remarkable progress during the last 15 years for enhancing contribution of forests towards poverty alleviation through empowering people with the ownership of NTFP (PESA, 1996).

India had taken a revolutionary shift in the approach of forest management from regulatory to participatory with the promulgation of National Forest Policy, 1988. This Forest Policy came four years before the Earth Summit which embodies all elements of sustainable forest management and India's forests are treated as social and environmental resources. India has shown sensitivity towards recognition of tenural rights of the tribal on forestland with the issuance of guidelines to the state governments in 1990.

The ecological security became the prime objective and focus was given for providing livelihood to the forest dependent communities. India has initiated the implementation of this policy in a big way to involve local communities in the conservation, protection and management of forests through JFM institutions in 1990 and expanded this programme to more than 22 million hectares of forests with the involvement of approximately 21 million people (Bhattacharya et al., 2008).

Another milestone was reached in 2006 by the enactment of a national level legislation, the Scheduled Tribes and Other Traditional Forest Dwellers (Recognition of Forest Rights) Act, 2006, for assigning habitation and occupation rights on forests along with responsibility of conservation of biological resource and maintenance of ecological balance to community. This law recognizes the rights of occupation of forests by tribes and forest dwellers and empowers them for management of forests used by them as common property resources. It is estimated that about 20% of the government controlled and managed forestland will come under the occupational titles recognized under this law. The recognition of right of common use conforms to the policy prescription of participatory forest management and also accepted principles of biodiversity conservation as well as community involvement in conservation (Malhotra and Bhattacharya, 2010).

NEW CAPACITY BUILDING CONCEPT

Forestry sector, like any other sector, requires new technologies, new approach and new management tools to keep pace with the developments at the local, national, regional and global levels. With the shift towards sustainable development and participatory mode of forest management, forestry facing new challenges which are fundamentally different from earlier ones. There is an increasing emphasis on conservation practices and collaborative management by involving the stakeholder communities and individuals. On the other hand, there is pressure for increasing the productivity of wood and other non-timber forest products (NTFPs) using modern technical inputs. Various important areas like biological record keeping/assessment, environmental impact indicators, biodiversity based ecosystem services and valuation principles, trade and intellectual property rights, biopiracy, gene pool management, protected area effectiveness, environmental economics, forest certification, monitoring indicators, carbon sequestration, sustainable development of forests and people, participatory natural resource management, new techniques for raising productivity of forests, bio-economic modeling, ecotourism, valuation of ecosystem services are gaining importance in the forestry sector. Though some of these issues are addressed partially in their forestry school under regular in-service courses offered by National Forest Academy, state training institutions and other organizations, there is a need for further consolidation and streamlining the capacity building regime to adequately address these emerging issues to achieve excellence in new forestry sector developments and project initiatives.

The traditional approach to forest management was mainly focused on technical aspects and scientific management practices such as nursery techniques, seedling demand analysis, nutrient analysis, tree growth modeling, monitoring systems for afforestation, application of geographical information systems (GIS), remote sensing (RS) technique, impact assessment methodology, holistic approach to watershed management, soil and water conservation techniques in watersheds, low impact harvest technology, controlled burning using fire modeling, wood technology, international trade in wood products, crossborder marketing of NTFPs including medicinal plants, forest research methodologies, protected area management, process documentation, forest management information system, etc. However, the present-day foresters along with the above practices, have to deal with a number of non-technical issues such as inter-sectoral policy and programme linkages, social mapping, community mobilization, Forests as tool for rural development including income generation, micro-credit, micro-enterprise management, alternate energy technology, quality control skills, inventory management, accounting and fund management skills.

Capacity Building in Forestry Sector

The duties and responsibilities of a present-day forester have become much more challenging today. From being an exclusivist profession, primarily charged with forest management for timber production, foresters are called upon to perform a wide range of roles varying from traditional forest protection and production to comprehensive biodiversity conservation and sustainable use to ecotourism promotion and so forth. In today's global scenario, when the world has begun to take note of a fast degrading environment, it looks towards professionals like foresters to provide innovative stewardship in tackling concerns like climate vulnerability, biodiversity conservation, food and water security, and poverty alleviation through sustainable livelihood programmes based on sustainable use of natural resources. Therefore, the expectations from the Forest Service Officers have grown by leaps and bounds. The recent recommendations of the National Forest Commission (2006) and Approach Paper on the Eleventh Five-Year Plan reflect these expectations in no uncertain terms. While societal expectations from the present-day forester to fulfill these changing roles is increasing, more often than not the forest manager is left to his/ her own devices to cope with these emerging challenges. The National Forest Policy (NFP), 1988, also lays emphasis on capacity building and enunciates, "specialized and orientation courses for developing better management skills by in-service training need to be encouraged, taking into account the latest development in forestry and related discipline."

Forest administration in most of the states in India is more than a hundred years old. During this period, conventional forest management was practiced throughout the country. A rigorous training process at the forestry training institutes as well as strict follow-up by the senior officers in their respective states is a common practice. The orientation process of frontline staff was rather slow and insufficient in most of the states as compared to the higher officials.

> "*The rapid growth of cross-disciplinary and interdisciplinary work would indeed argue that new knowledge is no longer obtained from within the disciplines around which teaching, learning and research have been organized in the nineteenth and twentieth centuries*" – Drucker, 1989

Present Status and Perceived Gaps in Knowledge in Forestry Sector

The forestry as profession has gained tremendous importance in view of recent societal and environmental developments, and the concerns expressed for protection of global environment. In addition to upgrading their technical knowledge and expertise required in the changing scenario, present-day foresters have to deal with many social and managerial issues. An All India Service constituted under the provisions of Article 312 of the Constitution of India for the forestry professional service is considered technical service. Selection process is conducted by the Union Public Services Commission every year, comprising written examination, interview and an endurance test. Educational requirement for Indian Forest Service is bachelor's degree in any of science/engineering disciplines. Majority of selected officers have Master's Degree in Botany, Chemistry, Physics, Math, Agricultural Science, Veterinary Science, etc. A good number of B.Tech/ Ph.D. candidates are appearing for the job.

Officials have been trained for several decades with the traditional mode of training package. In the forest resources and areas, serious changes are happening on ground situation with the development of Indian economy and policies with focus on revenue orientation. As a consequence, there is a need for modification in the country's forestry training by pooling resources not only from within the country but also sharing them at global level. State level central forest schools as well as other forestry training schools in the state take due care of the above-mentioned issues while imparting training to the forestry personnel. However, any training institute has certain limitations. The technological and educational changes in the fast changing world require new forms of learning resources and collaborative learning at international as well as regional levels. With the increasing integration of the Indian economy with the world economy and rapid changes in technology, the new demands are created on the state apparatus and officers need to understand fully the implications of the economy and what measures can be taken to safeguard the interest of the forestry sector. Another area is that of understanding how to use information technology to simplify internal processes, build data and information bases for better decision making (remote sensing, geographic information system) and create innovative ways of delivery of services. Some exposures to forest officers would enable them to efficiently follow policies, plan and implement projects and thereby help to improve the performance of government and delivery of services to the public.

Capacity Building Gap and Future Scope of Improvement

Training and education are tools for transferring information and are essential for maintaining effective and innovative management and operational functions. Educational programmes focus on general principles and deal with all information about a subject. In contrast, training teaches specific information, functions, and skills to enhance a person's ability to perform management, operational, or research functions related to a particular discipline or subject. Training usually involves some type of practical exercise or practice that enhances a person's skills. Training should build upon trainees' existing knowledge and strongly encourage them to participate so they can benefit from the experience of others. Often our education is associated with formal, broad based, degree-granting programmes. Training, however, more closely resembles concentrated activities designed to provide new information or skills over a short term:

- Forest officials at different levels not adequately exposed to tools/techniques and principles of modern management.
- Inadequate perception of social behaviour/dynamics, conflict resolution, communication and media management.
- Higher technological inputs on forestry/wildlife management not integrated into working systems.
- Less exposure to economics of forestry management including business enterprises, marketing and sales, market research, etc.
- Limited skills in policy planning, policy research and analysis.

A JICA sponsored loan project to Government of India was implemented in the year March 2009 to March 2014 for the state forest departments, to improve the skills of the frontline staff of various Indian states. This project was assumed that local front-line staff will be in daily contact with local village communities. It received effective and well-structured training on the latest skills and techniques covering various aspects of the forestry sector, including forest management methods based on a new innovative approach, which will improve the functioning of the department. However, the frontline staff currently lacks such skills and techniques, and does not yet have adequate experience in sharing these techniques with the local community. Thus, this was a project to improve the training at the national level under the Directorate of Forest Education at CASFOS (Central Academy for State Forest Service), for the state forest officers who are the leaders and trainers for training the frontline staff.

This technical cooperation project is linked to the ODA Loan Project "Capacity Development for Forest Management and Personnel Training" (a project to provide funds for improvement of training environment for frontline staff through the rehabilitation of state forest training institutions and through capacity development of frontline forestry staff in about 10 Indian states). To assist this ODA Loan Project, in addition to long-term experts, short-term experts are also appointed. Their technical inputs extend to activities of ODA Loan Project such as:

1. preparation of syllabus and materials for master training and for model teaching note for frontline staff training;
2. implementation of master training at CASFOS, Dehradun; and
3. monitoring of the training of trainers implemented by master trainers in the targeted states.

New Thrust of Training and Objective

- To enhance professional knowledge about application of modern technology/skills in forestry management.
- To facilitate better comprehension of far reaching developments/changes taking place in the sector at all levels including national, regional and international levels.
- To ensure a good grasp and better understanding and integration of issues related to social welfare, poverty alleviation, integrated development approaches for marginalized communities and coherence in policies/programmes related to forestry development in the country.

With the new thrust of participatory forestry, other than timber and wildlife, management forestry organization started understanding human component in the system and adopting the strategy of joint management of resources.

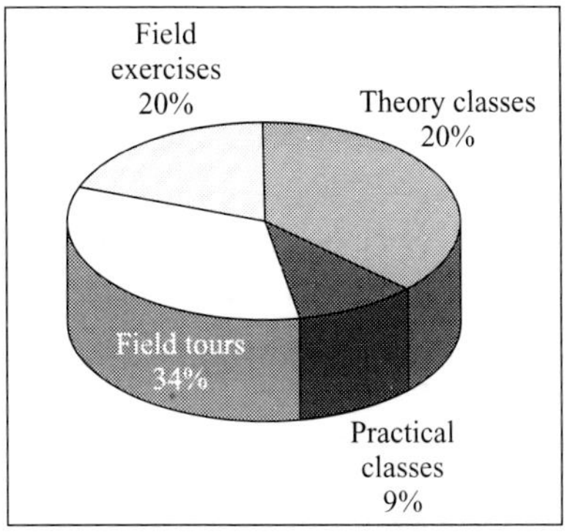

Fig. 1.1 Distribution of training content IFS traning

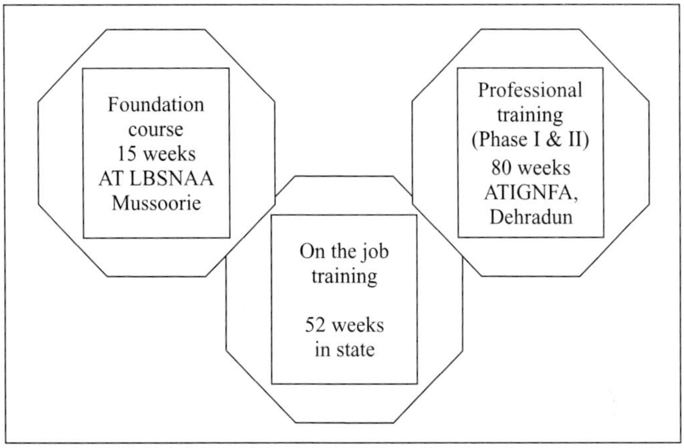

Fig. 1.2 Training schedule of IFS officers

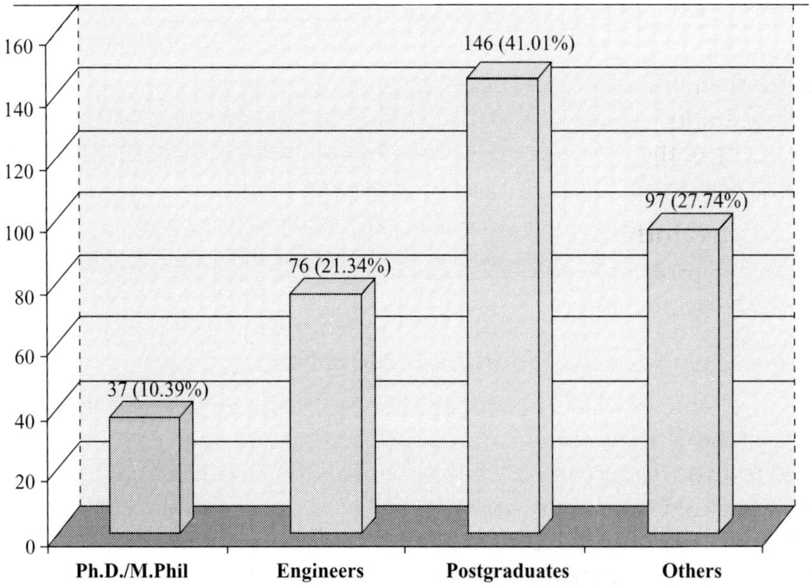

Fig. 1.3 Profile of IFS probationers (1995-2005)

Table 1.1 Syllabus for IFS professional training (Phase 1 & II).

Subject Area	*Sessions*
Core Forestry	530
Ecological Sciences	110
WL and Biodiversity Sciences	100 +70 (Spl. Mdl.)
Engineering Sciences	115
Basic Sciences (Geology, Soil Science, Botany, etc.)	140
Applied Sciences (GIC & RS, Computer Application, etc.)	115
Others (Qualifying Subjects)	140

Table 1.2 Job responsibilities of IFS officers.

Post	*Area of Responsibilities*
Dy. Conservator of Forests (DCF)	• Executive head of territorial/Social forestry/Production/ Research/Working Plan divisions • Responsible for general management/protection of forests in the divisions and financial matter related to revenue/ expenditure • Regulation and enforcement to protect forests and wildlife • Implementation schemes/programmes related to forests
Conservator of Forests (CF)	• Holding change of territorial circle covering 3-5 divisions • Exercising supervisory control over divisions • Dealing with PCCF/CCF on all related matters • Coordination with other department & report to higher ups in the dept.
Chief Conservator of Forests (CCF)	• Holding charge of territorial zone covering 2-3 circles • Heading functional cells at HQ, e.g., administration, budget & planning, research and management, etc.
Addl. Principal Chief Conservator of Forests (APCCF)	• Assisting the PCCF of the state • Holding charge of various wings of dept. at state level independently, e.g., wildlife, monitoring & evaluation
Principal Chief Conservator of Forests (PCCF)	• Head of department in state • Advisor to the govt. on technical and policy matters • General administration and control of forest department

Source: IGNFA, Dehradun, 2013

Training Requirement for the Frontline Forestry Staff

Divisional Forest Officers (DFO) are kingpins in the sector, though they come in the senior category. But if they are encouraged to work with the frontline players. It will provide better results for the capacity building of all the frontline staff, i.e., forest guards, foresters and range officers of the state, in order to ensure adequate representation of diverse groups of range officers, foresters and forest guards involved in diverse activities related to forest management. Other than their long-term two years training programme in their respective forestry college and school, there is a need for refreshers and orientation courses for the frontline staff as it is being practiced for the DFO and above staff through IFS compulsory training (Figs. 1.1, 1.2 and 1.3).

(i) Training need assessment (TNA)

Defining and listing the tasks to be performed by the target population:

Training needs assessment of the target group of learners for whom training is intended, and their expected tasks to be performed is important in designing the training manual. This exercise involves:

1. consultation with technical experts who can accurately describe the job;
2. referring documents and manuals that accurately describe the job;
3. need for capacity building to fulfil the requirements of sustainable forest management among different levels; and
4. best practice documentation (good performers).

(a) Training Need Assessment

- Defining and listing the tasks to be performed by the target population
- Listing the skills and knowledge required to do the task
- Selecting the skills and knowledge to be taught (training objectives and contents)

(b) Designing and Developing the Training Manual

- Developing of training manuals on the basis of subject relevance and identified topics through training need assessment
- Realistic examples and information for use in exercises
- Field-test the training materials

(c) Revise and finalize training materials based on the field test

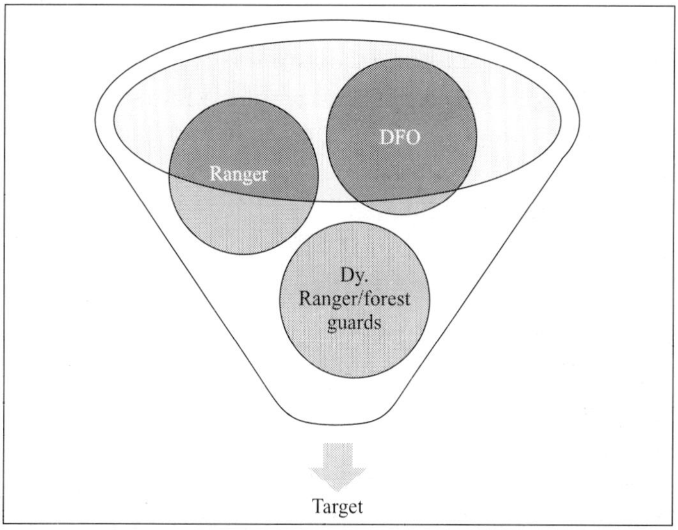

Fig. 1.4 Primary data collection through interview and focused group discussion

In addition, focused group discussions will also be conducted with range officers, foresters and forest guards separately so as to know and gauge skill gaps among them on various aspects of forest management.

- *Desk review of previous training and feedback of participants*

  Along with the primary data and information collection from the frontline staff, a desk review is done for training courses (and participant's feedback) conducted by the training institutions so as to gauge the demand for capacity building overtime.
- *Listing the skills and knowledge required to do the task*

  After analysis of the data and facts gathered from the field, the training modules are developed based on the skills and knowledge required to perform the task of forest management and related issues. Skills are generally actions such as measuring, mixing, recording, calculating and communicating, or making decisions. Required knowledge is the information needed to do a task correctly.
- *Selecting the skills and knowledge to be taught (training objectives)*

  The training objectives may be formulated further based on the skills and knowledge required for particular activities. This will help in harmonizing the information and avoiding irrelevant.

(ii) Designing and developing the training courses

Based on the TNA, the target group/s, number of programmes, and the themes of training are identified and training modules are prepared. A training toolkit consisting of pre-prepared printed flip charts, CDs on important themes of training, trainer's manual, FAQs, relevant training feedback formats is prepared by a team of experts and is given to all the trainers. The training manual is structured in such a way that the participation of the trainees is ensured in the learning events for this participatory teaching methodologies, and tools are prepared keeping in the roles and responsibilities as well as level of education of different categories of the frontline staff.

Realistic examples and information for use in exercises

The training should be based on the realistic examples and information need to be used, which can be used in the roleplays, case study probing and audio and visuals.

Course kit may include:

- articles on relevant/burning issues;
- case studies/success stories from the field;
- simulation games;
- few important standard formats;
- latest information on gps, other equipment, etc.
- The emphasis will be on orientation of the participants for the changed role and responsibility.
  - More group activities, discussions and presentations by participants will be ensured to achieve greater participation and better reception.

  - Personality development and general management will also be a major component to make the course more meaningful and fruitful.
  - Case studies and panel discussions will provide a broader platform to participants who are expected to play a role on higher levels.
- Field visit will be theme-based, e.g., watershed management/community participation/department management/NTFP based livelihood activities.

Field-test the training materials

The modules designed based on need identification are piloted. The duration of the training course varies from 2 to 3 days. Based on the suggestions of the participants of the pilot training programme, the training modules and materials may be modified and finalised at state level.

(iii) Revise and finalize training materials based on the field testing

To build capacity of the frontline forestry personnel at different levels enabling them to deal with the emerging issues efficiently, it will be imperative to pool forestry resources and establish linkages with other training and academic institutions not only from within the country but also with international organisations. The forestry training institutions in the country have been taking due care of the above-mentioned issues while imparting training to the forestry personnel. However, every training institution has limited capabilities. The resources available to support forestry training are generally insufficient to sustain the requisite expanse and quality of training. It is important that the forestry personnel are provided opportunities to present their views at the international events related to forestry and environment. This will help in capacity building of not only the officer as an individual but also the forestry sector as a whole. Exchange of faculty among the forestry training institutions and other academic institutions, sponsoring/organizing training of trainer (ToT) courses for the faculty of forestry training institutions, and encouraging forest officers to undertake higher studies in the institutions within the country and abroad, would be a right step for capacity development of the forestry sector.

After analysing the feedback, the material needs to be revised and final training module may be implemented at the ranger's college and forest guard school as per the requirement of inputs number of session should be developed.

Table 1.3 Proposed subject areas suggested for the front line staff

Subject	*Areas*
Forest Management	1. IITO guidelines on sustainable forest management 2. Regeneration 3. Rehabilitation 4. Sustainable production 5. Forest certification
Silviculture — Nursery, Plantation	1. Ecological and silvicuttural characteristics of major tree species 2. Tree species identification, biosystematics 3. Forest soils and tree nutrition

Contd...

	4. Silviculture 5. Forest ecology 6. Forest genetics 7. Nursery management 8. Methods of tree propagation
Forest Products	1. Production, types of NTFPs and timber 2. Sustainable harvesting 3. Storage 4. Value addition 5. Enterprise development
Mensuration	1. Biometrics 2. Forest inventory 3. Modelling stand and forest dynamics 4. Remote sensing, photointerpretation and photogrammetric 5. Geographic information systems (GIS)
Research Methodology	1. Sampling techniques 2. Research methodology for social, ecological and economic study in forest areas
Forest Protection	1. Fire ecology and fire management 2. Forest pest management
Forest Law	1. Forest policies of India and respective states 2. Biodiversity Conservation Act, 2002 3. PESA Act,1996 4. Forest Rights Act, 2006 5. JFM
Forest Accounting	1. System of national accounting 2. Inventorying
Wildlife Management	1. Habitat mapping 2. Policies related to wildlife 3. Conservation of wildlife 4. Ecotourism
Information Technology	1. Operating System –Windows Vista 2. Microsoft Office
Joint Forest Management Components	1. Organization behaviour 2. CPR management 3. Forest and livelihood 4. Conflict management 5. National forest policy 1988
Micro-planning and PRA	1. Land use planning 2. Village resource mapping

Physical Fitness and Self Defence	1. Physical fitness 2. Self-defence
Use of Firearms	1. Manual for using firearms
First-Aid	1. Forest safety and survival
Syllabus for Refresher Courses	1. Refresher course material based on module focus
Syllabus for Induction training for promoted ranks.	1. Syllabus for induction training and promoted ranks

Special Importance of Topics at Range Officer Level

1. *Forest Resource (Production and Demand Mitigation)* - timber, fuel wood, fodder (production and gap)
2. *Ecological rehabilitation* – degraded forests, grassland, mangrove, mined out area
3. *Plantation* – high-tech plantation, special innovation clonal
4. *NTFP* - Sustainability, harvesting, resource assessment, marketing/value chain improvement
5. *Joint Forest Management* – Case studies, innovative approach
6. *Poverty alleviation through forestry* – Case studies, PPP
7. *Agroforestry/outside forest* - Innovative model case studies, unique urban forestry efforts
8. *Forest Based Microenterprise* – Case study, types of forest products, role of microcredit
9. *Biotechnological Applications on Forestry*
10. *Forest Assessment* – Practical assessment methodology, field practices
11. *Biodiversity Conservation* – Field level innovation
12. *Ecotourism as forest conservation and livelihood generation, cases, experience sharing*

Theme-Wise Distribution of Sessions in Forestry Stream

- **Main Issue: How to manage forest resources : Issues, challenges and approaches**
  - Planning for management of forests and natural resources aimed at sustainable development
  - Ecosystem approach to resource management
  - Public participation, co-management and partnerships in natural resource management: Conceptual and case studies
  - Reproductive biology and NTFP availability/conservation.
  - NTFP certification, value chain analysis and its improvement.
  - Sustainable forest management (SFM)
  - Techniques of economic assessment of forest resources – Case studies of two forest divisions

- Poverty alleviation and addressing livelihood concerns through forestry
- Micro-enterprises functioning, micro-credit and institutional finance — Case studies
- Forestry extension
- Trees outside forests, agro-forestry models — Case studies
- Urban forestry — planning and management — Case studies
- Landscape planning and management

- **Main issue: How to increase productivity and planting stock improvement in forestry**
  - Productivity enhancement in natural forest and plantation forestry
  - Scope and need for biotechnological interventions (clonal) in forestry
  - New trends in seed technology for better forestry
  - Cell/tissue culture techniques
  - Role of genomics and proteomics, bioinformatics and eco-informatics
- **Main issue: How to assess forest resources through improved scientific techniques and providing new skills to foresters for inventory management through geomatics**
  - Changing perspectives and challenges in natural resource management - information needs
  - Contribution of geomatics technologies, e.g., GIS, remote sensing in natural resource management — concepts and case studies
  - New trends in assessment of forest cover/forest type
  - Rapid inventory technique for working plan
  - Use of GIS methods to visualise and analyse spatial data related to NRM
  - Biodiversity characterization in relation to GIS
- **Main issue: For improved forestry—Successful linkages, integrated planning and management**
  - Intersectoral policy and planning approach in forestry and rural development
  - Phases of project cycle - the project format and its limitations
  - Objective tree analysis - hierarchy of project objectives - analysis of strategies
  - Logical frameworks - project logic - understanding and formulating log frames coordination and control of interrelated activities
  - New scientific skill for protected area management
  - Monitoring and evaluation, tools and techniques in forestry/environmental project
  - Project development concerns - common reasons for project failure
  - Legal aspects of Forest Right Act (FRA 2006) and its implementation issues
  - Community mobilization and case studies
  - Forests and livelihood
  - Forestry governance - concept of forest governance/tools and techniques for better governance/social justice through forest governance
  - Writing of scientific paper - finalizing and dissemination of a paper

Various states were considered for the present study, where the concentration of tribal people high at the same time climatic condition, forest products, system of market functions are different, but some common parameters were analyzed and recommendations were framed. Several reports were reviewed, which highlighted that in recent past agriculture became difficult in the tribal areas due to competition and climate vulnerability. Forest produce collection is still a popular livelihood activity amongst the tribals for which they face hardships. There is a need for NTFP database development for making proper livelihood planning on various aspects like seasonality of collection, processing, storage, trading and marketing information for training. Orientation requires policy changes towards liberalization and deregulation which reflects the good intention of the government, but needs to be strengthened in several ways for boosting the sector. Protected area management in India is suffering due to lack of human, financial and material resources. Given with the situation of undertrained staff on wildlife and PA management became further difficult. State wildlife boards (through nodal department) with wildlife division of MoEF, in association with academic/training institutions such as WII, IIFM, IISc, FRLHT, CEE, and state forestry training institutions, and conservation NGOs like WWF, ATREE, BNHS, CPREEC, Kalpavriksh, etc., may cater long-term responsibility for the sector.

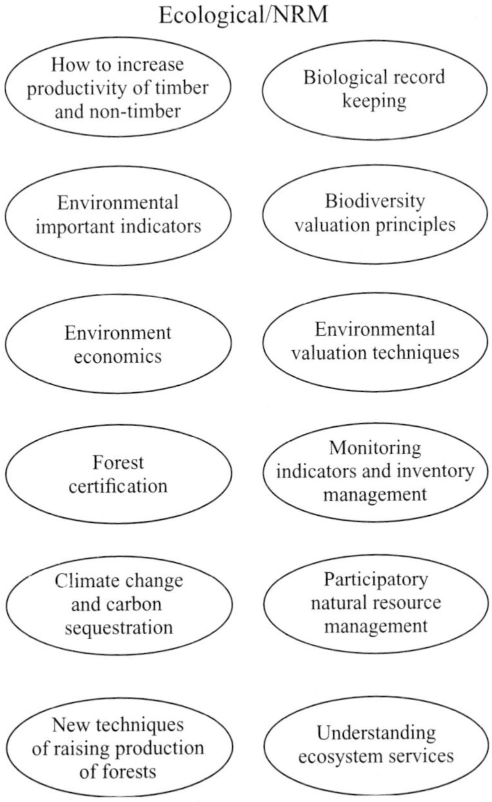

Fig. 1.5 Knowledge gaps and challenges for future

Fig. 1.6 Social and developmental issues and challenges for forest officials

Types of Effective Learning Methods

Sessions can employ several basic learning methods, grouped by subject matter into modules. A training activity can use a combination of the following learning methods:

Lectures: Oral information from trainer to trainees usually accompanied by view guides, slides, and other visual aids. Trainees with limited background on the subject should gain new information.

Lecture-discussion: Oral information to trainees, providing interaction between trainer and trainees, and using trainees' collective experience to confirm information presented on a subject.

Panel discussion: Group interaction among trainees, directed by the trainer towards a specific goal. It can involve problem solving, consensus building, and decision making by the group.

Small group discussions: Group interaction among trainees but less formal than the panel discussion. The trainer provides varying content and leadership aiming at problem solving, consensus building and decision making.

Case study: Report of a problem or situation by a designated presenter, who usually has direct knowledge of the situation. It is usually a detailed, in-depth discussion based on experience that considers all pertinent aspects and attempts to point out lessons learned.

Examples: Illustrative discussions that point out methods or situations without the in-depth detail and all-inclusive considerations of a case study.

Problem-solving exercise: Application of knowledge to a given hypothetical situation to gain practical acquaintance or skill in a subject. These are usually working sessions to practice applying methods or knowledge gained in the training activity.

Structured role playing: Individuals in trainee groups assume role playing with different levels of direction and background information. Its purpose is to make individuals more aware of how they or others would react in new situations.

Game or simulation: Extended, structured role playing with specific rules and a larger trainee group. It helps identify behaviours and attitudes in different settings.

Demonstration: Physical actions, which describe, illustrate, or explain a subject by example.

Field trip: Excursion to a different location for demonstration purposes. It often complements other learning methods in information delivery.

Trainers can use one or more learning methods to develop modules for specific topics. Sometimes emphasis can be on local or regional conditions and viable approaches, for example, to achieve sustainable watershed management. These modules would rely heavily on examples and case studies from the targeted region. There are many ways to package modules depending upon the training purpose, the audience, and time and resources available to conduct the training activity. The section, "Designing the Training Activity," discusses the use of modules and sessions for different training purposes and audiences.

Developing State Level Training Improvement Plan

The approach to training improvement plan for capacity building is to prepare training in a systematic way, using the SAT (systematic approach training) tenets - analysing tasks, assessing training requirements, stating training needs, developing course design and training improvement plan.

Foresters today are required to play multifarious roles to deal with a variety of externalities besides coping with traditional forestry management practices. It is, therefore, essential to develop expertise in the field of forestry and wildlife management as well as to create awareness among the personnel of other services and all other stakeholders who directly or indirectly influence the development and management of forests and wildlife ecosystems.

Training and skill development programmes to improve the capacity of producers in establishing new enterprises need to be carried out. Most of the government programmes in rural areas are for individuals and cover areas such as poultry, food processing, piggery, diary, handlooms and handicrafts, pottery, beekeeping, etc. They provide technical skills, but do not assist in marketing the products; here the not-for-profit sector is still in an embryonic stage. Most of these training programmes are not related to NRM and community based enterprises are not included in most of these programmes; implementation may be a challenge and new learning for foresters.

The duties and responsibilities of a present-day forester have become more challenging today. From being an exclusivist profession, primarily charged with forest management for timber production, foresters are called upon to perform a wide range of roles varying from traditional forest protection and production to comprehensive biodiversity conservation and sustainable use to ecotourism promotion and so forth. In today's global scenario, the world has begun to take note of a fast degrading environment and it looks towards professionals like the foresters to provide innovative stewardship in tackling concerns like global warming, biodiversity conservation, food and water security, and poverty alleviation through sustainable livelihood programmes based on sustainable use of natural resources.

Forestry schools have been trying to adjust their curricula to produce graduate foresters with necessary "core competence", augmented by the study of allied disciplines. This is commendable for tertiary-level education. However, universities, as leaders of society, can no longer ignore global forestry problems and concentrate only on producing graduates trained to deal with local and national issues. Global issues are too serious to be left to diplomats and politicians. It is high time that leading forestry education institutions pool resources and coordinate activities in order to offer training in international forestry, at least in some continuing education programmes (El Lakany, 2004).

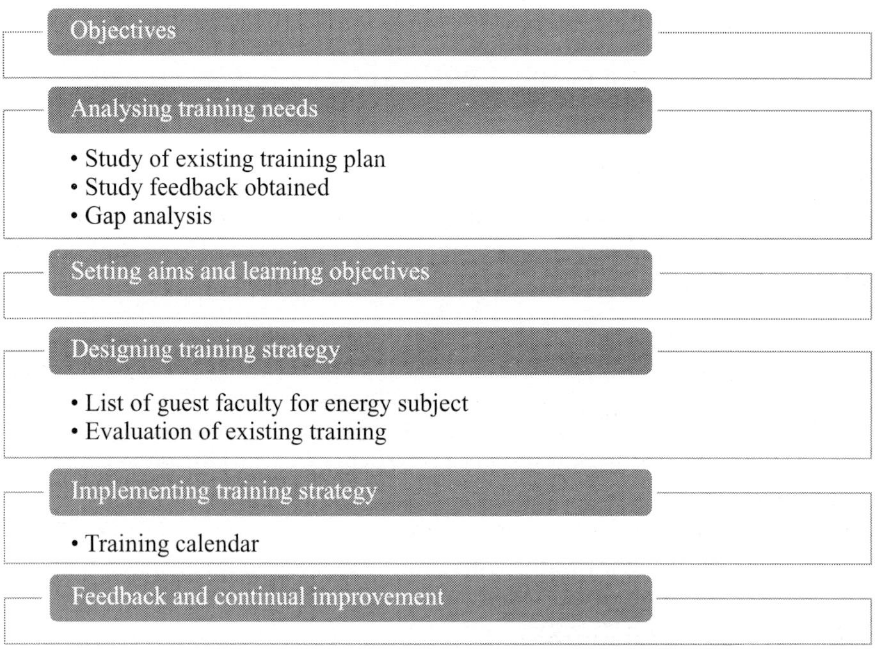

Fig. 1.7 Flowchart for organising training programme

Objectives

Understanding the organisation's aims and needs and how the aims are met by range of jobs that exist in the organisation.

Analysing training needs

Finding out what people need to learn. This is done by:

- analysing the knowledge, skills and attitude/behaviour that each job requires; and
- assessing the degree of competence needed for job-holders to meet these requirements.

Study of Existing Training Plan

Need for study on the existing training plan, physical achievements, effective utilization of master trainers and facilities.

Study of Feedback Obtained

Feedback on the existing trainings by participants and trainers are usually referred for the corrective action and improvement of the existing trainings.

Gap Analysis

Understanding in-service departmental system of training and identifying the gaps that need to be addressed in the new training plan.

Setting Aims and Learning Objectives

Specifying what trainees should be able to do as a result of training.

1. Expected role to be played by trainees
2. Achieving the training gaps

Designing Training Strategy

Deciding on a strategy to meet training needs, e.g., by designing courses/modules, suggesting various methodologies, deciding key learning points trainees must grasp and also sending some learning material (preview) to trainees.

List of guest faculty for every subject

The list of guest faculty for each subject will be identified. This exercise will help in the number required and expertise available for training session.

Evaluation of existing training

An evaluation of training impacts on the participant will be conducted to know the applicability of training and scope of implementation of the trainings. This will help in devising strategy for the future.

Implementing Training Strategy

Strategy for putting the training into practice in the three forestry training centres will be finalized. Training programme will be formulated in a way that the three forestry training centres conduct the training throughout the year.

FEEDBACK AND CONTINUAL IMPROVEMENT MECHANISM

Continual improvement and feedback mechanism will be placed in the training plan to improve and continuously monitor the training progress. Continual improvement and feedback mechanism should be in place under the training plan to improve and continuously monitor the training progress. This should be done by placing a system of revision of training syllabus and making the training more contextual.

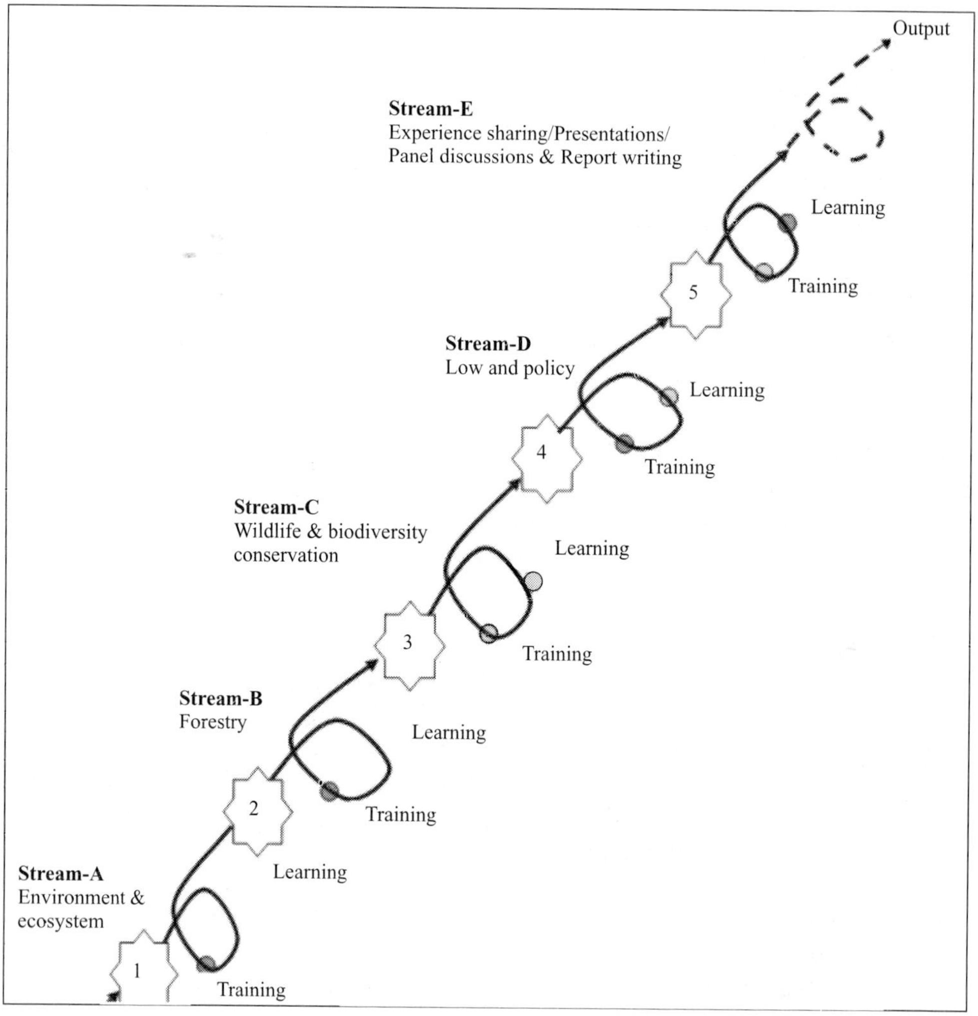

Fig. 1.8 Sequencing of training inputs

PLANNING STATE ToT PROGRAMME

A state training-of-trainers (ToT) programme may be planned to build capacity for quality improvement among existing trainers and incubating new trainers. The programme will be designed to build strong experiential background in forest management and provide participants exposure to adult learning theories and available training resources.

The overarching goals of the ToT programme include:

- enlarge the pool of qualified trainers on forestry;
- further spread of quality management concepts by increasing the number of training opportunities for trainers;

- increase the ability of the ToT trainers on innovative training methods; and
- training calendar

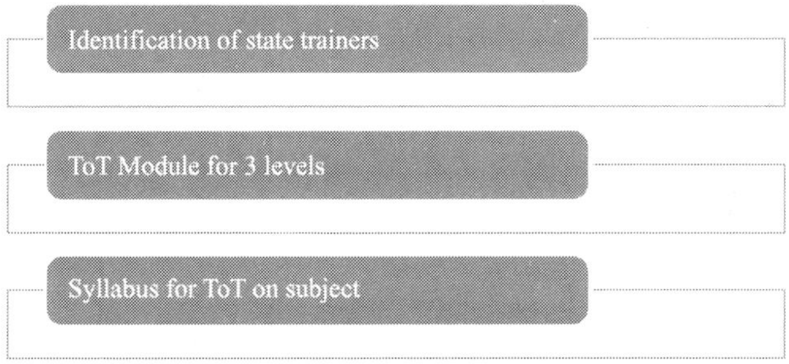

Identification of state trainers:

State trainers may be selected from the in-service department persons, retired personnel from the forest department, NGOs, universities and other departments on continuous basis to create a pool of trainers for training of all three levels of the front staff, viz., range officers, foresters and forest guards.

Considerations for selecting trainers include:

- technical expertise;
- familiarity with the geographical, cultural, and political region;
- appropriate language capabilities;
- necessary presentation skills, specifically at the level of the planned training activity;
- availability for the preparation time and presentations at the training activity; and
- ability to become part of a team-taught effort.

Table 1.4 Syllabus for ToT on subject: The syllabus of ToT module changes with the officer level

Range Officer ToT	*Foresters and Forest Guards ToT*
Forest Management	Protection
Joint Forest Management	Law
Forest and Livelihood	Micro-planning and PRA
Forest Law	Preparation of vouchers and payment documents.
Micro planning and PRA	
Information Technology	
Accounting	

Study of Syllabus

Study of Syllabus for Frontline Staff: For review purpose, the study syllabus of forestry school is important and needs to be updated for better performance of our staff.

Device model syllabus in Hindi for frontline staff: The training modules need to be translated into Hindi.

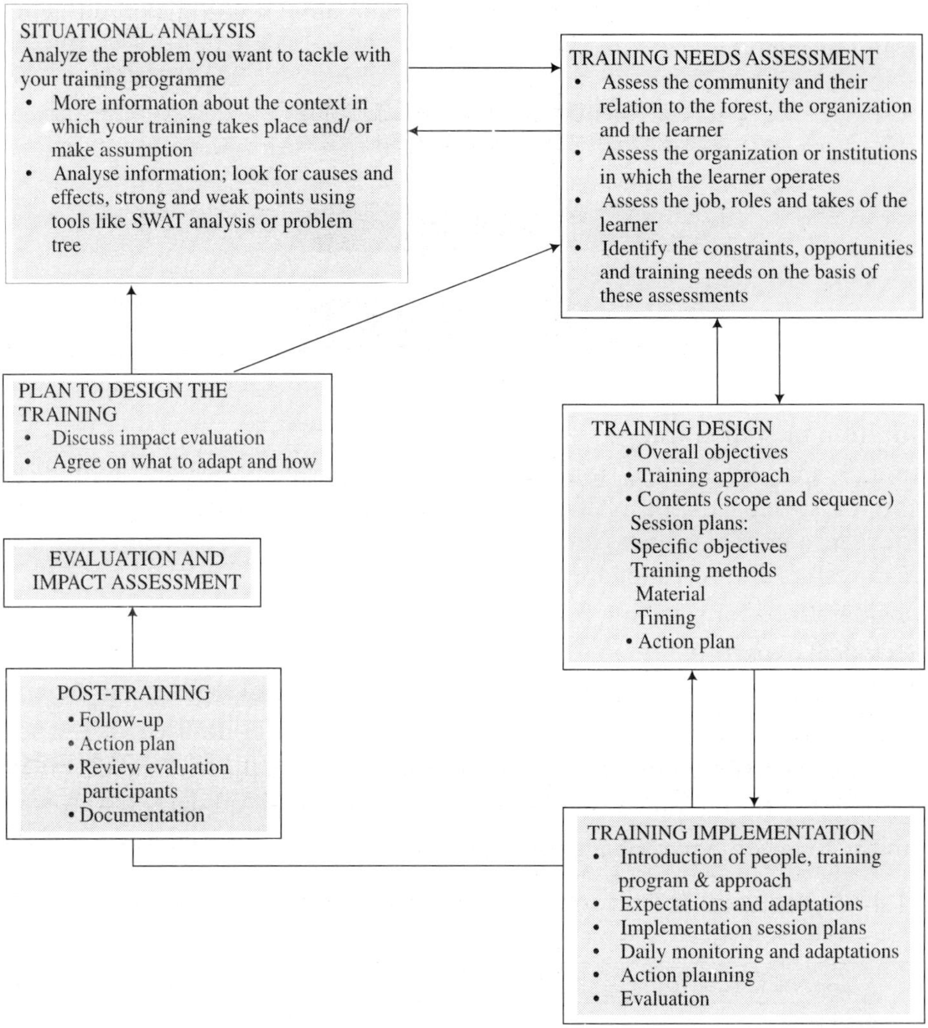

Fig. 1.9 Training design cycle

For Master Trainees

Master ToTs will be trained in engaging and conducting effective workshops, while linking them with experts and peers facing similar challenges. Before attending the rigorous face-to-face ToT session, a participant is asked to complete the pre-work, which includes online learning, reading of available quality publications, and participation in two-conference calls.

Preparing necessary training material is also a critical task. Training material must be technically correct, pertinent to reflect the region, and be in the language and at the educational and skill level of the trainees. Training materials can include a training guide, a training manual, case studies and examples, problem-solving exercises, instructions for

Supervise ToT

Supervise training of Frontline staff

Collect feedback from course

Modified ToT module as per requirement

Suggest modified training of frontline staff per needs.

Give advise to trainers for improvements

games and simulations, demonstration descriptions, software, and field trip itineraries. In most instances, regardless of the type of training, we need to prepare both a training guide and a training manual.

1. Monitoring Training of Forestry School

Regular feedback from participants and trainers will be collected and the physical target achievement of training as per the training calendar will be reported quarterly. This will also include the physical target of achievement and effectiveness training. The quarterly report will contain following things:

(1) Training material used by trainers.

(2) Training delivery on

- Knowledge (contents)
- Skill (state-of-the-art)
- Attitude (conceptual)

(3) Training acceptance by trainees

(4) Feedback based modifications

- **Training material used by trainers:** The training modules need be updated as per the needs of trainers and participants.
- **Training delivery:** The training delivery on knowledge (contents), skill (state of the art) and attitude (conceptual) will be monitored continuously. This will include surprise visits to the training halls and discussion with trainers on effective delivery of training to the participants.
- **Training acceptance by trainees:** The training acceptance level of trainees is also constantly monitored with mock tests and their performance in the exams.
- **Feedback based modifications:** The feedback from the participants needs to be monitored and modification if done may be enacted immediately.

We will also use the well known Kirkpatrick's training evaluation methodology for judging training programmes, which is known as Kirkpatrick's Four Level Evaluation Model.

Four levels of evaluation include the following:

- Reaction - how the learners react to the learning process
- Learning - the extent to which the learners gain knowledge and skills
- Behaviour - capability to perform the learned skills while on the job
- Results - includes such items as monetary, efficiency, moral, etc.

2. Assessing Training Performance

Evaluation of trainee participants

In each module it is necessary to work out the requirement of trainees in order to seek out their interest areas as well as field-based needs. Accordingly, a three-tier evaluation of trainee officers is proposed to be conducted, which is given below:

1. Punctuality and discipline (20%)
2. Class participation and attitude towards learning (20%)
3. Subject knowledge (60%)
   (i) Quiz – 20%
   (ii) Presentation – 20%
   (iii) Assignments – 20%

   Total marks in the form of percentage = 100%.

All participants will be evaluated in each step of the module, i.e., environment, forestry, wildlife and biodiversity conservation, management module and experience sharing/presentation/panel discussion and report writing. Evaluation of resource persons is an important component of the training programme. Like evaluation of the trainees, evaluation of faculty resources is also a continuous process and would be helpful in constant upgrading of content as well as the pedagogy of the training courses.

Evaluation of resource persons will be done by both informal as well as formal methods. Informal method constitutes every day feedback/suggestions received from the participants and the output is mostly internal. These are helpful for self-assessment of resources persons and enable them to come up to the level of trainees' expectation. It will also help the resource persons and programme organizer to make immediate mid-term corrections in the course.

Feedback Card/Post-its will be given to participants, and a board will be placed outside the training hall. After a session, trainees can reflect some constructive feedback to the resource person for subsequent session. Such methods include reflections on day's proceedings at the end of each day or module peer group evaluation, mood reflection of the trainees through moderator, brief questionnaire, etc.

Formal Feedback Form

However, in order to make major improvements/modifications in the overall content, structure and delivery method, formal evaluation methods will be followed. These formal methods are also useful to assess the relative strength and weakness of different resource persons. Though it is a conventional method, using a multidimensional feedback questionnaire is still a useful method to evaluate the different resource persons either at the end of each module or a particular phase of the training programme.

However, care will be taken to make the feedback, questionnaire, particularly that of resource persons a multidimensional one so that it helps different aspects resource persons. To ensure a successful training activity, the design process should include constructing a timeline, showing when to attain specific planning milestones. Selecting appropriate trainers and preparing necessary training materials, such as training guides, training manuals, case studies and examples are also important tasks. Managing the training activity efficiently with an appointed member having the required authority is necessary. Timely feedback to administrators and trainers must evaluate the training activity and its subsequent impacts to improve training activity quality and ensure the training focuses on trainee needs. Training materials and methods must be appropriate for trainees, skill levels, and the means by which they learn best. Trainers should use a

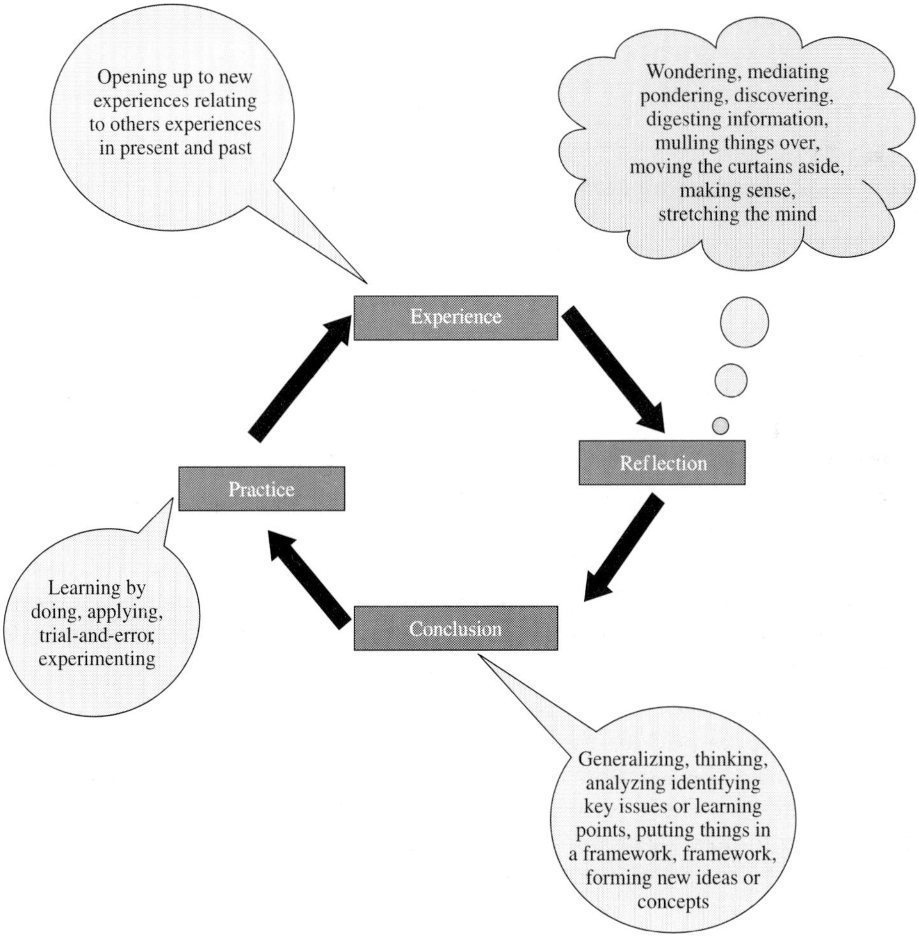

Fig. 1.10 Experimental learning cycle

standard professional method for student trainees. If trainees are experienced managers, trainers might find a more participatory, adult education method more appropriate. There needs to be effective communication between sponsors and trainees to determine their satisfaction. After getting feedback, trainers should change the training activities accordingly.

CONCLUSION

Hence, to build the integrated capacity of the forestry personnel at different levels and enabling them to deal with the emerging issues efficiently, it will be imperative to pool forestry resources and establish linkages with other training and academic institutions not only within the country but also with international organizations, universities or institutions. The forestry training institutions in the country have been taking inadequate care of the above-mentioned issues while imparting training to the forestry personnel. However, they have limitations in terms of capabilities and facilities, and insufficient availability to support qualitative forestry training. In addition, it is important that the forestry personnel are provided opportunities and exposure to confidently present their views at the international events, bilateral and international agreements, treaty and negotiation/governance processes related to forestry, climate change and environment. However, governing forests and forestry demands good and interconnected decisions to be made by many people. But many people are often denied access to key decisions about forests, while others who have the access do not have the ideas or incentives to change things, or are actively preventing change (Mayers, 2014) .

Some of these strategies include making appropriate changes in the curriculum to encompass the broad disciplines of 'new forestry', improvement in teaching methodologies, development of research partnerships with industries and networking with other research and education institutions including universities. Most often these are inadequate and too late, and their impacts are so short-lived that institutions need to make continuous adjustments to survive.

As per the latest changes in legal aspects in India, various local institutions were given powers and responsibility in forests, biodiversity, NTFP, and wildlife management. There is a need to have close relationship of forest official with JFM committees, Biodiversity Management Committees (BMCs), Forest Right Committees (FRCs), Village Eco-development Committees (VEDCs) and Gram Sabha for implementation of various forestry activities including the National Biodiversity Action Plan (NBSAP) implementation.

The importance of policies that support forest resources and reforms to markets has been recognised as essential to enable more effective contributions from forests for poverty reduction. There is a need for recognition of such policy reforms, market, tenure and poverty to be addressed by the forestry curriculum so that such connects are properly appreciated. To familiarize participants with the sound principles of general management/administration including HRD aspects, good governance, etc., are critical for the training:

- to expose them to the best practices in the field of forestry/environmental management during foreign exposure visit,
- To assimilate the processes involved in policy planning/programme formulation and learning from the experiences of other countries

For taking feedback from trainees subject expertise, delivery (presentation), pedagogy used communication skill, involving the trainees, motivational / in-spiritual skills, time management, interactive skills, session control, use of training material like audio visual aids, body language, personality, usefulness of content, etc. Training curriculum should be simple to understand and appreciate for the forest department and the curriculum should be such that people can adopt in varied situations. There is a need to strengthen regular programmes of government agencies and make them less dependent on project financing. Institutional strengthening should include training and manpower development and an orientation towards programmes of peoples' participation.

REFERENCES

1. Bhattacharya, P., Kandya, A.K., and Krishnakumar, K.N. (2008). JFM at Crossroads: Future Strategy and Action Programme for Institutionalizing Community Forestry, in Bhattacharya P., Kandya, A.K., Krishnakumar, K.N., edited Joint Forest Management in India Vol I, Avishkar Publishers, Distributors, Jaipur, India. pp. 3-31.
2. Drucker, P.F. (1989). The new realities: in government and politics/in economics and business/in society and world view. New York , USA, Harper & Row.
3. Mayers James (2014). Social justice in forests: gains made and tactics that work. A report from the Forest Governance Learning Group. IIED, London.
4. Malhotra, K.C. and Bhattacharya, P. (2010). Forest and Livelihood. Published by RULNR. Centre for Economic and Social Studies, Hyderabad. pp. 1-246.
5. Matta, J.R., and Kerr, J. (2007). Barriers beyond the partners: bureaucratic and political constraints to implementing Joint Forest Management in India. *Environment, Development, and Sustainability*, 9(4): 465-479.
6. M., Hosny, El-Lakany (2004). Looking outward: Incorporating international forestry in higher forestry education and research. Unasylva No. 216
7. Nair, C.T.S. (2003). Forests and forestry in the future: What can we expect in the next 50 years? Presented at the XII World Forestry Congress, Quebec City, Canada, 21-28 September 2003.
8. Nair, C.T.S. (2004). What does the future hold for forestry education? Unasylva, No. 216.
9. Van Lierop, P. (2003). Trends in forestry education. Presented at the XII World Forestry Congress, Quebec City, Canada, 21-28 September 2003.

2

Forest Certification: Benefits and Issues of Sustainable Forestry

Prodyut Bhattacharya* **and T.R. Manoharan****

**University School of Environment Management*
GGS Indraprastha University, New Delhi - 110078
*** Forest Stewardship Council*
Email: prodyutbhattacharya@yahoo.com & tmanoharan@in.fsc.org

ABSTRACT

Forests are considered an important resource for coping mechanism to local communities as well as high value internationally traded material, a carbon sink and home for biodiversity of flora and fauna. The protection, utilization, conservation and monitoring of forests are being considered at national and international levels to ensure its long-term sustainability and to minimize adverse social and ecological impacts.

Forest certification is a voluntary market tool which attracts "green consumerism" by ensuring sustainability of the resources. However, it is still very recent and largely untested except in a few cases, but it has great potential for global business. Certification is done on the basis of developed principle, criteria and indicators towards achieving sustainable forest management (SFM). Greater challenges are there for the development of any generic standards for certification, even though many of the principles, criteria, indicators and verifiers are universally applicable for certification. One possible benefit of certification is that certified producers receive a higher price premium for their products.

The harvest of forest products is coming under increasing scrutiny from certification programmes because of the key role it plays in the sustainable forest management and community benefit worldwide. Forests have two kinds of products—timber and NTFPs. Traditionally, NTFPs are collected by local communities and the dynamics of indigenous NTFP management systems, capability of communities to manage and control of NTFPs are major determinants of sustainability of the resources and are still not very clear to the major certification schemes. On the other hand, NTFP based certified products market is still to develop and tend to occupy the niche market for high quality products, particularly for medicinal plants, honey and other food items.

The emergence of forest certification as a conservation strategy has triggered a debate concerning the role of certification and sustainable forestry versus the role of traditional approaches to forest conservation. The purpose of this paper is to examine the role and contribution of forest certification, its present status in India and how the government, local communities and industries can get benefits out of it.

INTRODUCTION

Certification is a market-based tool that is becoming a hot topic in many natural resource sectors. A certification logo or label enables potential customers to differentiate products, based on the social and environmental 'qualities' of the commodity they to buy. This market opportunity is motivating many producers to adapt their management practices to meet certification criteria (Mallet, 1999).

What is Certification?

Certification is a process of confirming that a system or component complies with its specified requirements and is acceptable for operational use. The process of certification involves an independent and accredited third party visits an organization that had applied for certification and audits their management system and certifies that a product, service, system, process or material conforms to specific requirements and show that the organization abides by the principles set out in the standard, so following industry best practice (BSI, 2008).

Forest Certification

Forest certification is a process of confirming that the forest product certified under the scheme has been harvested from a well (sustainably) managed forest and that the organization abides by the principles set out in the standards following the best management practice. Certification of sustainably managed forests is mainly based on the principles for sustainable forest management (Bhattacharya et al., 2008).

Forest certification involves certification of the source of origin (forest) and also the certification of the chain of custody (CoC) through which the forest products are transported and transferred from the source of origin through various custodies (timber depot, processing units, etc.) till the finished product is delivered in the market.

Forest certification takes place at two seperate process:

Forest management unit (FMU) certification is a process that leads to issuing of a certificate by an independent party, which verifies that an area of forest/plantation is managed to a defined standard.

Chain of custody certificate (CoC) is a process of tracking wood products from the certified forest to the point of sale to ensure that the products originated from a certified forest.

The steps followed in the process are (FSC, 2002):

- Development and approval of principle(s), criteria, indicators verifiers and standards of certification for forest resource (point of origin) and for the chain of custody (CoC) at national and regional level as per the accreditation policy of an international certification council.
- Development of pilot testing of draft standards.
- Preliminary accreditation of national/regional certification standards and certification body by the international certification council.
- Development and approval of standards as per social and environmental international standards.
- Adoption of management principles and standards at forest management unit level (origin of source) and its various chains of custody.
- Documentation of the conformity of management system as per the certification standards at the source of origin and various chains of custody.

- Visit and auditing by the certification body to the FMU (source of origin) and various chains of custody.
- Issue of the certification and logo for a specific time period.
- Monitoring and evaluation of the conformities from time to time and renewal of the certification.

In comparison to the timber certification, NTFP requires more intrinsic and detailed methodology, research and planning (Pierce and Laird, 2003; Pierce et al., 2003). However, despite having all these complexities and hurdles, attempts towards NTFP certification are increasing (Brown, 2002). Although there are only a few officially certified NTFP products to date, an increasing number of initiatives are being taken to develop standards for NTFP certification. It is interesting to see that timber and NTFP certification (Table 2.1) have quite different approaches.

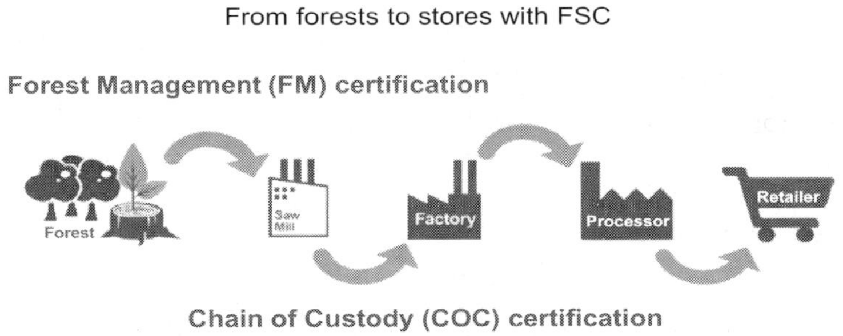

Fig. 2.1 Schematic view of two processes of forest certification

Why NTFP Certification

As per an estimate of EXIM Bank, international trade in medicinal plants is worth US$ 60 billion per year and is growing at the rate of 7% annually (Pandey and Mondal, 2008). At present almost 90% of the collection of herbal raw materials is from the wild. Most of the raw materials are not only unsustainably harvested, but are also socially and economically depriving the benefits to the local communities. The reason for huge amount of herbal raw material coming from the wild is that domestication of wild medicinal plants in cultivation often requires huge financial resources, time and intensive research activities, which many times restrict its economic feasibility (Bhattacharya et al., 2004). Another important fact is about the active constituents, which are often believed to be present in higher amount in the wild materials than in their cultivated counterparts (Scrippmann et al., 2006). This has resulted in destruction of the habitat of the species, resulting is low availability and poor quality of life of the local inhabitants (Miteva et al., 2015). In India there are around 8000 species of medicinal plants, and around 1000 species are estimated to be facing various degrees of threat-across different bio-geographic regions in the country (Ved, 2007).

Presently, there is an international demand of medicinal herbs used for herbal tea. Large Indian tea companies are planning to develop herbal tea in line with China that has huge demand in Europe, but this requires certification of herbal composition used with tea. NTFP certification does not have government policy and guidelines, in true sense now it is totally a market-driven system.

Table 2.1 Comparison between forest (timber) and NTFP certification

	Forest (timber) Certification	*NTFP Certification*
1.	A few important primary timber products are the focus as general trend, development of timber harvesting protocol is standardized.	Multiple products, huge diversity, development of harvesting protocol is difficult.
2.	Bulk collection on particular age and yield is almost predictable. Harvesting cycles are known.	Seasonality, production, collection, pattern vary a great deal. Harvesting cycle is still emerging. Sporadic harvesting is being followed, since NTFP does not grow uniformly and are available in pockets of good growth.
3.	Follows principle of SFM and sustainable yield. Since it is timber based yield, growth, regeneration, plantation data are quite predictable and long term. Silviculture systems are standardized.	Biology, ecology and unsustainable harvesting parameters are more complex and unpredictable. Silviculture systems are not developed.
4.	Standard chain of custody developed for timber products.	In general the chain of custody is very long.
5.	Standards are developed; already well established system and many certifiers are working globally.	Standards in most cases follow FSC standards, national standards not developed in various parts of the world.
6.	Timber certification market base is very big; timber demand is high as well as certified products are available in the market easily. The timber producers are much more powerful and organized.	NTFP certification is mostly important for tropical forest areas. Not much demand for premium products as compared to timber. The NTFP producers are unorganized and less powerful and often work in poor working conditions.
7.	Timber certification responsibility is with the Govt. or private parties, certifier industries are more resourceful and standardized in different forest certification schemes.	Group certification is more relevant as collectors form their groups but paying high certification cost is difficult for the groups.
8.	Comparatively less time consuming as timber assessment is a bit straightforward method and standard methodology is available.	NTFP certification is more time consuming because auditing requires more time for critical review of products and species.
9.	Area harvested for timber can be easily monitored for post-harvest impact.	NTFP post-harvesting impact assessment does not give an accurate picture of resources until there is a gross change.

The primary goal of certification is to bring about positive environmental and social change in resource stewardship (Mallet and Karmann, 2001). Since, certification of NTFP is a market-based tool to ensure sustainable harvesting and attract "green consumerism". It has many benefits: (i) price premium for the collectors/harvester and producers; (ii) social sustainability; and (iii) sustainability of the resources through responsible collection practices and resource generation.

By purchasing the certified product, a distant consumer, who is not directly involved in the sustainable forest management (SFM), contributes in SFM as a token of respect to the natural products. Forest certification promotes the following aspects, if implemented in responsible manner:

- Supports legal and customary rights
- Social-impact-secure contracts and local employment
- Workers safety and well-being—training and empowerment
- Biodiversity conservation
- Reduced impact on felling coup
- Improved organizational management
- Market access and price premiums
- Raising awareness in civil societies
- A useful instrument for government

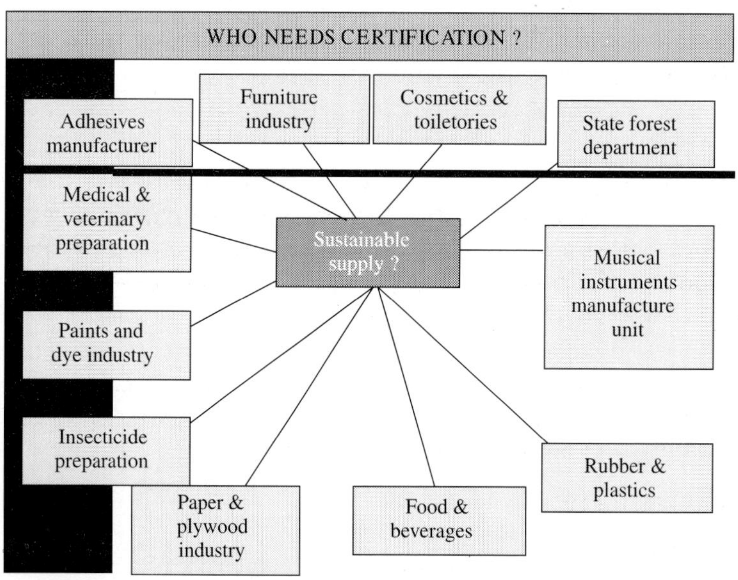

Fig. 2.2 Various stakeholders who need certification for their business promotion

Does NTFP Certification Reduce Unsustainable Harvesting?

Certification can be considered one of the tools among many more methods and approaches used for sustainable harvesting of NTFPs and to move towards more sustainable production systems. It is an innovative tool which links our local common people and international consumers who use the products but hardly know about massive unsustainable harvesting existing in this sector. It ensures that the NTFPs are coming from legal sources and that the sustainability of the resource has not been compromised. However, it will take further refinement of certification programmes to meet local realities (Brunori, 2007).

Once, more producers and harvesters are willing to test the certification market and there is an increase in the demand by consumers for certified products, the full benefits of NTFP certification will be realized. Nonetheless, the principle, criteria, indicators and verifiers of sustainable harvesting and NTFP certification can be used by producers and harvesters anywhere as a model for best practices.

Industries mostly use raw products for making drugs which are procured from forests (about 90%) by many people but generally industries are least bothered about the consequences of unsustainable harvesting as they get products from any place where-ever it is available. The crisis of raw drug and quality raw material availability they have started facing (Bhattacharya, 2004). NTFP certification promotes tracking, the illicit nature of industry and lack of information about harvest sustainability.

A set of criteria and indicators (C&I) has been developed under national standards for NTFPs/medicinal plants certification, which shall be used for assessing sustainability of NTFP production for that area.

There are no specific management plans for NTFP management, and the forest working plans are yet to incorporate NTFPs in their prescription. The micro-plans prepared at JFMC level also lack specific management plan for the NTFPs. However, there is a general list of NTFPs available in the regions, and in some cases broad thumb rules for sustainable harvesting are mentioned in the plans.

In case if certification process is to be initiated, there must be a proper species specific management plan for the NTFP as per its set of principle and criteria. In view of this, ICCF-IIFM has initiated the process of incorporation of management of NTFPs in the National Working Plan Code so that the working plans prepared have explicit planning for the management of NTFPs. This will not only strengthen the process of NTFP certification but will also improve the management of these resources, and special emphasis will be provided to the NTFP management. This will also help in regulation of the harvest and improving the production system. Figure 2.3 shows the ways certification contributes to sustainable harvesting and biodiversity conservation.

What does Certification Involve?

Certification Systems

- Certification is envisaged as a market-driven mechanism that promotes sustainable forest management in three main ways:
    - by establishing standards for forest practices and management that guarantee a certain level of management performance;
    - by enhancing market opportunities for products from sustainably managed forests; and
    - by promoting public education about improved forest management for both producers and consumers.

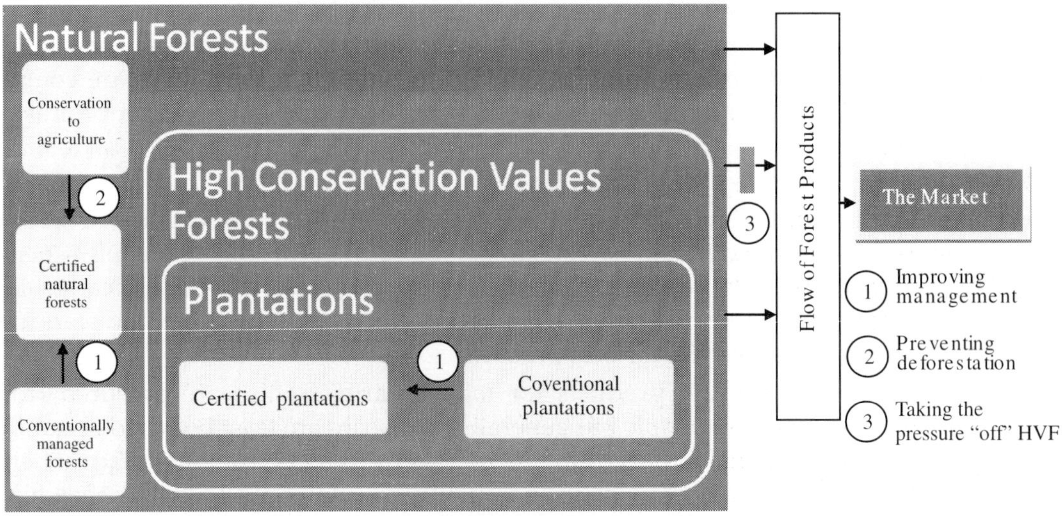

Fig. 2.3 Schematic view of three possible areas where forest certification contributes to biodiversity conservation (Gullison, 2003)

There are, in general, three types of certification initiatives in NTFP certification (Mallet, 1999):

1. Well managed forests (sustainable forestry) by applying criteria that address ecological, social and economic issues.
2. Organic certification
3. Fair-trade aystem.

Under NTFP certification, fair trade certification promotes and assures the benefits to the collectors. It tries to redistribute some of the benefits back in the production chain, to increase the profitability of NTFP collection to local people. Quality is a major concern in this sector. The international "good collection practices" are part of the certification process, which assure the sustainability of the products in nature. In this reference "good manufacturing practice" (GMP) requirement (under the Drugs and Cosmetic Act, 1940) that also tries to assure the products originate from sustainable and legal sources.

Since 1998, the FSC has permitted certification of NTFP management systems on a case-to-case basis. Smartwood, rain forest alliance, and others are involved in NTFP certification (Smartwood 2002). One issue on which all certification programmes can agree is that certification specific to NTFPs is still very recent and largely untested. However, the recent work of the FSC NTFP working group has gone a long way to refining certification for NTFPs. In another case, Smart Wood, an FSC-accredited certifier has developed a generic addendum to their criteria that will be used as the basis for all their future NTFP assessments. The NTFP Working Group of the FSC has been undertaking field trials and interpreting the FSC principles and criteria to make them more appropriate for the harvest of NTFPs (Smith, 2006; Mallet and Karmann, 2001). There are a set of 10 principles and 54 criteria under its system. The area certified under FSC has now 185 millions Ha. (Fig. 2.4).

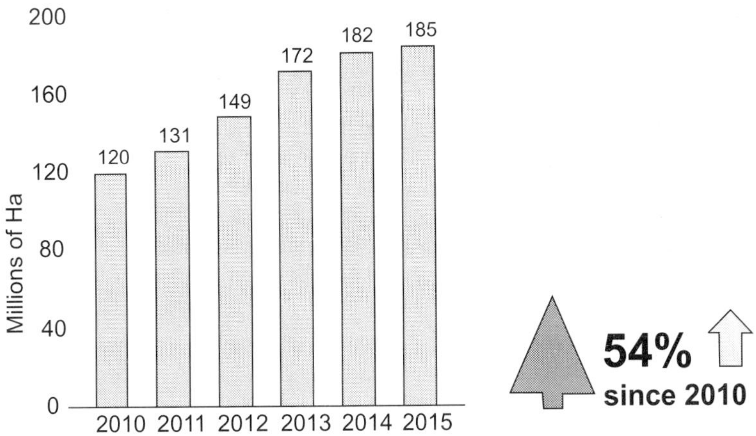

Fig. 2.4 FSC certified forest area growth (*Source:* FSC Certificate Database, data from January 2010 to January 2015)

WWF with IUCN under its TRAFIC network has developed International Standards for Sustainable Collection of Medicinal and Aromatic Plant (ISSC-MAP). The ISSC-MAP has a set of 6 principles and 18 criteria indicators for sustainable collection of MAP resources. The project is implemented in the Himalaya region (Nepal), greater Mekong region (Cambodia), Tropical Amazon Region (Brazil), Southern Asia (India), South Africa, and South East Europe.

WHO has also developed guidelines for good collection practices (GCPs), which are being widely discussed for their implementation by various NTFP producing countries (WHO, 2006; Karki and Rawat, 2004). Recently, National Medicinal Plant Board (2008) in India also set up a committee involving experts and practitioners to develop India specific guidelines for good collection practices other than the WHO standards for cultivated MAP resources.

IMAFLORA, an organization in Brazil, is assessing the possibility of certifying a forest region from which a wide range of medicinal plants are harvested, based on the individual management plans developed for each species.

In India, national standards for sustainable collection are as per Indian situation for development of national standards for NTFP Certification by ICCF-IIFM. ICCF-IIFM has adopted a set of 4 principles and 16 criteria for development of national standards for sustainable collection of MAPs adopted from IISC-MAP, 2007. A comparative analysis of different certification schemes, as applicable in India, is given in Table 2.2. Once these processes are concertized, they can be forwarded to International Certification Councils approval and accreditation of the standards, so that the certifier can follow these standards for NTFP certification.

Table 2.2 Comparison among different sustainable NTFP/Medicinal plants harvest standards

Criteria Organization	*Emphasis*	*Weakness*	*Inference*
ISSC MAP	Best practices in all the three strata viz. ecological, social and economic	The most comprehensive till date involving all the stakeholders but still the implementation agencies are not specified	Most relevant
FSC principles	Forest management units (FMUs) and chain of custody (CoC)	Absence of policy environment in India	Await recommendations of certification committee
GACP, WHO sanitation and handling of herbs	Little or no concern for ecology or sources	Largely applicable to cultivated MAPs and not the ones from the wild	

Source: Shanley et al., 2002

Many countries do not have national standards for certification; in that case the criteria are developed by the accredited certifiers who take the responsibility of actual certification under FSC- endorsed national standard.

Certified Products

Globally under the forests certification was primarily developed for the timber certification.

NTFP certification is experienced globally on species like chicle gum, cork, Brazil nuts, and chestnuts, pine nuts, rattan and rubber, Canada maple syrup, Nepal aromatic and medicinal oils, forest tea, etc., has been done by SmartWood-Rainforest Alliance following the FSC principles and Criteria (Rainforest Alliance SmartWood NTFP certification addendum). The total certified area under all schemes is given in Fig. 2.6 shows the main suppliers of certified forest products.

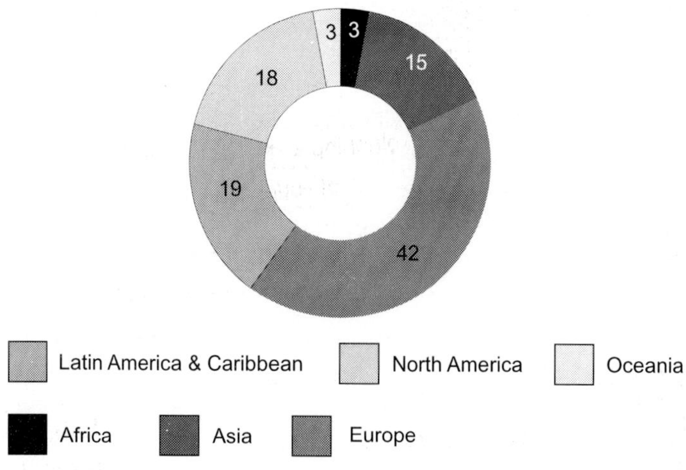

Fig. 2.5 Global distribution of FSC FM certificates by region (%)
Source: FSC Certificate Database, January 2015

Note: EU25. EECCA represents Eastern Europe, Caucasus and Central Asia (CIS). Certified forest area includes area certified by FSC, PEFC, CSA, SFI, ATFS (forests which are not independently certified, and process certification schemes such as ISO 14001, are not included in this analysis).

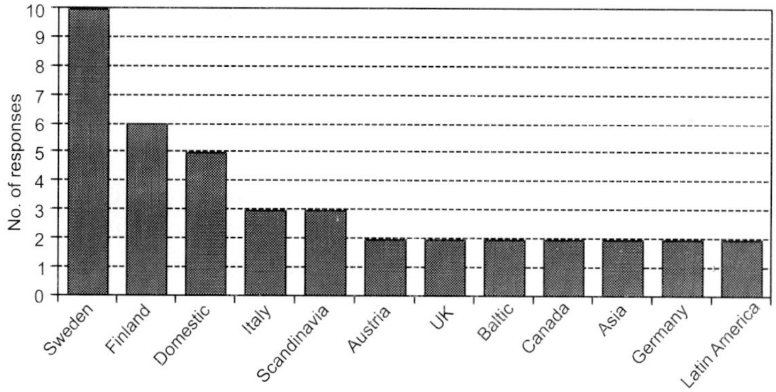

Fig. 2.6 Suppliers of Certified Forest Products.
Source: Forsyth, 2002

In India there is a huge certification potential for a wide range of NTFP products. These products are harvested and sold for export in huge quantities from natural forests; a few of them are mentioned here:

Amla	(*Emblica officinalis*)
Arjun	(*Terminalia arjuna*)
Atis	(*Aconitum rophyllum*)
Baheda	(*Terminalia bellirica*)
Beladona	(*Atropa belladonna*)
Nux vomica	(*Strychnos nux-vomica*)
Kullu gum	(*Sterculia urens*)
Neem seed	(*Azadirachta indica*)
Chiraita	(*Swertia chirayita*)
Gudmar	(*Gymnessa sylvestre*)
Salai and gugal	(*Broswellia serrata* and *Commiphora wrighti*)
Berberis	(*Berberis aristata*)
Taxus	(*Taxus wallichiana*)
Kokum	(*Garcinia gummi-gutta*)

FSC Certification

FSC is an international organization that provides a system for voluntary accreditation and independent third-party certification. This system allows certificate holders to market their products and services as a result of environmentally appropriate, socially beneficial and economically viable forest management.

Mission of FSC is to promote environmentally appropriate, socially beneficial, and economically viable management of the world's forests.

FSC provides setting standards for responsible forest management and valuing ecosystem services; is an accreditation/certification system to monitor adherence to FSC standards and a product labeling system that rewards responsible forest managers. Based on these standards, FSC provides a system for certification of organizations seeking to market their products as FSC certified.

FSC product label is affixed by certificate holders on products to illustrate the FSC status of the forest-based content, and it allows the end consumer to distinguish products from FSC sources. The four categories of labels are: (1) **FSC 100%:** All materials are from FSC-certified sources; (2) **FSC 100%: Small and Community Label Option (SCLO)** All materials are from FSC-certified sources from small holders (this is a new label); (3) **FSC MIX:** At least 70% of the materials contain a mix of FSC-certified sources and eligible recycled fibre, remaining are from FSC Controlled Wood; (4) **FSC Recycled:** The product is made from recycled materials.

Globally, the area under FSC Certification is 184.940 million ha, and the number of chain of custody certificates reached 29,458; and the nature of forest type (natural, plantation, semi-natural and mixed plantation & natural forest) and area are given in Fig. 2.7 and 2.8 (As of 7 August 2015)[1].

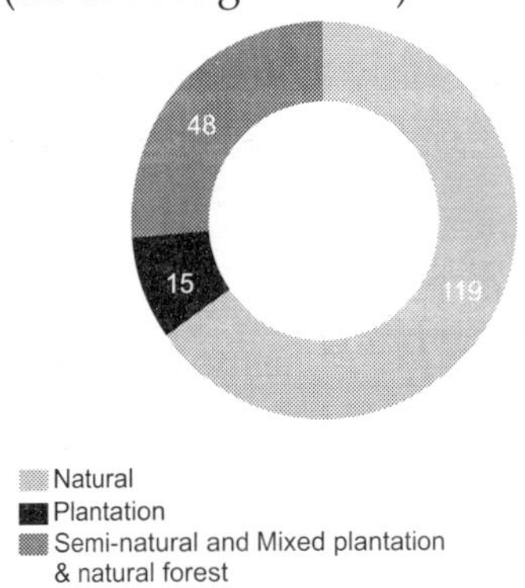

Fig. 2.7 Global FSC-certified area by forest type (Mha)
Source: FSC Certificate Database, January 2015

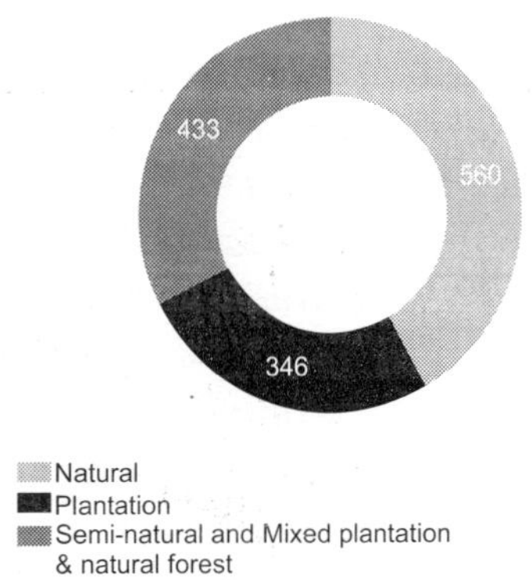

Fig. 2.8 Global FSC FM certificates by forest type (number of certificates)
Source: FSC Certificate Database, January 2015

Growth of FSC in India

The number of companies having FSC Chain of Custody Certificates in India has increased from 5 in 2005 to 258 in 2015. This includes small and medium enterprises, and government companies. The area under FSC certification has increased from 644 ha in 2007 to 811,815.80 ha (0.81 million ha) in 2015. This includes farm forestry/agro-forestry areas and forests managed by the state governments. FSC Certification of Bamragarh Forest Division (Maharashtra) and Uttar Pradesh Forest Corporation (UPFC) with 13 Forest Divisions in Uttar Pradesh have generated a strong potential for further growth of FSC Certification in India. The support from Indian paper industry has contributed significantly to the growth of FSC in farm forestry/agro-forestry areas.

Areas under FSC Forest Management Certification cover both FM and FM/CoC certificates.

[1]www.fsc.org

Table 2.3 FSC certified areas in India

S. No.	*Certificate Holder*	*Certified Area (Ha)*	*Species*	*State*	*Issue Date*	*Certificate*
1.	New Ambadi Estate (P) Ltd.	688	*Hevea brasilensis (Rubber)*	Tamil Nadu	12-09-2012 (644 ha certified in 2007	FM/CoC
2.	Patneswari Agri. Co-Operative Ltd. (PACL)	2,304	*Eucalyptus camaldulensis; Acacia mangium; Eucalyptus tereticornis*	Odisha	03-11-2011	FM/CoC
3.	Society for Afforestation, Research and Allied Works (SARA)	15,091	*Eucalyptus tereticornis; Eucalyptus* spp.; *Acacia harpophylla; Eucalyptus pellita; Dalbergia sissoo; Leucaena leucocephala; Eucalyptus urophylla; Eucalyptus urograndis*	Karnataka	13-12-2011	FM/CoC
4.	ITC Limited - PSPD Unit, Bhadrachalam	22,804	*Casuarina* spp.; *Eucalyptus* spp.;	Andhra Pradesh	22-02-2012	FM/CoC
5.	Tamil Nadu Newsprint and Papers Limited	15,304	*Eucalyptus* spp.; *Gmelina arborea* Roxb.; *Dalbergia sissoo; Acacia mangium; Melia azedarach* L.	Tamil Nadu	06-07-2012	FM/CoC
6.	Bhamaragarh Forest Division, Maharashtra Forest Department (Govt. of Maharashtra State)	3,67,732	*Tectona grandis;Terminalia* spp.; *Bombax ceiba L; Dalbergia sissoo; Pterocarpus marsupium* Roxb.; *Dendrocalamus strictus*	Maharashtra	12-12-2012	FM/CoC
7.	Lok Vaniki Kisan Samiti	175	*Dalbergia sissoo; Tectona grandis; Terminalia* spp. *Dendrocalamus strictus*	Madhya Pradesh	17-10-2013	FM/CoC
8.	International Paper APPM Ltd., Andhra Pradesh	28,635.80	*Casurina equisentifolia*	Andhra Pradesh	28-03-2014	FM/CoC

Contd...

S. No.	*Certificate Holder*	*Certified Area (Ha)*	*Species*	*State*	*Issue Date*	*Certificate*
9.	JK Paper LTD, Unit:JK Paper Mills	2,616.00	*Eucalyptus camaldulensis; Eucalyptus deglupta* Blume; *Eucalyptus* spp.; *Eucalyptus urophylla*	Rayagada, Odisha	23-12-2014	FM/CoC
10.	Kasu Brahmananda Reddy National Park* Wildlife Management Division, Hydrabad	142.50	Neem, Teak, Rosewood, Sandal Wood. *Anogeissus latifolia, lagerstromea parviflora, Albizzia amara, Cassia fistula, Sapinalus, Butea monosperma, Bamboo, Ficus* sp., *Dalbergia paniculata, Accacia sundra, Chloroxylon swietenia, Grewia teliafolia*, etc.	Telangana	01-12-2014	FM
11.	UP Forest Corporation, Govt. of Uttar Pradesh	3,49,296.00	*Acacia catechu* var. *sundra Duthie; Aegle marmelos* (L.) Correa; *Albizia lebbeck* (L.) Benth.; *Albizia procera* Benth.; *Alstonia scholaris* (L.) R.Br.; *Barringtonia acutangula (L.)* Gaertn.; *Bauhinia purpurea* L.; *Bombax ceiba L; Bridelia retusa* (L.) Spreng.; *Buchanania* spp.; *Cassia fistula* L.; *Cordia dichotoma* J.G. Forst.; *Dalbergia sissoo; Dillenia pentagyna* Roxb.; *Diospyros ebenum; Ehretia laevis* Roxb.; *Eucalyptus tereticornis; Ficus* spp.; *Grewia tiliaefolia* Vahl; *Haldina cordifolia* (Roxb.) *Ridsdale; Holoptelea integrifolia* (Roxb.) *Planch.; Hymenodictyon orixense* (Roxb.) *Mabb.; Kydia calycina* Roxb.; *Lagerstroemia parviflora* Roxb.; *Lannea coromandelica* (Houtt.) Merr.;	Uttar Pradesh	19-02-2015	FM/CoC

			Leucaena leucocephala; Litsea glutinosa (Lour.) C.B. Rob.; *Litsea* spp.; *Miliusa velutina* Hook.f. & Thoms.; *Mitragyna parviflora (Roxb.)* Korth.; *Phyllanthus emblica* L.; *Schleichera oleosa* (Lour.) *Oken; Semecarpus anacardium* L.f.; *Shorea robusta Gaertn. f.; Sterculia villosa Roxb.; Stereospermum chelonoides* (L.f.) A.DC.; *Streblus asper* Lour.; *Syzygium* spp.; *Tectona grandis; Terminalia* spp.; *Toona ciliata; Ziziphus mauritiana* Lam			
12.	Tripura Forest Development Plantation Corporation Ltd. Govt. of Tripura	7,027.00	*Hevea brasilensis* (rubber)	Tripura	23-02-2015	FM/CoC
	Total	811815.3				

Source: FSC (www.info.fsc.org)

Challenges

Cost of certification is still higher, in particular, among small holders/small companies; there is lack of awareness on the benefits of forest certification, and identifying the domestic markets for certified products.

Certified products need to be described as such on sales document (e.g., invoices) with an accompanying FSC COC certificate number of the supplier.

Opportunities

Opportunities for FSC certification in India include the demand for green public procurement, private sector initiatives, FDI in retail sector by MNCs (many of them have committed to FSC globally), legality requirements in international markets: (e.g., US Lacey Act, EU Timber Regulation, similar legislations in Australia and Japan), demand for green buildings, and increased recognition of voluntary sustainable standards (VSS).

Table 2.4 The 10 FSC Principles (Revised in 2012)

P1	**Compliance with Laws** The organization shall comply with all applicable laws, regulations and nationally-ratified international treaties, conventions and agreements.	P6	**Environmental Values and Impacts** The organization shall maintain, conserve and/ or restore ecosystem services and environmental values of the management unit, and shall avoid, repair or mitigate negative environmental impacts.
P2	**Workers Rights and Employment Conditions** The organization shall maintain or enhance the social and economic wellbeing of workers.	P7	**Management Planning** The organization shall have a management plan consistent with its policies and objectives and proportionate to scale, intensity and risks of its management activities. The management plan shall be implemented and kept up-to-date based on monitoring information in order to promote adaptive management. The associated planning and procedural documentation shall be sufficient to guide staff, inform affected stakeholders and interested stakeholders and to justify management decisions.
P3	**Indigenous Peoples' Rights** The organization shall identify and uphold indigenous peoples' legal and customary rights of ownership, use and management of land, territories and resources affected by management activities.	P8	**Monitoring and Assessment** The organization shall demonstrate that progress towards achieving the management objectives, the impacts of management activities and the condition of the management unit, are monitored and evaluated proportionate to the scale, intensity and risk of management activities, in order to implement adaptive management.

P4	**Community Relations** The organization shall contribute to maintaining or enhancing the social and economic well-being of local communities.	P9	**High Conservation Values** The organization shall maintain and/or enhance the high conservation values in the management unit through applying the precautionary approach.
P5	**Benefits from the Forest** The organization shall efficiently manage the range of multiple products and services of the management unit to maintain or enhance long-term economic viability and the range of environmental and social benefits.	P10	**Implementation of Management Activities** Management activities conducted by or for the organization for the management unit shall be selected and implemented consistent with the organization's economic, environmental and social policies and objectives and in compliance with the principles and criteria collectively.

Certification and Market

Like organically certified products market, the progress of NTFP certification is growing at a slow pace. As per the concept certified, products receive a higher price for sustainable/responsible management. But available evidence is somewhat contradictory, and generally suggests that buyers are unwilling to pay a price premium for certified products (Bass et al. 2001; Teisl et al., 2001; Gullien, 2003) only a very small premium (Stevens et al., 1997). The conscious willingness on the part of consumers to pay or sustainability of production system is still to come.

The certified product market for high value medicinal plants is increasing in North America and Western Europe. Main market is Europe (although low interest in Southern and Eastern Europe), but the bulk purchase in Asia has less demand for certified NTFP items (Fig. 2.9).

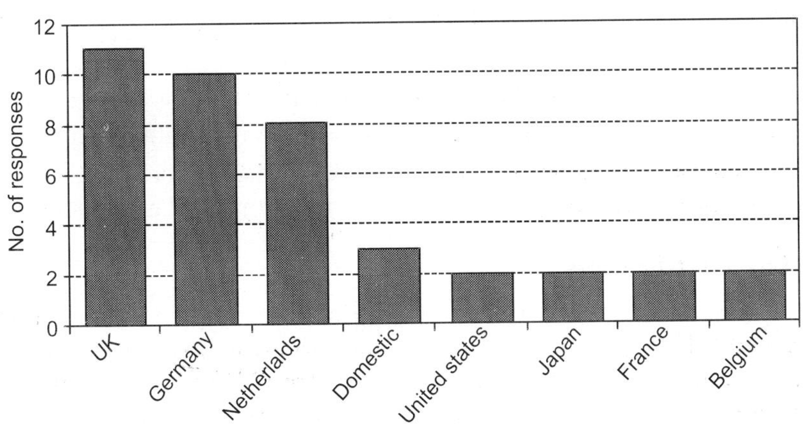

Fig. 2.9 Important markets for certified products
Source: Forsyth, 2002

The market demand in other countries is less as compared India's domestic demand. There is a great scope in India where nearly 9500 registered herbal industries and a multitude of unregistered cottage-level herbal units depend upon the continuous supply of medicinal plants for manufacture of different herbal formulations based on Indian system.

Due to an increase in the purchasing power of Indian middle class section, and interest for quality products, the domestic market of NTFPs for certification product has good future opportunity. Pharmaceutical companies are also trying to understand the prospects of certification and its scope, as in case of ISO, Bureau of Indian Standards, and other food standards, understanding has helped by marketing quality products.

Benefits and Cost of Certification

The costs of certification are of two types. The direct costs of certification are the cost of the certification process itself. Direct cost may vary with size of the area of the enterprise and distance travelled by the certifiers. While been made in the indirect cost of the certification process adopts change of management at field to meet the sustainable forest management standards.

At present, whatever the progress has certification at global level, it is a considered a costly affair. In the value chain, cost of certification is borne by the retailer as well as consumer, but initially the producer has to pay for certification. While FSC is closely associated with forest, it is also the most expensive certification programme to implement.

The FSC system is difficult to apply to the vast majority of informal community-based NTFP operations that constitute the bulk of NTFP harvesting worldwide. FSC is beginning to look at new models of community-based certification where a number of harvesters are certified as a group, or where a resource manager is certified to oversee multiple harvesting operations. Despite this recent progress, FSC certification is probably still most appropriate for large scale industrial NTFP operations (Mallet and Karmann, 2001).

The impact of FSC certification on developing countries has been described by CIFOR (2005):

- Acted to reduce social conflict in and around the certified forest.
- Helped in securing land tenure and usufruct rights in certified community forests.
- Helped promote sustainable forests management more generally through dialogue between the private sector, government organizations and civil society.

Future of NTFP Certification

At present the FSC certification system is difficult to apply to the varied informal community based NTFP operation, which is a common way throughout the globe. There is a need for some innovative and new model of community based certification where a number of collectors/harvesters are certified as a group as proactive at Chhattisgarh State, Jaributi Samiti under group certification by CGCERT.

The International Standard for Sustainable Wild Collection of Medicinal and Aromatic Plants (ISSC-MAP), published in 2007, is currently being implemented by WWF, TRAFFIC, IUCN and their partners in six wild collection projects worldwide. Within these projects models of adaptive ISSC-MAP implementation are developed to demonstrate effective management and sustainable use of wild-collected plants, ensuring thereby the long-term survival of the natural populations and contributing substantially to local livelihoods (ISSC-MAP, 2007).

Progress of NTFP Certification in India

India, although having a rich tradition of medicinal plant usage as well as its conservation through socio-cultural practices (viz., offering prayers to plants and animals, sacred groves and mythological concepts, etc.), represents a gloomy picture of the forest based MAPs. On one hand, there is indiscriminate collection of medicinal plants from wild sources leading to rapid depletion of natural resources, disturbance of ecosystem and environmental services, on the other hand, privately driven exploitative marketing system deprives the MAP gatherers from their due remuneration (Verma, 1998). Although forest plays a vital role in national economic scenario, until recently, NTFP/ MAP was a meager part of forest-based strategic planning and development. The problems, beginning from MAP procurement to marketing, were never addressed with due attention and seriousness which therefore worsened the condition day by day. The idea of MAP certification gained importance due to direct and indirect benefits it offers. Direct benefits cover sustainable collection of material which secures future availability, improved quality value added material, ensuring price premium as well as market expansion through international acceptance. Indirect benefits cover conservation of biodiversity by following environmental standards, respecting traditional rights and practices, improvement in social setup, economic development and benefit sharing. In a nutshell, MAP certification not only secures MAP population in forest area, it also helps to address other important issues which are linked with its stakeholders without whom MAP sector is incomplete (Sindhi and Choudhury, 2003).

MAP certification standards can be categorized as generic and specific. Generic standards may cover good collection practices with consciousness in conservation of biodiversity, ecological monitoring, impact analysis, management plan, social issues as well as economic development. Species, specific standards mainly deal with scientific harvesting methodology, storage and maintenance details, processing and value addition ideas, ecological importance, religious and cultural importance, marketing values, trade details, etc. Comparing the present scenario of MAP sector in India certification may sound ambitious, but step-wise advancement is necessary towards adopting "good practices" for wild medicinal plant collection, procurement and its management.

Attempts towards MAP certification have been started recently. World Wild Fund (WWF) India initiated a study in Himachal Pradesh on three medicinal plants in forest land to evaluate the principles of Forest Stewardship Council (FSC) in 2001 (Rastogi and Pant, 2004). Another state government led attempt was started in Chhattisgarh by forming Chhattisgarh Certification Society (CGCERT, Raipur Report, 2003). Although CGCERT covered NTFPs other than MAPs, it was accorded priority due to MAP richness and its economic potential for the state. A project was initiated on medicinal plant certification in Dhamtari district of Chhattisgarh sponsored by NMPB-IDRC. Under this project, a set of generic standards were devised covering good collection and other practices. Along with this, a set of species-specific standards were developed for 10 species namely, Aonla (*Emblica officinalis*), Harra (*Terminalia chebula*), Kalmegh (*Andrographis paniculata*), Baibiding (*Embelia tsjeriam-cottam*), Tikhur (*Curcuma angustifolia*), Baichandi (*Dioscorea hispida*), Satawar (*Asparagus racemosus*), Anantmool (*Hemidesmus indicus*), Nagarmotha (*Cyperus scariosus*), Dhawaiphool (*Woodfordia fruticosa*) (Katiyar, 2007). However, the project is under the implementation phase.

Similarly, in Uttarakhand, MAP certification work has been started by Winrock International India on 5 medicinal plant species (including lichen) with an aim to develop some species-specific standards for certification (Chaterjee et al., 2008). In 1991, the Government launched the 'Ecomark' scheme aimed at certifying and awarding a special product label to goods meeting specific environmental standards. This scheme has been unsuccessful, with very few companies participating, the Ecomark label having very low consumer recognition, and consumer demand for 'environment-friendly' products currently thought to be low (Jain, 2004).

The International Standards for Sustainable Collection (ISSC-MAP), developed by IUCN, TRAFFIC, WWF are also being jointly implemented at Western Ghats by FRLHT and TRAFIC India.

International Centre for Community Forestry (ICCF) at IIFM, Bhopal has recently started a project on "MAP certification standard development" sponsored by National Medicinal Plant Board (NMPB) covering the states of Madhya Pradesh, Chhattisgarh, Odisha and Uttarakhand. This project is designed to evaluate the potential for MAP certification and to develop generic standards for MAP certification with special emphasis on raw material collection to marketing. This paper presents some preliminary findings related to MAP certification possibilities in Indian background as well as a draft of MAP certification standards (Bhattacharya et al., 2008, Bhattacharya et al., 2009).

Standards are benchmarks meant to provide a comparison with the existing situation. It helps to provide a tool to resource manager as to whether it conforms to the standards/ benchmarks, and if not, how far or close it is with reference to benchmark. Development of standards is a continuous ongoing process as for many aspects the standards/benchmarks are not available and need to be developed taking into consideration the use of different stakeholders. In view of the multiple stakeholders, interest as well as their role in MAP sector development, local and regional level meeting/workshop as well as internal assessment have been organised for sharing their experience on different issues of MAP certification. The process of standard development has been started from reviewing and accepting (national and international) existing guidelines on standards as per the Indian conditions permitted and has gone through several phases of modification and refinement as and when required according to stakeholders opinion, field survey and assessment at different levels. To most of the stakeholders, the costs of certification as well as its painstaking documentary process are the major constraints in achieving the goal. Therefore, certification standards should be based on real scenario rather than idealistic or prescriptive situation, and therefore emphasis should be on field inspection/ verification rather than cumbersome documentation.

Considering the present scenario of Indian MAP sector and its varied range of problems, attempts have been made to evolve the standards in a way that its implementation could be possible at ground level along with options for further improvement and development. However, the most recent version of framework of standards is presented here with an aim to disseminate the standard to wider range of stakeholders for their critical review as well as constructive feedback.

A 4-tier framework has been developed consisting of principles, criteria, indicators and verifiers. The following are the 4 principles :

1: Legal and policy framework (4 criteria 13 indicators)

2: Wild area conservation and management plan (4 criteria 13 indicators)

3: Responsible collection and use practices (4 criteria 15 indicators)

4: Economic development and benefit sharing (4 criteria 12 indicators)

These set of principles have related set of criteria, indicators and verifiers which have been the outcome of a number of field survey, discussions, interaction and feedback from all possible stakeholders, varying from gatherers to foresters (Bhattacharya, et al., 2008).

It is a common experience that NTFP certification is developing as synergic system of international scientific and bureaucratic norms and rural resource use and practices and cultures. In most cases certification guidelines are a concept imposed by the developed countries; approaches are often top-down with inadequate community incentives or participation in setting the standards and actual field compatibility. In addition there are wide administrative burdens, financial costs, which only a few can afford. To date, the most successful NTFP certification has either been subsidized by donors or sales of certified timber under forest (timber) certification.

The process of third-party certification for ensuring sustainability, quality of product and benefit sharing to collectors is no doubt an important step in NTFP sector. However, at present there is no clear process, and no certifying agency exists in India, except organic certifiers who are mostly involved in agriculture commodities.

MoEF and NMPB have initiated processes for grounding the concept in real sense with Indian conditions. There is a need for wide awareness generation and stakeholder consultation for understanding the positive and negative factors of certification of NTFPs, before it is implemented.

The possible threats of certification visualized by stakeholders of NTFPs are on small collectors and sustenance users, which includes their requirement to be proven and formalized as customary NTFP users, including their tenure and access rights who until now enjoy free collection.

The critical question before us is to motivate industries to use sustainable harvested NTFP material for their production.

In the process of certification, there is a need to develop strong local institution that will follow the simple rules and robust guidelines for sustainable collection of NTFPS as per their own decisions at local government with political will (Panchayat/FPC level). Then only certification will really benefit the community as well as the sustainable resource management.

CHALLENGES OF CERTIFICATION

There are a number of challenges facing certification. The most critical are the following:

- Unavailability of funds for meeting direct and indirect cost of certification. There is also difficulty in reporting and administrative procedures, which creates heavy workload for unprepared communities (Tesil, 2001).
- Lack of capacity (technical and ecological knowledge about individual species).
- Lack of baseline data, sustainable harvesting levels and resiliency levels.
- Non-equitable benefit sharing for the local communities. Unclear tenure rights and access.
- Complex chain of custody.
- Difficulty in realizing market benefits from certification.
- Lack of standard policies and difficulty with integrating timber and NTFP certification (Durst et al., 2006).

Certification standards development is an important and responsible task for a nation, which requires system integrity and credibility, and should follow the following aspects:

- multi-stakeholder supports;
- accreditation of independent third party certification bodies;
- genuine environmental and social protection;
- performance (not simply plans drawn up);
- governance and membership balance;
- transparency; and
- impact on the resource and socio-economic condition.

Opportunities of Forest Certification

- Green public procurement
- FDI in retail sector by MNCs (many of them have committed to FSC globally).
- Legal requirements in international markets: US Lacey Act (amended in 2008), EU Timber Regulation, effective from 3 March 2013, initiatives in Australia, Japan.
- Linking with programmes and projects (both govt. and corporate) to support sustainable standards.
- FSC established Standard Development Group in India in 2012 to develop national standards/local indicators for India to the FSC international standards.
- Policy guidelines on forest certification on government forests/plantations.
- Incentives for forest certification, support to small holders, farm forestry/social forestry schemes at various levels.

- Promote green public procurement to generate demand in the domestic market.
- Create awareness on the benefits of certification.
- Support the development of national standards, indicators with a view to adapt certification in the national and local contexts.

Political willingness and administrative capabilities are key to implement forest certification in India. It also requires incorporating reporting system, governance principles, transparency, accountability and involvement of local people in decision making for innovation. There is a need to allocate significant fund for research, documentation and information dissemination, as the programme needs further popularization looking at the dynamic nature of the system. Further, procedure required in Indian situation is information system development for effective M&E using modern technology of GIS/RS, science and technology intervention, quality testing, storage efficiency, harvesting tools and value addition option (Mass et al., 2001).

REFERENCES

1. Bhattacharya, P., and Hyat, S.F. (2004). Sustainable NTFP management for rural development: a case from Madhya Pradesh, India. *International Forestry Review* 6(2): 161-168.
2. Bhattacharya, P., Prasad, R., Bhattacharya, R., and Ashokan, A. (2008). Forest Certification: Development of National Standards for Sustainable Collection of MAP from Forest Area in India. Project Report Submitted to NMPB. ICCF-IIFM, Bhopal.
3. Bhattacharya, P., Prasad, R. Bhattacharya and A. Asokan. (2008). Towards Certification of Wild Medicinal and Aromatic Plants in Four Indian States. *Unasylva,* 230(59): 35-44.
4. Bhattacharya, R., Asokan, A., Bhattacharya, P. and Prasad, R. (2009). The Potential of certification for conservation and management of wild MAP resources. *Biodiversity Conservation* 18: 3441-3451, DOI 10.1007/s10531-009-9653-z (Springer).
5. Bass, S., Thornber, K., Markopolous, M., Roberts, S., and Grieg-Gran, M. (2001). Certification's Impacts on Forests, stakeholders and supply chains. *Instruments for Sustainable Private Forestry Series.* International Institute of Environment and Development, UK.
6. British International Standards (2008). www.bsi-global.com/.../management-systems/at-a-glance/what-is-certification/
7. Brown, L., Robinson, D., and Karman, M. (2002). The Forest Stewardship Council and Non-Timber Forest Product Certification: A discussion paper. Forest Stewardship Council (Secretariat) Avenida Hidalgo 502, Oaxaca 68000, Mexico.
8. Brunori, A. (2007). NWFP certification. Non-wood News No. Forest product and their marketing. http://www.unece.org/trade/timber/docs/fpama/2007/fpanr2007.pdf.

9. Chaterji, S. (2008). Developing standards for certification of medicinal and aromatic plants in Uttarakhand. Project Report, Winrock International India. New Delhi.
10. Durst, P.B., Mckenzie, P.J., Brown, C.L., and Appanah, S. (2006). Challenges facing certification and Eco-labelling of forest products in developing countries. *International Forestry Review* 8(2): 193-200.
11. Forsyth, K. (2002). Certified Forest Products Markets. http://www.unece.org/trade/timber/docs/tc-60/presentations/2-2-forsyth.ppt (accessed on June 27, 2008).
12. FSC (2002). FSC Policy Document for Preliminary Accreditation of National/Regional Standards. www.fsc.org.
13. FSC (2007). FSC Presentation April 17, 2008-9. http://www.fsc.org/fileadmin/web-data/public/document_center/powerpoints_graphs/facts_figures/08-04-01_Global_FSC_certificates_-_type_and_distribution_-_FINAL.pdf?PHPSESSID=e97ba6d4c5f100ac808688d16b162af1 (accessed on June 27, 2008).
14. Gullison, R.E. (2002). Does forest Certification Conserve Biodiversity? *Oryx*. Vol. 37. No. 2. April 2003.
15. International Standard for sustainable wild collection of medicinal and aromatic plants (ISSC-MAP). Version 1.0 Medicinal Plant Specialist Group, 2007.
16. Jain, P. (2004). Certifying Certification: Can certification secure a sustainable future for medicinal plants, harvesters and consumers in India? Traffic International, U.K.
17. Karki, M., and Rawat, R.B.S. (2004). Definitions, good practices and certification. Sustainable Forest Management, 1357-1367.
18. Katiyar, A. (2007). Group Certification for organic NTFPs : Dhamtari approach. IDRC - Canada (ICIMOD), NMPB-New Delhi, India and CGMFP Federation - CG.
19. Maas, Jelle., and Mirjam, A.F., Ros-Tonen NTFP Certification: Challenges for Research.. European Tropical Forest Research Network News. No. 32. Winter 2000-2001. pp. 71-75.
20. Mallet, Patrick, and Marion, Karmann. (2001). Certification of NTFPs : An Emerging Field. European Tropical Forest Research Network News. No. 32. Winter 2000-2001. pp. 68-71.
21. Mallet, Patrick (1999). Certifying the Harvest: Developments in NTFP certification. Forest communities in the third millennium: Linking research, business, and policy toward a sustainable non-timber forest product sector. Proceedings of a meeting held October 1-4, 1999 Kenora, Ontario, Canada.
22. Miteva, D.A., Loucks, C.J., Pattanayak, S.K. (2015). Social and environmental impacts of forest management certification in Indonesia. PLoS ONE 10(7): e0129675. doi:10.1371/journal.pone.0129675.
23. Pandey, A.K., and Mondal, A.K. (2008). Sustainable development of medicinal plants through non destructive harvesting practices. Community Forestry. RCDC. 2008.

24. Pierce, A.R., and Laird, S.A. (2003). In search of comprehensive standards for non-timber forest products in the botanicals trade. *International Forestry Review*. 5(2): 138-147.
25. Pierce, Alan., Shanley, Patricia., and Laird, Sarah (2003). Certification of non-timber forest products : Limitations and implications of a market based conservation tool. Paper presented at the International Conference on Rural livelihoods, Forests and Biodiversity, Bonn, Germany.
26. Rastogi, A., and Pant, R. (2004). FSC certification feasibility assessment report, Himachal Pradesh, India. Technical Report, WWF-India.
27. Scrippmann, Uwe., Leaman, Danna and Cunningham A.B.A. (2006). Comparison of cultivation and wild collection of medicinal and aromatic plants under sustainability aspects. R.J. Bogers, L.e. Craker and D. Lange (Eds.) Medicinal and Aromatic Plants, Springer. Chapter 6, pp. 75-95.
28. Sindhi, S., and Choudhury, P.R. (2003). NTFP certification: possibilities galore, Community Forestry. 3(1): 24-28.
29. Smart Wood (2002). Non-timber forest products certification standards addendum. Guidance Notes. November 2002. http://www.rainforest-alliance.org/forestry/documents/ntfp-addendum-generic-guidelines.pdf
30. Smith, M.W., Bowyer, J., Fernholz, K., and Howe, J. (2006). Combining organic and FSC certification of non-timber forest products reducing costs, increasing options. Dovetail Partners Inc.
31. Stevens, J., Ahmed, M., and Ruddell, S. (1997). Forest products certification: A Survey of manufacturers. http://www.certwdmkt.com/mrktStd/fpjfin95certman.coc (accessed on 2008)
32. Tesil, M.F., Platinga, A.J., Allen, T.G., and Field, D., (2001). Funding Forest Certification. Choices, 7, 1-8.
33. The Raipur Report. 2003. Chhattisgarh Forest Department led Non-Wood Forest Produce Certification Initiative - http://cgforest.nic.in/nwfp_certification_1.htm
34. Ved, D.K. and Goraya, G.S., (2008). Demand and Supply of Medicinal Plants in India, Foundation for Revitalisation of Local Health Traditions, Bangalore, p. 127; 139; 195
35. Verma, S.K. (1998). Evolving mechanism for NTFP oriented forest management. *Forest Usufructus*, 1(1&2): 1-22.
36. World Health Organization (WHO). (2002-2005). WHO Traditional Medicine Strategy [online]. Geneva, 2002 [cited 10 October 2003]. Portable Document Format. Available from Internet - http://www.who.int/medicines/library/trm/trm_strat_eng.pdf.
37. UNECE/FAO. 2006. Forest Products Annual Market Review, 2005 and 2006 editions. http://www.unctad.org/infocomm/anglais/timbertrop/quality.htm (accessed on June 27, 2008).

3

REDD+: Indian Experience Towards Sustainable Forestry

Subhash Chandra

Forest Policy & Forest International Cooperation, Ministry of Environment, Forests & Climate Change, Indira Paryavaran Bhawan, Jorbagh Road, Jorbagh Government of India, New Delhi – 110003
Email: subhaash.chandra@gmail.com

ABSTRACT

Degradation and deforestation of the world's tropical forests are cumulatively responsible for about 10% of net global carbon emissions. Therefore, tackling the destruction of tropical forests is core to any concerted effort to combat climate change. Traditional approaches to halting tropical forest loss have typically been unsuccessful, as can be seen from the fact that deforestation and forest degradation continue unabated.

REDD (reducing emissions from deforestation and forest degradation) incentivises a break from historic trends of increasing deforestation rates and greenhouse gases emissions. It is a framework through which developing countries are rewarded financially for any emissions reductions achieved associated with a decrease in the conversion of forests to alternative land uses. Having identified current and/or projected rates of deforestation and forest degradation, a country taking remedial action to effectively reduce those rates will be financially rewarded relative to the extent of their achieved emissions reductions. This paper highlights the REDD + initiatives in India, its thrust areas, and the challenges faced by the country so far.

REDD provides a unique opportunity to achieve large-scale emission reductions at comparatively low abatement costs. By economically valuing the role forest ecosystems play in carbon capture and storage, it allows intact forests to compete with historically more lucrative, alternative land uses resulting in their destruction.

INTRODUCTION

Forests are one of the most valuable renewable resources of the planet. They provide multiple benefits to the society in the form of various products and services. In India, forests are particularly important for livelihood and sustenance of forest dwelling communities, whose socio-cultural and economic life revolves around forests. The scientific forest management in India dates back to mid 19th century. India has a comprehensive policy and legislative framework for guiding forest conservation and management.

Climate change is one of the most serious threats faced by our generation, which has the potential to change the way we live on the earth. The climate change is impacting forests, which, in turn, also influences climate change. Emissions from deforestation and forest degradation constitute around 17% of the global GHG emissions. Reduced emission from deforestation and degraded forests (REDD) is an international effort to create financial value for the carbon stored in forests and to provide incentives for developing countries to reduce emissions from forested lands for mitigating climate change and invest in low-carbon paths to sustainable development.

Though, REDD+ benefits are co-benefits in addition to the vast products and ecological services from forests which include both monetized as well as non-cash benefits. In Indian context, the financial benefits arising from implementation of the REDD+ programmes would help the local communities in their efforts in conservation and management. Though, internationally REDD+ architecture is still evolving, efforts must be directed to prepare the country for implementing REDD+ programmes effectively so that in future, India can expect from REDD+ international mechanism for its pro-conservation and pro people policies.

SUSTAINABLE FOREST MANAGEMENT

The definition of Sustainable Forest Management (SFM) as adopted by the Food and Agriculture Organization (FAO) is *"The stewardship and use of forests and forest lands in a way, and at a rate, that maintains their biodiversity, productivity, regeneration capacity, vitality and their potential to fulfill, now and in the future, relevant ecological, economic and social functions, at local, national, and global levels, and that does not cause damage to other ecosystems."*

SFM aims to bring a balance between society's increasing demands for forest products and benefits, and the preservation of forest health and diversity which is critical to the survival of forests, and to the prosperity of forest-dependent communities. It means that the goods and services derived from the forests meet present-day needs while at the same time securing their continued availability and contribution to long-term development. It encompasses the administrative, legal, technical, economic, social and environmental aspects of the conservation and use of forests. The seven thematic elements or key components: are: (1) extent of forest resources; (2) biological diversity; (3) forest health and vitality; (4) productive functions of forest resources; (5) protective functions of forest resources; (6) socio-economic functions; and (7) legal, policy and institutional framework.

A BRIEF HISTORICAL PERSPECTIVE ON REDD+

The UN's Intergovernmental Panel on Climate Change (IPCC) AR4, 2007 estimated that the forestry sector and other sectors that impact land use through deforestation, forest degradation and other changes in forests contribute approximately 17% of GHG emissions or approximately 5.8 Gt of CO_2 equivalent per year, mainly in tropical developing countries. REDD+ as a concept has generated lot of interest and possibilities on climate change mitigation and adaptation. In COP 11, Montreal in 2005, a coalition of nine rainforest nations, headed by Papua New Guinea and Costa Rica, sought attention to the growing threat of deforestation, and introduced the concept of reducing emission from deforestation (RED). At COP 12, 2006 at Nairobi, India proposed a new policy approach namely, 'compensated conservation' intended to compensate the countries for maintaining and increasing carbon pools of their forests as a result of conservation and increase/improvement in forest cover backed by a verifiable monitoring system. This approach was finally recognized at COP 13, Bali in 2007 Action Plan1/CP (Para 1(b) (iii).

The COP 16, Cancun, 2010 marked the official entry of REDD as a UNFCCC-enabled mitigation mechanism linking deforestation, degradation, conservation and enhancement of forest carbon stocks and sustainable management of forests with reduction of GHG emission. The COP 19 at Warsaw, 2013 agreed on a Warsaw REDD+ framework. It reaffirmed that results-based new, additional and predictable finance, from a variety of sources, public and private, bilateral and multilateral, including alternative sources, to be provided to developing countries for the implementation of REDD+. It emphasized that the result-based progression is to occur for all phases of action and activities of REDD+ as agreed by COP16 at Cancun. The result-based actions have to be fully measured, reported and verified. This will include development and implementation of national strategies or action plans, and capacity building in a phased manner. All REDD+ elements have to be in place in a developing country that undertakes result-based actions. A country has to address safeguards and provide information on them for receiving result-based payment.

The CoP 19 at Warsaw has given positive signals for taking forward the REDD+ action at global level. There is a need to analyse several critical aspects of REDD+ including forest governance, technical aspects of measurement, reporting verification, institutional arrangements, livelihoods, carbon stock assessment, biodiversity and ecosystem services to operationalize REDD+ in the country. Capacity building of forest officials and communities is important for the implementation of REDD +, and it is important that such ideas are ingrained in planning and implementation processes. Forthcoming CoP 21 at Paris is expected to give further fillip to REDD+.

STATUS OF INDIA'S FOREST RESOURCES AND MANAGEMENT PRACTICES

India holds 10th position in terms of forest area in the World as per GFRA, 2010, and is represented by 16 major forest types and 251 sub-forest types. As per India State of Forest Report, 2011, the total forest cover of the country is 692,027 km^2 which works out as 21.05% of the geographical area of the country. The total forest and tree cover constitute 78.29 m ha, i.e., 23.81% of the geographical area which includes 2.76% of tree cover including 21.4 mha protected forests. In terms of density classes, area covered by VDF is 83,471 km^2 (2.54%), that by MDF is 320,736 km^2 (9.76%) and that by OF is 287,820 km^2 (8.75%).

Forests are managed by State Forest Departments. India is one of the 17 mega-diverse countries with 4 global biodiversity hotspots. The protected area network includes 102 national parks, 515 wildlife sanctuaries, 47 conservation reserves, 4 community reserves, etc., extending over 16 million ha 4.9% of geographical area of the country. Total growing stock of India's forest and ToF is estimated as 6,047.15 cum. Total bamboo bearing area in the country is estimated to 13.96 mha. Total carbon stock of the India's forest is estimated to be 6663 million tonnes. India is one of the few countries of the world to have a robust and scientific system of periodic forest cover assessment.

An analysis of forest cover data of India shows that a significant part of the forest cover falls in the open to medium categories in various degree of degradation. There is great scope for improvement of forest cover by addressing drivers of degradation with people's participation and cross-sectoral support for sustainable forest management. Forests neutralize 11% of India's GHG emissions. India added around 3 mha in last decade 1997-2007). India is world's 4th largest economy and 5th largest GHG emitter, accounting for about 5% of global emissions. Its emissions increased 65% between 1990 and 2005 and are projected to grow another 70% by 2020. On a per capita basis, India's emissions are 70% below the world average and 93% below those of the United States. A number of policies and programmes in the country contribute to climate mitigation by reducing or avoiding GHG emissions. Basic information about India's forests is given hereunder.

FOREST MANAGEMENT IN INDIA

India's scientific forestry dates back to 1860's with one of the oldest forest services in the world. The emphasis on management of forests has been based on the principle of sustained yield effectively implemented through the instrument of working plans. The forestry sector has over the years developed considerable institutional strength and technical expertise. Country's forests have been managed by professionally trained foresters supported by scientific and research institutions. Forest Survey of India carries out biennial assessment of country's forest cover since 1987. Presence of various forestry institutions and enabling policy framework will give India an early advantage in implementing REDD+ programmes.

India's forest and tree cover has increased in recent years and the total forest and tree cover amounts to 24% of the geographical area of the country. Over the past two decades progressive national forestry legislations and policies of India have transformed India's forests into a net sink of CO_2. With its focus on sustainable forest management, afforestation and regulating diversion of forest land for non-forest purpose, India plans to increase its carbon stock. Government of India's long-term goal is to increase its forest cover through a planned afforestation drive which includes a number of programmes and initiatives like Green India Mission, Green Highways Policy, Financial Incentive for Forests, Plantation along Rivers, REDD-Plus and other Policies and Compensatory Afforestation Fund Management and Planning Authority. For the first time devolution of funds to states from the federal pool would be based on a formula that attaches 7.5% weight to the area under forest. It takes into account the changing realities in order to rebalance the fiscal system of the country in a way that will incentivize greener distribution of resources. This initiative will give afforestation a massive boost by conditioning about USD 6.9 billion of transfers to the states based on their forest cover, which is projected to increase up to USD 12 billion by 2019-20.

PRESENT POLICY ENVIRONMENT AND ITS APPLICABILITY FOR REDD+ FRAMEWORK

The forestry subject is in the concurrent list of Schedule-VII of the Constitution of India, therefore, both State and Central Governments can make laws. The Central Government deals with policy and regulatory issues at broad level, which in advisory capacity provides policy directions, guidance to the states. The day-to-day functions of protection, management and conservation of forests lie directly with the state government. The Indian Forest Act, 1927 provides for law relating to consolidation of forests, transit of forest produce, duty leviable on timber and other forest produce, empowering forest officers with certain powers to ensure forest protection. The Wildlife (Protection) Act, 1972 provides for conservation, protection and management of Wildlife. The Forest (Conservation) Act, 1980 regulates diversion of forest land for non-forestry purpose. Recently enacted "the Scheduled Tribes and Other Traditional Forest Dwelling Communities (Recognition of Forest Rights) Act, 2006", aims to recognize and vest the forest rights and occupation in forest land of forest dwelling Scheduled Tribes and other traditional forest dwellers, who have been residing in such forests for generations, but whose rights could not be recorded. This legislation has also empowered forest dwelling communities in managing and conserving forest resources.

The National Forest Policy, 1988 (NFP) aims to ensure environmental stability and maintenance of ecological balance which are vital for sustenance of all life forms—human, animal and plant. The policy makes derivation of direct economic benefit subordinated to this principal aim. The policy aims at maintaining one-third of the country's geographical area under forest or tree cover, and calls for massive afforestation and social forestry programmes for increasing the forest/tree cover in the country and also towards meeting the domestic requirements of forest products including fuel wood, fodder, minor forest produce and small timber of the rural and tribal populations. NFP suggests increasing the productivity of forests to meet essential national needs and encouraging efficient utilization of forest produce through the application of scientific and technical inputs. All forests are to be managed as per approved working plans.

The policy provides protection of rights and concessions of tribals and other poor living in and around forests while making their domestic requirements of fuel wood, fodder, minor forests produce and construction timber to be the first charge on forest produce, within the carrying capacity of the forests. The policy recommends to closely associate the tribal people in the protection, regeneration and development of forests as well as to provide gainful employment to people living in and around the forests.

REDD+ AND ITS RELEVANCE TO INDIA

REDD+ goes beyond REDD by including financial incentives for positive elements of conservation, sustainable management of forests and enhancement of carbon stocks. It provides for flow of positive incentives for demonstrated reduction in deforestation or for enhancing quality and increase of forest cover and incorporates important benefits of livelihood improvement, biodiversity conservation and enhancement in environment and ecological services, etc. The REDD+ architecture is still evolving, REDD (Reducing Emissions from Deforestation and Forest Degradation) is the global endeavour to create an incentive for developing countries to protect, better manage, and save their forest resources, thus contributing to the global fight against climate change. REDD+ goes beyond merely checking deforestation and forest degradation, and includes incentives for positive elements of conservation, sustainable management of forests, and enhancement of forest carbon stocks (FCS).

Livelihood and Sustenance from Forests: There are around 300 million forest dependent people, including around 87 million tribals, deriving their livelihood and sustenance from forests. Joint Forest Management (JFM), a partnership between local communities and the Forest Department, which started in India in the nineties, has been a successful model under the "Care and Share" principle with over 100 thousands JFM committees involving around 20 million people managing over 22 million hectares of forest area (Bhattacharya et al., 2010).

Scope of Forests for Mitigation of Climate Change: There is significant scope for improving quality of forest cover by addressing drivers of degradation, as a significant part of the country's forest cover falls in the open to medium categories owning to various drivers of degradation.

In its recently submitted Intended Nationally Determined Contribution (INDC) to UNFCCC, India has shown its initiative by setting fair and ambitious targets for itself to work towards a low carbon emission pathway while pursuing the goals of socio-economic development. In spite of the huge tasks of attaining faster, equitable and sustainable economic growth for generating employment, poverty reduction, food security and health for all, India has declared a voluntary goal of reducing the emissions intensity of its GDP by 20–25%, over 2005 levels by 2020, despite having no binding mitigation obligations as per the Convention. United Nations Environment Programme (UNEP) in its Emission Gap Report 2014 recognized India as one of the countries on course to achieving its voluntary goal.

India has a good network of national institutions in the field of forestry and wildlife. Indian Council of Forestry Research and Education (ICFRE) with its eight forestry research institutes in different parts of the country provides forestry research support. Indira Gandhi National Forest Academy imparts training to Indian Forest Service Officers, while Central Forest Academy/State Forest Service colleges provide training to State Forest Service Officers. There are training colleges for field frontline staff like Forest Rangers and Forest Schools for field foresters. Wildlife Institute of India carries out research in wildlife. Indian Institute of Forest Management imparts training in forest management, education and research. These national level institutes have played an important role in capacity building, research support and providing technical capacity for forest management in the country.

India accords equal emphasis on economic development, social inclusion and environmental sustainability as equal and critical components of sustainable development. There are more than 300 million forest dependent people including around 87 million tribals, deriving their livelihood and substantial part of their income from forests. India has a long history of forest partnership with local communities who have contributed significantly in conservation efforts. Joint Forest Management (JFM), a partnership between local communities and the Forest Department started in India in the eighties, has been a successful model under the "Care and Share" principle. There are around 116 thousands JFM Committees involving around 20 million people managing approximately 22 million hectares of forest area. JFM has contributed significantly in stabilizing forest and tree cover of the country.

REDD+ has specifically opened the possibilities for India to expect compensation for its pro-conservation approach and sustainable management of forests which will lead to increase in forest cover and thereby its carbon stock. India's national position on REDD+ underlines that carbon service from forest and plantations is one of the co-benefits and not the main or the sole benefit. This position is similar to what is stated in the National Forest Policy, 1988, which aims to ensure that environmental stability and maintenance of ecological balance are vital for sustenance of all life forms. The policy states that the derivation of direct economic benefit must be subordinated to this principal aim. National Forest Policy envisages greater involvement of people in sustainably managing forest resources to strengthen their sustenance and livelihood security. REDD+ offers an additional incentive to the community for working towards the conservation of forests. We should aim to facilitate passing on of the financial incentives under REDD+ directly to the community which would motivate community for conservation of forests as co-benefits. These benefits will further encourage the local communities which are partners in conservation and sustainable management of forests.

TOWARDS INDIA'S PREPAREDNESS FOR REDD+

USAID Forest PLUS Programme

Under the bilateral programmes between India and the USA, a Partnership Agreement for Sustainable Forests and Climate Adaptation Programme, Forest PLUS (Partnership in Landuse Programme) is being implemented since 2011. The Forest-PLUS programme aims to strengthen technical capacity of forestry professionals in sustainable forest management and implementation of REDD+. The programme provides an opportunity to bring together Indian and US expertise to develop technologies, tools and methods of forest management to meet the technical challenges of managing forests for ecosystem health, carbon stocks, biodiversity, and livelihood. The programme is divided into

2 components. **Component one** focuses on US-India collaborative scientific research and exchanges that explore methods and approaches to implement REDD+. This will be vital for India's own programmes and will further demonstrate India's advances in science and technology to develop lessons that may be applied in other developing countries. Component one has focus on three tasks: (i) improving forest management practices to increase sequestration; (ii) developing methods to conduct carbon inventories and create carbon stock baselines; and (iii) analyzing social and economic incentives for improved forest management. **Component two** will share the results from component one with relevant stakeholders for feedback and refinement. Under this component, cost-effective, geographically relevant and culturally appropriate methods and approaches will be piloted in the field at scales that lend the results and learning to replication at higher levels. Component 2 focuses on three tasks: (i) establishing government and stakeholder dialogue and communication processes; (ii) engaging stakeholders constructively in REDD+ implementation; and (iii) strengthening human and institutional capacity and the enabling environment. The activities under this agreement have been organized in two phases.

Phase 1: Conducting training sessions to bolster Indian expertise in completing carbon estimations, forest inventories, and related analyses.

More specifically, Indian forestry professionals will be trained on:

(i) remote sensing applications to map and estimate forest cover, and estimate burn severity;

(ii) forest inventory and monitoring; and

(iii) establishing baselines/reference levels.

Phase 2: Set up pilot projects to demonstrate methodologies, tools and technologies in the Indian context.

Activities in phase 2 will build upon momentum generated during phase 1 to implement and improve data collection and decision-making tools for forest resources (including carbon) assessments, and monitoring and management at multiple scales. The following activities would be undertaken:

(i) pilot project for carbon estimations in mangrove systems of india; and

(ii) pilot testing carbon estimation methodologies.

Under the Forest PLUS Programme, four pilots projects are being undertaken in the states of Himachal Pradesh, Karnataka, Madhya Pradesh and Sikkim. The programme is helping REDD+ preparedness in the country through capacity building of forestry professionals in forest resource assessment, growing stock and forest carbon estimation, exposure to new remote sensing techniques including SAR, LIDAR, etc., ecosystem restoration, and ecological serves valuation.

India has recently joined UN REDD Programme to learn from international development in REDD+ and build capacity for implementation of REDD+.

Green India Mission

India has recently initiated National Mission for Green India (GIM). The programme highlights the significance of forests in ensuring food and water security along with livelihood security. Green India Mission considers holistic greening in the context of climate change adaptation and mitigation, improvement in ecosystem services and biodiversity conservation at a landscape level, such as CO_2 sequestration to the tune of 50-60 m tonnes/annum.

GIM aims to achieve afforestation and restoration of identified vulnerable landscapes of 10 million hectares into sustainable ecosystems and building resilience to tribals and rural poor inhabiting these vulnerable landscapes to climate change effects. GIM is pro-active to ensure convergence with other natural resource and socio-economic development programmes of the country with local communities, as the main stakeholders, at the centre, based on participatory approach.

The REDD+ offers an opportunity to manage forests sustainably for multifunctional role based on scientific approach to take leading role in addressing the challenges faced in view of threats due to climate change, increasing vulnerability of the poor due to degradation of forests, deforestation and depletion of natural resources. The enormity of challenges is quite daunting in the face of growing population, increasing prosperity, rapidly increasing demands for forests products and lands for meeting other demands of society aspiring faster development. REDD+, also highlights close interrelationship of sustainable forest management and economic development. The REDD+ as an added carbon incentive, is expected to create more economic opportunities and harnessing the true potential of the forestry sector through appropriate interventions like increasing productivity of forests, afforestation, investment in Tree outside Forests (ToFs) like agroforestry, enhanced availability of forest products, value addition, creation of green jobs in forestry activities including processing of forest products, research, etc., thus increasing greening of economy.

REFERENCES

1. Bhattacharya, P., Pradhan, L., and Yadav, G. (2010). "Joint forest management in India: Experiences of two decades" in *Resources, Conservation and Recycling* 54(2010) 469-480.
2. Global Forest Resource Assessment, 2015, food and Agriculture Organisation of United Nations.
3. National Forest Policy, 1988.
4. India State of Forest report, 2013.
5. Reference Document for REDD+ in India, 2014. Government of India.
6. The Scheduled Tribes and Other Traditional Forest Dwellers (Recognition of Forest Rights) Act, 2006, Government of India.

7. The Indian Forest Act, 1927.
8. The Forest (Conservation) Act, 1980.
9. National Agroforestry Policy, 2014, Government of India.
10. UN REDD Programme National Forest Monitoring Systems: Monitoring and Measurement, Reporting and Verification (M&MRV) in the context of REDD+ Activities.

4

Sustainable Harvesting of Non-Timber Forest Products: A Difficult Subject to Achieve

Prodyut Bhattacharya

University School of Environment Management
GGS Indraprastha University New Delhi–110078
E-mail: prodyutbhattacharya@yahoo.com

ABSTRACT

A sustainable system for extracting non-timber forest resources is one in which leaf, roots, fruits, gums, barks, nuts, and other products can be harvested indefinitely from a defined area of forest with negligible impact on the species being exploited. Harvesting of NTFPs, therefore, is a key area of management in the tropical forests. Sustainable harvest of a resource ensures an undiminished flow of benefits from the resource to its users over time so that it does not have any deleterious effect on regeneration of target population or any other species in the community or on ecosystem structure and functioning. This paper focuses on sustainable harvesting practices of various NTFP species of ecological and economical importance. How the management of NTFPs can be integrated within the working plan of forest division is also described. Information on sustainable NTFP resource management needs much attention at national level for the products available in different forest types in tropical, subtropical, temperate and alpine zone. It focuses on urgent need for development of baseline data and for large-scale plantation under Green India Mission to support NTFP sock improvement for next 10 years. Integration of NTFP in forest management is necessary to achieve under sustainable forestry and the same is highlighted in the 12th Plan of Planning Commission of India. Forestry technologies are developed with timber values uppermost, and now it is vital to develop forest management technologies that take into account both timber and non-timber values.

INTRODUCTION

A healthy forest can provide many benefits to people, more than timber/woods—a most common product of forest. Initially, in India the NTFP collection from the wild was less in quantity up to 1990, but gradually the market demand forced excessive resource extraction. Foresters initially were confident about natural replenishment of this stock in forests, but in reality sustainable harvest of NTFP requires quite a bit of careful step-by-step management, controlled harvesting and period monitoring for sustainability of resources. The controlled exploitation of NTFPs holds great potential as a method, for integrating the use and conservation of resources. Though there are more than 800 types of NTFPs harvested in India, reliable harvest data of only a few are available, which are often in the category of nationalized by the state governments. Because of this harvest/collection of large number of NTFPs by local people go unrecorded in government data bank (Forest Sector Report India 2010).

NTFP plays the crucial role of safety nets for the poor who rely on forest for their dependence. It contributes to food security, health cover, cash income and also provides dependable wage labor to local people in remotest areas where income opportunities are almost nil. In rural areas, marginalized, landless and very poor people in different states of India meet their livelihood requirement from NTFP's which varies from 35-80% based on several factor (Bahuguna, 2000; Gadgil 2003; Bhattacharya and Hyat 2004). Women are mostly dependent on NTFPs due to the nature of its production, quantity, collection procedure, processing and local selling. They use most of NTFPs for their own consumption and they have better knowledge on resources. Few research based studies give the information, like in South India Kerala, Karnataka and Tamilnadu, women use 62 different NTFP items in their food as vegetable and regular medicine for healthcare. Similarly, in North Eastern part like in the state of Tripura, Arunachal and Meghalaya, women regularly go to forest areas and wetlands for collection of leafy vegetables from the range of 85 items in the list, surprisingly all of which are uncultivated sources. Many primitive tribes of India still make their huge share of nutritional sources from range of NTFPs in different forest types. Most NTFPs are seasonal and we observe that there are various cultural rituals attached in society, which correspond with the availability of these NTFPs, either fruits, leaf, flower, twigs , roots/rhizomes and various other animal NTFPs.

Historically, the NTFP sector was neglected for many decades from the mainstream forestry and they were considered as 'minor', till the importance was recognised. Though commercial dimension of NTFP is very impressive, traditionally that serves meaningful manner in various parts of India. Export of NTFPs and its products contributes 68% out of the total export from forestry sector.

Herbal product raw materials from NTFP source contribute 90% of the supply for the industry which are practically sourced from natural forests. About 7000 plants are used in Indian system of Medicine; 960 of these items have been recorded in trade, and 178 are traded in high volumes in quantities exceeding 100 MT per year. According to a study, a total annual demand of botanical raw drugs in the country for the year 2005-06 was estimated as 3,19,500 metric tonnes (MT) with corresponding trade value of 1069 crores, while according to another study the Indian share of global medicinal plants trade is increasing at an annual growth rate of 23%. India stood 3rd biggest exporter of medicinal plants during 2009 after China and Canada respectively (NMPB, 2011).

Due to various reasons like overharvesting, premature collection, and climate change (extreme drought and flood situation), production of several NTFP has gone down. Excessive demand by the herbal industries forced premature product collection from various forest ranges through local villagers. Local people are engaged by traders and suppliers for collection of selected products. Due to destructive collection practices their regeneration is going down and plants are coming under threat. Studies from original field survey give a picture of reduced production from different forest division from 30 to 85% in last 30 years time. Biodiversity experts observed the population size of such trees is coming down rapidly. The associated animals, honeybees and bird species and their population have also gone down drastically, which has created negative impact on reproductive biology of several NTFP tree species (Bhattacharya, 2013).

All the forests in India are managed under the prescription of the working plan prepared on the basis of principles of sustainable forest management and recognized innovative silvicultural practices, which is recognized by National Working Plan Code, MoEF (Anon, 2004). Indian forestry has an established system of planning in the form of working plan; this system commenced during the period 1871-1900. A working plan is a written scheme of management aiming at continuity of policy and action and controlling the treatment of a forest (BCFT). This document is a means of enforcing systematic obligatory and mandatory regulations for continuous management of a given forest property. NTFP harvesting in various forest divisions now started bringing under working plan, tries to answer four fundamental questions in forestry, viz., how much, where and when to satisfy the given objective of management and direct the forest development towards normalcy and tries to achieve sustained production and yield from the forest under the plan. The working plans are prepared within the framework of national policy and five-year development plans and other considerations of national and regional level planning.

It broadly involves the following:

- Survey and assessment of the past results, present facts, resources and constraints.
- Analysis of the facts with respect to prospects leading to conclusion on policy, objects of management.
- A plan for future action in definite period in conformity with the long-term plan.
- Provisions for control of prescriptions for maintenance of records and collection of new facts by research and new techniques (Prakash, 1986).

NTFP based benefit sharing models initiated in five states initiated of India which started in 1970s, these states are namely Madhya Pradesh, Chhattisgarh, Andhra Pradesh, Odisha and Uttarakhand, but they yet have to establish their sustainable harvesting models. Various activities of micro-enterprise effort based on NTFPs are available in the respective selected states. Different stakeholder groups of NTFP (primary collectors, traders, processing units, and FD staff) primarily observed that institutional innovation has changed the trade situation in a big way and further manifold improvement is possible. Primarily, state governments of the above mentioned five states have established separate organizations which are dedicated to the procurement, primary processing, storage and further marketing of raw products for NTFPs, and recently government of India has started the Minimum Support Price (MSP) procurement for 12 items in 8 states which aims to benefit 100 million people. This step is a long desired policy step in NTFP sector, but there is a chance that without having a sustainable harvest model, MSP may lead to more unsustainable collection of NTFPs in respective areas.

The demand for certified products is increasing in European and American markets, and India being one of the major exporters of NTFP to these markets, the need for certifying its products cannot be ignored. The demand for certified products calls for scientific certification system to ensure that the resources are managed sustainably (Parrotta 2002, Bhattacharya and Prasad 2008). In spite of this increasing demand, the medicinal plants are still not given due recognition in Forest Working Plan in the country which is still timber-oriented, as a result of which there is a threat for sustenance of the NTFPs in view of the ever increasing demand in the national and international market, exploitative and unsustainable harvesting practices and discrepancies in the management plan.

NTFPs are collected from the wild without any awareness of the stage of maturity and parts required to be collected. It is essential to ascertain the quality of a plant material before it is employed, which is governed by the following factors:

1. Place from where the plants are collected.
2. Climate, soil, habitat.
3. Ignorance about genuine raw drug, its appropriate time of collection, etc.

As the price paid to the dwellers tends to be low, they apply destructive methods to generate their income. A critical factor in the wild harvesting is the non-availability of skilled labour for good collection of herbs. Unmanaged collection practices result in depletion of medicinal plants population, and the outcome of drugs is reducing day by day.

State-wise, there are certain norms for collection of medicinally important plants. There are also restrictions on collection of some threatened medicinal plants species, collected from the wild sources. These species must be protected in nature through in situ conservation and ex situ cultivation, and existing laws. However, unmanaged and over-harvesting/urbanization, habitat loss and increasing demand have caused considerable threat to large number of NTFPs. This has compelled all stakeholders and government to give a serious thought to evolve methods of sustainable harvesting practices to ensure sustainability. Looking at the concern for its conservation and sustainable management is the purpose of writing this paper.

Objectives of Sustainable Harvesting

The main objectives of sustainable harvesting of NTFPs are as follows:

1. To ensure conservation of species through sustainable harvest.
2. Long-term availability of various products in natural ecosystem, making balance between different biodiversity components.
3. To enhance earning of local people in a sustainable manner.
4. To eliminate drudgery to women and children mostly engaged in the collection of NTFPs.
5. Capacity building among the local people towards conservation of natural resources and organizing themselves for better marketing access and to eliminate the possibility of distress selling of NTFP.

CONCEPTS OF SUSTAINABILITY, SUSTAINABLE FOREST MANAGEMENT AND SUSTAINABLE HARVESTING

Concept of Sustainability

Sustainability is the ability of the system to maintain a certain well-defined level of performance over time, and if required, to enhance the same in response to changing needs, either by itself or through linkages with other system without damaging the long-term productivity of its resource base and essential ecological integrity of the system (Jodha, N.S., 1995).

DEFINING SUSTAINABLE FOREST MANAGEMENT

International Tropical Timber Organization (ITTO, 1997) defines Sustainable Forest Management (SFM) as "The process of managing forest to achieve one or more clearly specified objectives of management with regard to the production of a continuous flow of desired forest products and services without undue reduction of its inherent values and future productivity and without undue undesirable effects on the physical and social environment".

Sustainable Management of NTFPs

The basic idea behind sustainable harvesting is that a biological resource should be harvested within the limits of its capacity for self-renewal. More than that, the manner of its harvest should be such as not to degrade the environment in other ways. The dire consequences of unsustainable harvesting of some biological resources are evident today on the global scale.

Harvesting of NTFPs, therefore, forms a key area of management in the tropical forests. Very little research is done in this area and hence data and information related to this area is lacking. Since this area is already much dented leading to the extinction of some of the valuable NTFPs, there is an urgent need to have some scientific basis of management of NTFP within a reasonable period of time to prevent further damage to these natural resources.

Defining Sustainable Harvesting

A sustainable system for exploiting non-timber forest resources is defined, as one in which fruits, nuts, latexes and other products can be harvested indefinitely from a limited forest area with negligible impact on the species being exploited.

Sustainable harvest of a resource ensures an undiminished flow of benefits from the resource to its users over time in such a way that the harvest of NTFP does not have any deleterious effect on regeneration of target population or any other species in the community or on ecosystem structure and functioning. A broad definition of sustainability includes inter-generational equity, i.e., maintenance of site values and opportunity for future options for use.

Types of Unsustainable Harvesting

1. Destructive harvesting – e.g., uprooting the whole tree/plant or devoiding the tree from bark, cutting, lopping, debarking, tapping for gum/ resin, latex, etc.
2. Premature harvesting – plucking of fruits before maturing, resulting in low regeneration.
3. Overharvesting – extracting more than the area's productivity.

The trees, shrubs, herbs and climbers yielding medicinal forest or seeds and found in natural forest are subjected to several problems, some of which problems are as follows.

Tendency to collect unripe fruits

In most states, collection of fruits and seeds from the forest is free to public; any person can go to forest and collect forest fruits. Therefore, a large number of people go to forest for collecting forest fruits. Further, these collectors are instigated by the pharmaceutical companies and other industries using these fruits for preparation of these products. So there is a competition among the collectors and also among the industries for the collection of maximum quantity of fruits, with the result that immature fruits are collected from the forest. For example, Aonla fruit ripens during February, but many times collection begins from October-November in central India. Similarly, other fruits such as bel, chironji, harra, bahera, etc., are also collected much before these are ripened. The situation is more serious in fruits which are in demand by pharmaceutical and other industries. The problem is not so severe in case of fruits collected and used locally. Even after the formation of JFMCs, the situation has not improved. The collection of unripe fruits is common in almost all important fruits (Prasad et al., 1999).

Tendency to damage the tree

It is seen that most trees are damaged by the collectors while collecting the fruits. The branches are lopped; sometimes even the tree is cut disregarding the fact that the tree, if cut, would not give any fruit from the next year. The collectors look for the easiest way of collection. They do not want to climb the trees as many trees are difficult to climb and their collection targets are high. Therefore, the collectors tend to fell the branches, and sometimes even the tree, unmindful of the damage being caused to the tree and the ecosystem. The damage is more pronounced in larger trees. In shrubs and herbs, which can be handled from the ground, the damage is not that significant. In herbs and annuals, the tendency is to uproot the plant along with stem and branches like for Anantmul, Asperagus, Kalmegh, etc.

New recruitments are poor / absent

In general, regeneration in forests is poor or absent in most forest types. The medicinal plants which occur in different forest types are no exception. New recruitments and younger plants of trees and shrubs are generally absent. Repeated fires, uncontrolled grazing, illicit removal, etc., cause serious damage to the new recruitment and younger plants. The density of herbs, creepers and climbers is gradually reducing. Several plants occurring as a ground flora are highly susceptible to fire and grazing. Firstly, the fruits/ seeds that are medicinally important are collected on a large scale and only a small percentage of fruits fall on the ground. These fruits, being unripe, have poor viability and germination capacity resulting in poor germination. Even if some seeds germinate, the biotic pressures like fire and grazing kill the newly germinated seedlings. The result is no recruitment and less regeneration. The situation, therefore, is serious, and after a few years when present fruit yielding trees die, there will be no tree to take their place and fruit yield will reduce and there will be no yield one day.

Determining Sustainable Harvesting Limits

The conservation of commercially important forest resources in state owned natural tropical forests is a challenging task because these forests constitute a common property resource and local people have the right to free access to collect NTFPs. Owing to increased commercial utilization of forest resources, local user communities are encouraged to over-exploit forest products, ignoring the traditional practices of sustainable harvesting of utilizable resources from natural forests. In the prevailing forest management system, extraction of NTFPs is not at all sustainable, either in ecological or economic terms. A participatory approach involving local forest dependent user community seems to be an inevitable tool for sustainable management and *in situ* conservation of indigenous forest resources.

Keeping this in mind, Madhya Pradesh State Forest Research Institute, Jabalpur, India, has taken the lead to determine the sustainable harvesting limits of over-exploited NTFPs in natural forests, employing an integrated participatory approach in the tribal dominated forests of Madhya Pradesh. It used different treatment plots with 20%, 40%, 60% and 80% levels of harvesting. The result of various levels of harvesting is given in Table 4.1.

Table 4.1 Various levels of harvesting of NTFP

S. No.	Treatment plots	Sustainability index		
		2005/06	2006/07	Average
1.	T_1 (20% harvesting)	3.54	1.91	2.723
2.	T_2 (40% harvesting)	2.09	2.15	2.121
3.	T_3 (60% harvesting)	1.95	2.24	2.094
4.	T_4 (80% harvesting)	1.193	1.783	1.488

From the table it is clear that 80% harvesting gives maximum productivity. Therefore it was recommended that full protection should be observed for having maximum productivity of 80% harvesting and maintaining sustainability. Whereas some studies recommend harvesting around 50% product, rest should be left for its ecosystem functions as thumb rule (Murali, et al.1996, Bhattacharya et al., 2008).

In theory, any harvesting plan should consider the availability of resources, the rate of use, and the renewal rate of the resources. Thus, only plans that promote harvest below the resource regeneration threshold and that do not modify the natural prevalence of the harvested species should be used (Lawes & Obiri, 2003). Seedling mortality is high and over 90% of the new seedlings usually die before becoming established in the understory. Only a small fraction of these seedlings (less than 1 in 100,000) will make it to the canopy and produce fruits, so there are many biological factors we need to consider to understand sustainable harvesting (Table 4.2).

Table 4.2 Overall framework of NTFP and conservation biology parameters understanding harvesting

	Low	*Medium*	*High*
Resource Group	Bark, stem tissue, roots	Some resins, fruits and seeds	Latex, fruits and leaves
Yield/Plant	Low	Medium	High
Species characteristics:			
Flowers	Few, large	Intermediate	Small, many
Fruits	Few, large	Intermediate	Small, many
Seed germination	Low viability	Intermediate	High viability
Sprouting capability	None	Low	Low
Population structure:			
Size-class distribution	Type III curve	Type II curve	Type I curve
Tree density/hectare	0-5 adults	5-10 adults	10+ adults
Spatial distribution	Scattered	Clumped	Homogenous
Regeneration guild	Early pioneer	Late secondary	Primary
Flower/Fruit Phenology	Unpredictable	Supra-annual	Annual
Reproduction biology:			
Pollination	Biotic, with specialized vector	Biotic, with generalized vector	Abiotic
Pollinator abundance	Rare; bats, humming birds	Intermediate; beetles, moths	Common; small insects
Seed dispersal	Biotic, with specialized vector	Biotic, with generalized vector	Abiotic
Disperser abundance	Rare; large birds, primates	Intermediate; small mammals	Common; bats, small birds

Source: Charles M. Peters (1994)

Sustainable Harvesting Rate

Maximum percentage of harvesting at which maximum value of regeneration is recorded to determine the capacity of regeneration of selected medicinal plants after harvesting (annual & biannual).

$$\textbf{Regeneration Index} = \frac{\text{No. of plants regenerated}}{\text{No. of plants harvested}} \times 100$$

Extent of regeneration can be an indicator of sustainable harvesting of NTFP.

Sustainability index can be defined as SI = (Nh + Nr)/ Nt

Nh = No. of plants harvested

Nr = No. of plants naturally regenerated

Nt = Total no. of plants before harvesting

For sustainability, the value of SI should be > 1

Difficulties in Sustainable Harvesting

There are a variety of different ecological characteristics that can make sustainable harvesting a very difficult objective to achieve. The major problem areas are:

- the high diversity and low population density of plant species;
- the irregularity of flowering and fruiting;
- the importance of animals for pollination and seed dispersal;
- the high mortality and low success rate during seedling establishment; and
- the sensitivity of population structure to changes in the level of natural regeneration

In order to conserve valuable species of medicinal value and other NTFP yielding species, harvesting of plants and their products have to be sustainable. This paper, therefore, focuses on sustainable harvesting practices of various species of NTFP importance, even for ecological and economical importance. The plant species have been categorized on the basis of the parts used for NTFP purpose which are as follows:

- Fruit
- Flowers
- Leaves
- Root/rhizome leaves and flower
- Bark
- Gum/resin
- All plant parts being medicinal

The complete process of sustainable exploitation may have various steps like: (i) species selection, (ii) forest inventory, (iii) yield condition, (iv) regeneration surveys, (v) harvest assessment, (vi) harvest adjustments.

Important Indicators of Sustainable Harvesting

- Size class distribution of trees is an important indicator of regeneration rate.
- Monitoring of seedling, sapling and adult tree densities help in appraising the sustainability of current harvest.
- While monitoring the size class distribution, if density drops below these values, the intensity of harvesting should be reduced.

PROTOCOLS FOR NTFP HARVEST

- Plant species should be harvested under the best possible conditions avoiding wet soil, dew, rain or exceptionally high air humidity. For plants with volatile oils leaves, stems or flowers should always be gathered on a sunny day after the dew has evaporated the plants but before the full heat of the day.
- Based on gatherer's perception, the following retentions of different plant parts are recommended during collection:

- For harvesting fruits, 25% have to be left for regeneration.
- For harvesting seeds, 30% must be left for regeneration.
- During the harvest of flowers, 30% of flowers of each plant and 20% of the population must be left untouched.
- In case of harvesting leaves, 70% of the plant leaves have to be left out.
- Similarly, for harvesting roots or bulbs, 80% of the plant population should be left untouched.
- For harvesting whole plant, at least 30 % of population should be left untouched.
- Leaves and roots of annual plants should be collected after flowering to ensure regeneration and better product quality unless otherwise recommended.
- For harvesting roots, a hole should be dug straight down and to one side of the roots. Gradually, remove soil on one side of the hole towards the root. Then simply pull the root sideways into the hole. This method will damage the roots less than the common practice of digging down.
- For harvesting bark from trees or shrubs, the entire tree or shrub should not be girdled by removing the bark all the way around the tree, unless the tree has been or is to be removed for other purposes, e.g., for timber harvest, or is otherwise to be destroyed.

❖ Rainwater washes away some of the aromatic oils from many herbs, so after a rainstorm, harvesting should be done after 2-3 days after rains to let the plants oil collect again.

❖ The collectors should select harvest sites where the target plant can be readily found and the plant materials are likely to be of good quality and free of contamination from pollution and other negative environmental influences. Picking of herbs alongside highways or farm fields, unless it is organic, must be avoided.

❖ If herb leaves are to be harvested, the harvested stems or branches must be hanged upside down in bunches for a few days. This will bring the sap present in the stems or branches into the leaves.

❖ Sharp cutting tools like chopper, sickle, secateurs, scissor, and hedge scissor must be used to avoid undue damage to the tree. With tender, non-woody stemmed herbs, gathering of leaves, stems, or flowers can be accomplished easily with the help of scissors or sharp knife. Cutting tools must be cleaned and sharpened between two collections to reduce contamination.

❖ For harvest that involves taking the entire plant (e.g., roots), the harvest should be limited in any population to leave a portion sufficient for regeneration of that population. Harvesting should be done only after the fruit has ripened and the seed has been released, if the species is seed propagated.

❖ Plant species listed under RET or the Convention of International Trade in Endangered Species (CITES) or those which are not allowed to be harvested under state regulations due to concern about over harvest should not be harvested (e.g., *Taxus baccata* in Himalayas, Sarpagandha in M.P. and Chhattisgarh).

- Collection must be made from the abundant stands. Harvesting from stands where the plant is sparse or that are outside the species normal range should be avoided. For harvest only well-grown and adult plants should be selected.
- Whenever possible and acceptable, for meeting quality standards the bark should be harvested from the branches rather than the tree trunk.
- While harvesting roots of perennial plants, some plants from each life stage should be left out. For species that regenerate from portions of roots or root crown, a portion of the root should be left in the ground or the whole or divided crowns should be replanted, as appropriate (e.g., Satawar).
- For propagation by seeds, the seeds of collected species that reproduce sexually in a suitable environment should be planted. For asexual propagation, the whole or divided root crown, as appropriate, or other asexual propagation material should be prepared and planted in a suitable environment.
- All containers used during harvesting must be clean and free of contamination from the previous plant material. When the containers are not in use, they must be kept under dry conditions free of pests and inaccessible to rodents, livestock and domestic animals.

Table 4.3 Examples of important commercial NTFP trees, that need immediate care before depleting from state forests

Trees	*States*
• *Aegle marmelos* (Bael)	Central India – MP, CG, OR
• *Terminalia arjuna* (Arjun)	Central India
• *Garcinia indica* (Kokum)	South India – Karnataka, TN
• *Saracca ashoka* (Ashoka)	Tripura
• *Litsea dlutinosa* (Maida)	Orissa and North East
• *Commiphera wrightii* (Guggal)	North Central India, Gujarat, Madhya Pradesh, Rajasthan
• *Emblica offinalis* (Aonla)	Central and South India
• *Buchanania lanzan* (Chironzi)	Central India
• *Aquilaria malaccensis* (Agarwood)	Tripura
• *Acacia catechu* (Katha)	Central India and Himalaya foothills
• *Cinnamonum tamala* (Tejpatta)	South India and Northern Himalaya
• *Sapindus mukorossi* (Ritha)	Andhra Pradesh
• *Taxcus baccata* (Thuner)	Uttarakhand and Northern Himalaya

BIG QUESTIONS ON NTFP SUSTAINABLE HARVESTING?

Author strongly considers that NTFP sustainable harvesting must address the concern like the real assessment of production, regeneration status, climate vulnerability effects, correlations of NTFP availability with wildlife, whether to support any NTFP based enterprise as per the threat status, and given such situation how long NTFPs will be available in India's forests. NTFP resource assessment in tropical regions is relatively new and received little scientific attention; consequently only local methodologies have been developed by individual researchers.

How often are we open to accept some of these methods incorporating existing local knowledge with inventory methods from wildlife management, horticulture and other disciplines? Only a limited extent of these methods is scientifically tested. On the other hand, the existing scientific inventory methods are not easily adjustable to local circumstances and are not easily understood and interpreted by local management. Therefore, research should relinquish the need for inventory methods, which include traditional knowledge as well as some extent of scientific rigor.

Production Picture

Some of the important NTFPs which were widely collected from tropical dry deciduous forests of central India have been almost wiped off; *Emblica officinalis, Chlorophytum borivilianum, Asparagus racemosous* and *Comifora wrightii* were widely found in Sheopur. Table 4.4 clearly reflects the declining trend of the NTFPs in this region. Information could be gathered through Participatory Rural Appraisal (PRA) methods as there was no base line data. This information is considered authentic, which was further cross-checked with Forest Department office and during observation in the forest survey.

Table 4.4 Change in volumes of NTFP collection in last 20 years

MAP species collected	*Place of collection*	*Quantity traded*			
		15 years before	*10 years before*	*5 years before*	*Present Rate*
Emblica officinalis	Sirgitti	10 trucks*	2-4 trucks	1-2 truck	Very low<1 truck
Sterculia urens	Rohini	4 trucks	2 trucks	1 truck	< I truck
Baswellia serrata	Khairi	10 trucks	5 trucks	2 trucks	1/2 truck
Aegle marmelos	Khori	15 trucks	10 trucks	6 trucks	5 trucks
Chlorophytum borivilianum	Khori	2 trucks	1 truck	1/2 trucks	20 kg
Commiphera Wighitti	Birpur	1-2 trucks	< 1 tuck	Not found	Not found
Asparagus Racemosous	Birpur	60-70 trucks	20-25 trucks	8-10 trucks	2-5 trucks

Source: PRA exercise with NTFP trader and local NTFP collectors conducted during a workshop at Sheopur

* One truck could transport 8-10 tonnes of NTFP

Regeneration Status of NTFP Bearing Trees in Forest

Case studies revealed that increased commercialization and trade in NTFPs is a global phenomenon which has significant implications for their conservation and sustainable use. There has been a marked change in the climate of the region due to different anthropogenic and biotic factors. The erratic rainfall and the change in the temperature has not only decreased the production and regeneration potential of the NTFP species which also affected the total biodiversity. The reproductive biology of many trees, responsible for seed dispersal and pollination, has been affected due to decrease in beneficial insect and bird population. This in turn, has affected the regeneration potential of many NTFP species. Regeneration of NTFPs, viz., *Commifira wrightii, Anogeissus latifolia,* which were occasionally found in the forest area has declined completely. Only a few, species like *Aegle marmelos* and *Baswellia serrata* are still maintaining their regeneration status, as was 15 years back (Table 4.5).

Table 4.5 Regeneration status of NTFPs in forest

Name of the MAP species	15 years before (%)	10 years before (%)	Present (%)
Emblica officinalis	100	1-2	0
Aegle marmelos	100	90	80
Asparagus racemosus	100	90	60
Boswellia serrata	100	80	75
Annnogeissus latifolia	100	10	0
'Kala Dhavara'	100	90	70
Chlorophytum borivilianum	100	2-5	0
Buchanania lanzans	100	10	0
Acacia catechu	100	80	50
Sterculia urens	100	20	5

Source: PRA data of workshop conducted in Sheopur, 2008 keeping 15 years base line.

There is a need to understand size class distribution of trees, few other important indicators like regeneration rate, monitoring of seeding, sapling and adult tree densities that helps in appraising the sustainability of current harvest, while monitoring the size class distribution, if density drops below the values, the intensity of harvesting should be reduced otherwise it will have impact on the regeneration of species in the wild.

Future Climate Vulnerability Effect

Climate vulnerability is more rigorous than it was predicted earlier. Climate variability and change may alter the productivity of forests and thereby shift resource management, economic processes of adaptation and forest harvests (Alig et al., 2004). Bhattacharya and Prasad (2008) paper narrates a case which amplifies the scientific and economic evidence by demonstrating the risk of climate impact to most vulnerable primitive tribal group of dry deciduous forest of Sheopur district in central India. The dry forest, covering 2592 sq km, having rich variety of commercially valuable non-timber forest products (NTFPs) as lifeline and first step survival options, is at stake due to gradual climate change in the area. The mean annual rainfall is 800 mm, but during the past 5 years, the people have experienced drought in successive years. The number of rainy days has drastically changed; rainfall ranging from 460-790 mm without any definite pattern resulted in agricultural and forest crop failure and lack of fodder for cattle, forcing them to migrate to other states for employment. The study was conducted in 12 sample villages, and in all villages, climate change response has impacted phenology of trees, their flowering, fruiting and seeding pattern as well as pollination pattern. In social dimension, the income level from such impact has reduced 20-60% in their economy due to low production. The annual average income of the households is INR, 4500 (approx. US $ 90) from collection and sale of NTFPs in the local market. Due to the resource scarcity and unsustainable harvesting of limited resources the regeneration of all important tree species has gone down drastically.

The other observation made by the local people is that the bird population and wild honeybee population has also reduced in recent past, which may have created impact on reproductive biology of tree species. The decline in production of NTFPs has been 10-90% in the past 10 years. Many NTFP items, e.g., *Chlorophytum borivilianum* (roots), *Buchanania lanzans* (kernals), are no more available for collection. The production of fruits of Aonla (Indian gooseberry—*Emblica officinalis*), *Asparagus racemosa* (roots), *Sterculia urens, Anogeissus latifolia* and other gums has also reduced drastically (50-75%). The study provides strategy regarding how to incorporate such initial research results and local knowledge system in planning and management of forestry so that government can take appropriate action for supporting local community to opt for suitable adaptation measures.

Correlations and Dependency of NTFP Availability with Wildlife

Majority of forest fruits available in natural forest ecosystem are eaten specially by large mammals like sambar (*Cervus unicolor*), gaur (*Bos gauras*), sloth bear (*Melursus ursinus*) and a large number of birds. Over-exploitation of such NTFP has profound effect on the relative abundance of the animals that feed on the exploited resource (Murali, et al. 1996). Several evidences have been found which suggest that high levels of NTFP harvest and enhancement plantings may alter the composition and diversity of bird or animal populations dependent upon them, which corroborates the hypothesis that exploitation can affect plant-animal interactions. Galetti & Aleixo (1998) illustrated that harvest of *Euterpe edulis* palm hearts negatively affected the abundance of two of 15 large frugivorous birds known to eat its fruits. Large mammals such as primates, bears, wild boars, etc., are also important seed dispersers and predators, and harvesting of their food resources may produce competition with human harvesters (Kinnaird 1992). Harvest and management of NTFP may also have effects on pollinator populations which are yet to be studied under thorough research projects that may provide new role of NTFPs for ecosystem services, so our silvicultural management must look into the relationship of NTFP other than human being.

- Species differ in their responses to harvest depending on the:
  - plant part extracted; and
  - harvesting techniques.
- Regeneration and population densities of medicinal plant species are adversely affected by intensity of extraction.
- Such adverse effects are attributed to:
  - damaging harvesting practices; and
  - accompanying anthropogenic disturbances.

How Long NTFPs will be Available in India's Forests?

There is a need to understand given such huge demand and exploitative harvesting pattern; it may not support the resource in nature for long time. Sustained and coordinated efforts are needed to transform currently unsustainable practices of NTFP "mining" from wild sources to more ecologically sustainable, socially acceptable and economically equitable production and utilization system (Parrotta 2002). Many NWFP species are highly sensitive to the level of harvest and fragility of the ecosystem. Probably NTFP resources will only be available in protected areas like national parks, sanctuary and biosphere areas in future. These resources are a critical lifeline for poor, if they go for more and more exploitative harvest when NTFP resources become scarce. Increased demands and scarcity of resources will ultimately lead to total non-existence, which may even end up in endangering the very existence of the resource, and only then the concept of domestication will be a welcome step for NTFP conservation.

PRESCRIBE SILVICULTURAL MANIPULATION FOR NTFP

Silvicultural manipulations aim at providing suitable conditions for the growth of the most desired species. Management of NTFPs through silvicultural manipulations is most difficult task because of multiplicity of species, their varied nature, and growth requirements.

Silvicultural Requirement

Different plant species have different requirement for growth. Some species are light demanders, while others may be shade bearers, and still some others may be shade demanders. Some species require more water for growth than others, while others can grow under dry conditions. According to the nature of growth of plants, they tend to occupy different layers and different kinds of habitats. Most trees occupy either top canopy or under storey; shrubs generally form lower canopy. The herbs and grasses form the ground flora. Most shrubs are shade bearers or shade demanders, while most trees are light demanders. Smaller trees are shade bearers. The nature of trees, shrubs, herbs and other plants vary considerably from the point of view of light, moisture and temperature requirements. The silvicultural manipulations may include the following operations:

- Manipulation of top canopy
- Manipulation of middle storey
- Manipulation of ground vegetation
- Seedling adoption and weeding around the desired plant

Manipulation of Top Canopy

Top canopy is generally occupied by large trees. When an area is allotted to a specific NTFP working circle, the area has to be managed keeping the requirement of that NTFP. Other species become subordinate to that species. Some important NTFPs species are located in the shrub canopy or the under storey or underground vegetation; it may require manipulation of the top canopy. The extent of manipulation may depend on the nature of under storey species. Many small trees, shrubs, herbs and annuals grow better under shade than under full light conditions. Therefore, it may not be necessary to manipulate top canopy extensively. For light demanding or to some extent shade bearing species, manipulation of top canopy may be required.

Manipulation of under storey

Dense understorey trees and shrubs often create problems for the ground vegetation. Manipulation of understorey may be required to remove intense competition in the understorey and create favourable conditions of growth for the ground vegetation. Care has to be taken to create judicious openings in the under storey to favour some important species of ground vegetation. Many shrub species such as *Lantana camera, Carissa spinerum, Dodonea viscosa*, etc., create dense undergrowth and cause intense root and light competition with the ground vegetation resulting in poor or no growth of underground vegetation. Under these conditions removal of some understorey shrubs becomes necessary for the regeneration of NTFP species.

Manipulation of ground vegetation

A large number of species having medicinal and other uses occupy the level of ground vegetation and remain in the form of herbs, grasses, annuals and creepers. In order to provide favourable conditions of growth for the desired species or individuals and to remove competition among the ground vegetation, manipulation of ground vegetation becomes essential. Many species of the ground level and shrub level are considered weed as their uses are not known, or some of them may be less useful. Such species like weeds and relatively less useful species may be removed to favour more useful species.

Seedling adoption and weeding around the plant

Natural regeneration in form of recruits takes place in many species of NTFPs, but these recruits do not develop further to become seedlings, saplings and mature plants because these recruits are generally smothered by weeds and other vegetation growths. The seedlings of important NTFPs, if available, can be adopted and future growth can be ensured, e.g., for big trees like Mahua, Bahera, Harra, Chironji, etc. The process of natural regeneration of selected species can be assisted and regeneration can be ensured. For this, the seedlings of one or several species can be marked on the ground and these seedlings can be adopted through weeding around the plant. The area has to be protected against fire and grazing in order to ensure natural regeneration of important NTFPs.

STRATEGY FOR SUSTAINABLE HARVESTING OF NTFPs

Enrichment Planting or Seeding

When an independent working circle for the conservation and management of NTFPs is constituted, the objective is to manage the area for the conservation and development of some specific medicinal plants or group of medicinal plants and other NTFP species. There are several species of NTFPs, which once occurred in some forests in higher densities, but have now become rare in these areas. These areas may be taken up for enrichment planting or seeding. Some of these species include *Chlorophytum*, spp., *Rauvolfia serpentina*,

Withania somnifera, Andrographis paniculata, Asparagus spp., *Aloe barbadensis, Centalla asiatica, Curcuma* spp. and a large number of others. The following points have to be kept in mind while enrichment planting and seeding.

- The ecological requirements of NTFPs have to be known. The most efficient way of regeneration has to be found out. Most species regenerate through seed; some species regenerate through cuttings or other vegetative parts, and some other species regenerate through nursery raised seedlings. The most efficient way of regeneration has to be used.
- The area included in the NTFP working circle can be of a few square km. A very large area is not required. The area selected should be such that it meets the ecological requirement of the species or group of species.
- Since most of the NTFPs grow in the understorey, it is therefore feasible to identify even good forest area for enrichment seeding and planting of MADPs. The climbers get support of the existing trees and no artificial support is required.
- The seed and other planting material collected for enrichment seeding and planting should be the one most desired, i.e., superior in quality.
- The species selected for enrichment planting or seeding should be one, which is in great demand or whose population has dwindled in forest due to over exploitation.

Prescribe Deferred or Rotational Collection

Continuous harvesting of NTFPs has resulted in dwindling of these plant resources. Sustainable harvesting is generally not practiced in case of most NTFPs. These being common resource, every collector tends to collect maximum quantity. The fruits are collected before they are ripe; the plant is totally damaged in case of root/rhizome collection. Even when only fruits or flowers are required, entire plant is cut. Collection of bark leads to girdling resulting in death of the plant. The gum and resin tapping cause serious injury to the plants. It is, therefore, suggested that some area be rested for some time whereas collection may continue in other areas. It is suggested that 5 years rest may help to rejuvenate the NTFPs resource. The period of rest can be decided keeping in view the condition of the resource. The JFMCs may also be motivated to practice rotational collection of NTFP in their area.

The Madhya Pradesh Biodiversity Rules 2004 formulated on the basis of Biodiversity Act, 2002 also prescribed closure of area for certain period. The closure can be for some seasons. The entire area can be closed for one or few seasons. Some areas can be prohibited for collection. This prohibition can be for certain years. The harvest limit can also be fixed, taking into consideration the requirement of the area and the species.

Control on Fire, Grazing and Shifting Cultivation

A large proportion of NTFPs are herbs, annuals, climbers and creepers. Fire and grazing in forests cause serious damage to these plants. The trees and shrubs, however, escape the damage to a certain extent. Repeated fires during summer cause irreversible damage to annuals, herbs and other ground flora. These plants do not withstand the intense heat and succumb to fires. Several herbs and ground flora have become locally rare due to repeated fire. Annual fires during summers create drier conditions resulting in loss of moisture loving species. The fires also alter soil properties resulting in poor growth of some of the species.

When a large number of livestock graze over the area, many species get trampled by their hooves; the tender parts are eaten by animals. The flowers, fruits, etc., are also eaten resulting in poor regeneration. It would therefore appear that fire and grazing together create serious adverse effects on the ground flora including herbs, annuals, climbers, etc. Similarly, shifting cultivation results in creating adverse condition for the growth of these species. These operations have to be regulated in order to protect fast dwindling ground vegetation.

Establish Preservation Plots in each Felling Series

In order to check rapidly decreasing forest biodiversity due to heavy biotic pressure and unsustainable harvesting, it is necessary to have at least one preservation plot in each felling series. These preservation plots may be protected areas from the point of view of harvesting. These plots may be of 5-10 ha to 50-100 ha depending upon the availability of forest area and will of the local population. Such preservation plots may be identified both in the map and also on the ground.

These preservation plots (PPs) should be of permanent nature. As far as possible, these PPs should be well marked on the ground, and if possible, the area should be fenced to avoid any damage.

Increase Research Efforts

There is a limited data on the management of these plants. It is necessary to document whatever data is available and to initiate research on NTFPs. This is necessary so that the resource can be managed scientifically. The area of research may include ecological, silvicultural, economical, production, processing, etc. The best management practices may be documented and circulated among the researchers and managers. Research should be focused on the assessment of various factors influencing NTFP management rather than only on factors influencing their collection and use. Up to the present, most typologies of NTFPs have been based on their use characteristics. Such classifications should be considered a first step in developing a typology of NTFP management. Little attention has so far been paid to elaborating these typologies and placing NTFPs within a typological framework that explains patterns of NTFPs management (Ruiz, 1995; Peters, 1996; Bhattacharya, 2013).

Include Management of NTFPs in Working Plan Code

Working plan is the guide for the territorial divisions for managing forest. Any activity which is prescribed in the working plan is to be implemented. Working plan is prepared as per the guidelines issued by the government of India. The present code does not prescribe detailed management planning of NTFPs. It is, therefore, suggested that working plan code may be revised to include all states, the provision for preparation of working plan of NTFPs rich areas in a manner it is done for timber species by constituting independent working circles or overlapping working circles.

Unfortunately, neither the Panchayat (new owner of the resources), nor the Forest Department is taking interest in extending conservation and regulatory services in most cases for NTFPs. At present there is no governance mechanism to ensure that harvesting is not unsustainable. Good collection practices (GCP) and good manufacturing practices (GMP) have been developed but are yet to follow at field level to control wastages. Protocols for harvest are developed and practiced for many commercial NTFPs in different locations, and good results are found by participatory involvement of local people. Methods for community based ecological monitoring and robust methods of sustainable harvest practices need to be followed. Suitable training is required for the harvesters/local collector, traders, forest managers and biologists so that all stakeholders understand the benefits of sustainable harvesting.

Some of the fundamental principles followed in barefoot siliviculture like group selection system, multiple products concept, species association, natural succession, vertical stand structures and even role wildlife have been developed over thousands years of trial and error (Ganz, 2000). The indigenous knowledge system has lots of insights towards sustainable harvesting of diversified products list. Barefooted silvicultural system still exist in Mexico, Peru, Bolivia, Thailand, Laos, Indonesia and India, that supports to incorporate people's wisdom. Before 200 years of scientific forestry history, local communities were using such practices and still they are available having strong cultural knowledge. The success of such silviculture and management not only promotes ecological and economic sustainability, but also cultural sustainability and globally we are following the concept of sustainable development.

CONCLUSION

The responsibility of maintaining and practising sustainable harvesting is considered a forest department's job, but there are various stakeholders involved in the affair which needs urgent celerity in the system. The accessibility of NTFPs for local communities is still not clear for the forest department as well as local communities in terms of collection, trade and manufacturing/processing related areas in reference to Biodiversity Act, 2002, FRA, 2007 and PESA, 1996 provision. The paper does not address the innumerable economic and social factors, which are also important in determining the overall sustainability of forest resource exploitation. Presently, there is no single NTFP policy at national level; there are several acts, policies, or administrative orders like JFM resolutions and other legislations which partially address NTFPs in reference to the ownership, benefit sharing, monopoly, transit rules, tax, conservation need, etc., at national level and at various states across the country. Information on sustainable NTFP resource management needs much attention at national level for development of baseline data and for large scale plantation under Green India Mission to support NTFP all improvement for next 10 years. Future forestry may take up various scientific projects for further work on sustainable harvesting of many more species and develop various refined protocol to address future conservation need and economic development, keeping in view of certification needs as well as for environmental audit.

In many states field realties conclude that wild population of NTFP species have severe fluctuations in the yield, and this impact is due to unsustainable harvest at individual level, population level, ecosystem level, and at the genetic level. Integration of NTFP in forest management is necessary in order to achieve sustainable forestry. Because forestry technologies are developed with timber values uppermost, it is vital to develop forest management technologies that take into account both timber and non-timber values. In certification process of NTFP sustainability parameters are considered important basis in the certification standard. Global examples show that traditional knowledge could play a vital role while developing silvicultural regimes of NTFP management, and *in situ* experimentation will strengthen the regimes.

REFERENCES

1. Alig R.J., Adams, D., Joyce, L., and Sohngen, B. (2004). Climate change impacts and adaptation in forestry: responses by trees and market choices, American Agricultural Economics Association, 11 pp. (http://www.choicesmagazine.org/2004-3/climate/2004-3-07.htm).
2. Bahuguna, V.K. (2000). Forests in the economy of the rural poor: An estimation of the dependency level. *Ambio* 29, 126–129.
3. Bhattacharya, P. (2013). Prescriptions for sustainable NTFP management in forest working plan: Addressing new market demand. *International J. of Usuf. Mngt.* 14(2): 31–50.
4. Bhattacharya, P., and Hyat, S.F. (2004). Sustainable NTFP management for rural development: A case from Madhya Pradesh, India. *International Forestry Review,* 6(2): 161–168.
5. Bhattacharya, P., and Prasad, R. (2001). Integrated options for forest management in India. In: Victor, M. and Barash, A. 2001. Overview of an International seminar on cultivating forests: Alternative forest management practices and techniques for community forestry, 23–25 September 1998. RECOFTC Report No. 17. Bangkok, Thailand. 49–57
6. Bhattacharya, P., Bhattacharya Prasad, R., and Asokan, A. (2008). Towards certification of wild medicinal and aromatic plants in four Indian states. *Unasylva,* 230, Vol. 59, 2008.
7. CGMFPCF (2006). Chhattisgarh Rajya Laghoo Vanopaj, Bazar Sarvekshana Prativedan, Chhattisgarh Minor Forest Produce Co-operative Federation, Raipur, India.
8. FAO (2000). Report of workshop on developing needs-based inventory methods for non-timber forest products. Food and Agriculture Organization, Rome, Italy.
9. Gadgil M. (2003). India's biological diversity act 2002: An act for the new millennium. *J. Biosci.* 28(2): 145–147.
10. Gadgil, M., Berkes, F., and Folke, C. (1993). Indigenous knowledge for biodiversity conservation. *Ambio,* 22, 151–156.

11. Galetti, M., and Aleixo, A. (1998). Effects of palm heart harvesting on avian frugivores in the Atlantic rainforest of Brazil. *J. Appl. Eco*, 35, 286–293.
12. Ganz, D. (2000). Barefoot silviculture in Oaxa, Maxico: The adaptation of standard Silvicultural system for NTFP and the integration of Indigenous knowledge. Paper presented in Seminar on "Harvesting of Non-wood Forest Products" Menemem-Izmir (Turkey), 2-8 October 2000.
13. ICFRE (2010). Forest Sector Report India (2010). ICFRE, Govt. of India.
14. Kinnaird, M.F. (1992). Competition for a forest palm: use of *Phoenix reclinata* by human and nonhuman primates. *Conserv. Biol.* 6, 101–107.
15. Krishna H Gautam, Teiji Watanabe (2002). Silviculture for non-timber forest product management: Challenges and opportunities for sustainable forest management. The Forestry Chronicle, 2002, 78(6): 830-832.
16. Murali, K.S., Shankar, U., Shanker, R.R., Ganeshaiah, K.N., and Bawa, K.S. (1996). Extraction of non-timber forest products in the forests of Bilingiri Rangan Hills, India. II. Impact of NTFP extraction on regeneration, population structure, and species composition. *Eco. Bot.*, 50, 252–269.
17. Parrotta A. John (2002). Conservation and sustainable use of medicinal plant resources-An International Perspective. Compendium of Invited lectures, of World Ayurveda Congress-2002.kochi. pp. 52-63.
18. Peters, C.M. (1994). Sustainable harvest of non-timber plant resources in tropical moist forest: an ecological primer. Biodiversity Support Programme, World Wildlife Fund, Washigton DC, p. 45.
19. Peters, C.M. (1996). The ecology and management of non-timber forest resources. World Bank Technical Paper No. 322, World Bank, Washington, D.C., USA.
20. Prasad R., Kotwal, P.C., and Mishra, M. (1999). Impact of harvesting *Emblica officinalis* (Aonla) on its natural regeneration central Indian forests, *Journal of Sustainable Forestry*, 14 (4): 1–12.
21. Ruiz, M.R. (1995). A conceptual framework for CIFOR's research on non-wood forest products. CIFOR Working Paper No. 6, Center for International Forest Research, Bogor, Indonesia.
22. Ticktin, T., and Johns, T. (2002). Chinanteco management of *Aechmea magdalenae* (Bromeliaceae): Implications for incorporating TEK and TRM in management plans. *Economic Botany*, 56, 43–57.
23. Wiersum Freerk (1999). Understanding diversity in NTFP management: A neglected issue in NTFP research. NTFP research in the Tropenbos programme: Results and perspectives.

5

Pulp and Paper Industry Raw Material Scenario: ITC Plantation—A Case Study

H.D. Kulkarni

Plantation, ITC Limited, Paperboards and Specialty Papers Division, Unit: Bhadrachalam, Sarapaka - 507128, Khammam District, Andhra Pradesh, India

E-mail: hd.kulkarni@itc.in

ABSTRACT

The pulp and paper industry is one of the key industrial sectors contributing to the Indian economy. There are 759 paper mills in India with an operating capacity of 12.7 million tonnes and consumption at 11 million tonnes with 9.3 kg per capita consumption of paper. The annual pulp production is 2.71 3.3 million tonnes using 9.83 million tonnes of wood. Under public private partnership, ITC promoted 140,989 ha plantation using R&D to increase productivity to 20-58 t/ha/yr compared to 4-6 t/ha/yr from seedling plantations favouring fast wood forestry.

Paper mills in India continue to face challenges with forest-based raw material. Out of the annual paper production capacity of nearly 10.11 million tonnes, around 31% (3.19 million tonnes) is produced by 26 major wood-based mills and the rest 69% by waste paper and agro-based mills. The present annual requirement of wood is 9.83 million tonnes. The projected demand for paper by 2025 is 24 million tonnes with indigenous production of 22 million tonnes leading to a shortfall of 12 million tonnes of wood. The strategy to be adopted by the paper industry, to meet its ever-growing demand of wood on continuous and sustainable basis is to enlarge social and farm forestry plantation apart from raising plantations by forest development corporations.

INTRODUCTION

Global Scenario

Paper and paperboard are an essential requirement of our life. The current global paper and paperboard demand is 402 million tonnes per annum and there are more than 7745 mills producing 192 million tonnes of pulp. The paper demand has doubled in 20 years from 242.79 million tonnes in 1990 to 402 million tonnes in 2011 (Fig. 5.1).

The forecast for paper production by 2021 is projected at 521 million tonnes per annum (RISI, 2010). Asia produces nearly 177 million tonnes (44%) while; rest of the world produces 225 million tonnes (56%). The demand of paper is strongly linked to GDP growth. The 6.84 billion people of our planet consume 56.7 kg paper. Forest products sector is estimated to contribute about 1.2% of world GDP and 3% of international merchandise trade. Industry annual turnover exceeds US$ 200 billion for round wood and sawn wood, panels, pulp and paper (Anon, 2006).

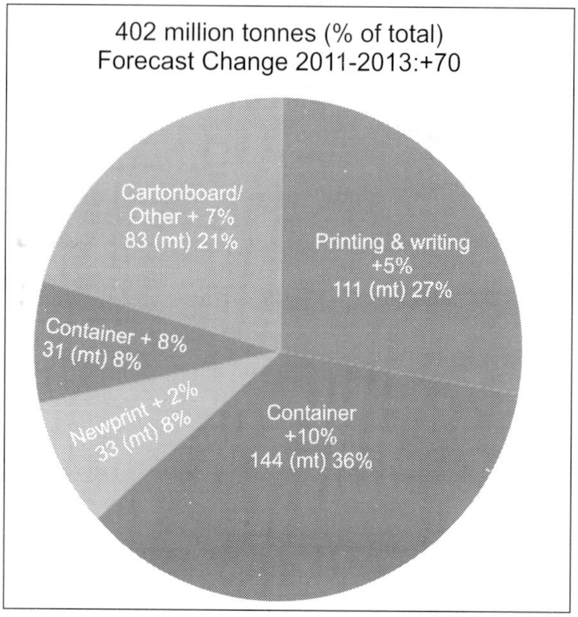

Fig. 5.1 Global paper and paperboard demand
Source: RISI Annual Review

Indian Scenario

Paper making in India is almost two centuries old with first unit commissioned in 1832 in Serampur, West Bengal (Ahuja S.P., 1980; Anon,1986). Rags and waste paper were the raw material used for producing paper in 1867. Commercial production started in 1882 with non-wood fibers. Bamboo pulping process was developed in Forest Research Institute, Dehradun during 1922-24, which provided impetus to pulp and paper industry in India. The growth of paper mills (Table 5.1) is quite rapid during the last two decades (GOI, 2011).

Table 5.1 Growth of paper mills in India

Year	*No of units*	*Installed capacity (million tonnes)*	*Production (million tonnes)*	*Capacity utilization (%)*	*Per capita consumption (kgs)*
1950	17	0.13	0.11	85	0.9
1970	57	0.77	0.75	99	1.9
1990	325	3.3	2.43	62	3.6
2000	380	3.94	4.87	99	5.5
2006	660	8.5	6.8	80	6.7
2007	667	8.5	8.3	100	8.3
2010	759	12.7	10.11	80	9.3

Today, there are 759 pulp and paper mills with an installed capacity of 12.7 million tonnes producing 10.11 million tonnes of paper and paperboards, which is 2.52% of the total world production of 402 million tonnes per annum. Apart from 10.11 million tonnes production, 1.04 million tonne is imported annually. Therefore, the present consumption of paper and paperboard including newsprint is at 11.15 million tonnes.

The per capita consumption in India is 9.3 kg as against 42 kg in China, 22 kg in Indonesia, 25 kg in Malaysia, 250 kg in Japan, 325 kg in USA and the world average of 56.7 kg. In India, the paper consumption is predominantly domestic and the demand is driven by GDP growth. The main growth drivers for paper demand include enhancement in government spending on education (6% of GDP), increase in literacy rate, improvement in standard of living, booming retail sector, construction boom, unprecedented growth in industries like food, pharmaceuticals and apparels, increase in packaging and advertising expenditure, etc. The printing and packaging industry is growing at 14% CAGR. The population of 1.2 billion and the changing demographic profile, which will put over 65% of the population in the working class and half of that is less than 30 years old, following western consumerism, will further fuel the demand for paper and paperboards.

With the GDP expected to average around 9%*, growth in consumption of paper and paperboard is expected to be about 7.8%. The Indian pulp and paper industry mirrors its impressive growth rates in the GDP at 8% CAGR (Gopalratnam, 2007). This translates into consumption of about 24 million tonnes and production of 22 million tonnes by 2025 resulting in per capita consumption of 17 kg. All this means rapid growth of pulp and paper industry in India that ultimately requires wood as raw material. Indian pulp and paper industry being the 15th largest industry in the world is an important industrial sector having a bearing on the socio-economic development of the country as it provides employment to nearly 1.5 million people (Suri R.K., 2007). The paper industries market size in 2009-10 was Rs. 317 billion (US $ 6.34 billion)** and is expected to touch Rs. 526 billion (US $ 10.52 billion) by 2014-15. Annually, India exports 0.527 million tonnes and imports 1.582 million tonnes of paper products (GOI, 2011).

Indian paper industry is highly fragmented. In terms of product-wise segmentation, the printing and writing is 38.58%, while packaging is 53.61% and newsprint is 7.81%. Depending upon the production capacity, paper mills are classified as large (33000 tpa), small (7500 tpa) and medium (between 7500 and 33000 tpa). The paper production from small and medium mills accounts for 60% while 40% comes from large mills. Out of 759 mills, 114 (15%) are large, 342 (45%) small and 303 (40%) medium. Further, the industry is divided in 3 sectors (Tables 5.2 and 5.3) based on raw material [8, 9] usage.

Wood and Bamboo: Mills using wood and bamboo contribute 31% production, which is 3.19 million tonnes, and there are 26 large integrated paper mills.

Agro-based: There are 150 mills using agro residues like bagasse, wheat and rice straw, etc., and produce 2.2 million tonnes, which is 22% of the total production. The agro-based fibre usage has decreased considerably.

* The GDP is now fixed at 8% by the Planning Commission

** 1 US $ = INR 50

Recycled fiber: Mills using waste paper contribute almost 47% of the country's current production which is 4.72 million tonnes, and there are 538 mills in operation. Over the years recycled fibre usage has considerably increased.

Table 5.2 Segment-wise composition trend in usage of raw material

Segment	Unit	1970	1995	2000	2002	2005	2009	2011
Wood based	%	84	39.18	29	36.8	35	39	31
Agro based	%	099	32.36	36	32.4	30	31	22
Recycled fiber based	%	07	28.44	35	30.8	35	30	47

Table 5.3 Total number of pulp and paper mills in India-2010 year

	Unit	Wood based	Agro based	Recycled fiber based	Total
Mills	No.	26	150	583	759
Production	Million tonnes	3.19	2.2	4.72	10.11
Share	%	31	22	47	100
Installed capacity	%	4.00	3.00	5.7	12.70
Capacity utilization	%	80	73	83	79

The agri waste is finding its way for power generation and availability is becoming tight and expensive. Moreover, the agro-based mills are shutting down on account of pollution related problems and non-availability of agri waste on sustainable basis as seen in Table 5.2. Waste paper quality and price has also put paper industry under pressure. Wood based mills have done some work on raising plantations but still the availability of wood is a major concern to the mills. As wood is a renewable resource to produce paper, it becomes necessary to look into the trends of its demand and ways and means to meet the raw material requirement to produce it.

Wood Requirement and Generation

India has a total forest area of about 75 million ha which forms only 22.8% of the total geographical area (328 million ha) of the country. Forests in India are fast disappearing. At the time of independence more than 22% of India's geographical area was covered by dense forests. Recent satellite surveys show that hardly 11% of the area now supports closed forests, i.e. forest with 40% crown cover, even though the National Forest Policy enunciated soon after independence in 1952 that at least 33% of geographical area of the country should be under forest cover. Further, the per capita forest area in India is only 0.064 ha against the world average of 0.64 ha (FAO, 2005). Out of 69.09 million ha of the recorded forest cover, nearly 28.84 million ha are degraded forest lands. The productivity of Indian forest is only 1.34 m^3/ha/yr against the world average of 2.1 m^3/ha/yr. Earlier forests were the main source for wood and bamboo based raw material for the paper industry. Depletion of forest areas and reduction in volume of extraction of wood and bamboo has hit badly the supply of raw material to the industry.

In recent times, the demand for the fire wood, timber and industrial wood in the country continues to grow because of increasing population and the growth of economy. The Indian Forest Act, 1927, the forest Conservation Act, 1980, the National Forest Policy, 1988 and the National Forest Commission, 2003 are the umbrella legislation and frameworks for forest protection and conservation. Under these Acts and Rules and Policy, the supply of raw material to wood based industries is phased out from forests. Participation of the private sector, even in reforestation of 28.84 million ha degraded forest lands and Joint Forest Management is not allowed as per the policy guidelines. In the National Forest Policy 1988, the wood based industries have been advised to encourage agro forestry for raising plantation to meet the raw material demand. Agriculture sector in the country covers about 143 million ha out of which 40 million ha is classified as degraded (Pandey M.C., 1992). Hence, non-forest lands such as private lands are explored for raising tree crops to augment the available wood resources. In course of time, social and farm forestry plantations emerged on a large scale for meeting the wood demand of the paper industry.

The wood based paper mills in India continue to face challenges with forest-based raw material. Pulp and paper industry consumes 3% of total national requirement of wood, while the major consumption being fuel wood (89.5%) and timber (7.5%). The annual pulp production is 3.03 million tonnes from 10 million tonnes of wood, agro-residue and waste paper. Nearly 20% of wood is procured from government sources, while 80% is from agro farm forestry sources. The bamboo and wood requirement is 0.82 and 9.27 million tonnes per annum respectively (Table 5.4). The strategy adopted by the industry to meet the ever growing demand of wood on a sustainable basis is to obtain wood from social and farm forestry plantations. Over a period of 23 years the paper industry has promoted nearly 657,093 ha plantations which is estimated to produce 39.42 million tonnes of wood at 60 tonnes per ha yield. However, if we consider last 5 years plantations of Eucalyptus, Casuarina, Leucaena and Acacia, it is 300,000 ha which can produce 18 million tonnes of wood. At the felling cycle of 4 years, the wood production annually is 4.5 million tonnes. The current level of planting by the paper mills is around 50,000 ha per annum. Apart from the industrial efforts, farmers are raising plantations on their own. In the recent years several private Eucalyptus clonal nurseries have sprang up and an additional 20,000 ha area every year is planted. This is adding to the general availability of wood to the industry. Hence, the industries' wood demand of 10 million tonnes annually is somehow met through farm forestry plantations.

In addition to pulpwood, the farm grown wood is required for mine props, scaffolding, plywood, particle board and biomass based energy industries, which is creating stiff competition to paper industry. There is drastic increase in pulpwood prices alone from 25 to 45 US $ tonne (Rs.1250 to Rs. 2250) in a span of 6 years. Prices of hardwood increased at a CAGR of 8% to around Rs. 3525 per tonne (US $ 70.5) in 2009-10 and from about Rs. 2450 per tonne (US $ 49) in 2004-05 (Figs. 5.2 & 5.3). Due to competition and high demand, the wood is transported from faraway places. Presently, the cost of wood-based raw material to Indian mills is US $ 70/t compared to US$17/t in Indonesia and US$25/t in Brazil. As one tonne of paper requires 4 tonnes of wood, there is a straight disadvantage of about

US$ 180 to 212 /t of paper produced. If this trend continues, India will have to import pulp and paper instead of making it domestically. Paper companies are aggressively looking at farm forestry to cut down on the landed cost of wood. With transportation cost accounting for nearly 30 to 50% of the wood cost, developing farm forestry plantations near the manufacturing units is being pursued vigorously by the mills.

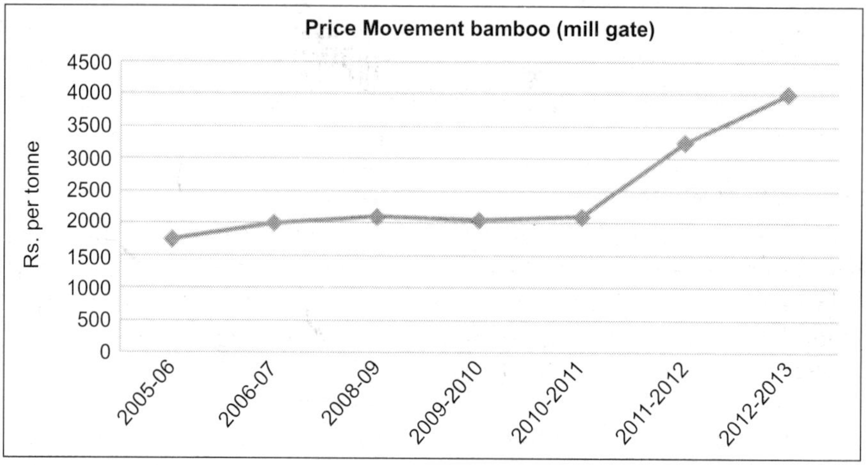

Fig. 5.2 Bamboo prices

Fig. 5.3 Hardwood prices

Andhra Pradesh State has four mega pulp and paper units and has become a wood surplus state in India. This state has brought in legislation wherein Eucalyptus, Casuarina and Leucaena wood are declared as agricultural produce and marketed through agricultural market committee (GoAP., 1999). Pulpwood from social forestry district, viz., Prakasham and Krishna in Andhra Pradesh meet the wood demand of mills not only of Andhra Pradesh but also of mills located in Karnataka, Maharashtra, Gujarat, Odisha, Uttar Pradesh, Uttaranchal and Madhya Pradesh (Saigal and Kashyap, 2002). The plantation efforts lead to net increase of 939 km^2 in forest cover in 5 years time including 3000 km^2 in non-forest areas and nearly 400,000 ha is with trees outside forests (FSI, 1999, 2000, 2000). The present consumption of wood is at 9.83 million tonnes. Considering the future demand of paper by 2025, an additional 12 million tonnes of wood is required from 1.2 million ha of pulpwood plantations and 0.36 million ha (30% extra) for fuel, fodder and local usage. The total land mass therefore required to raise pulpwood plantation will be around 1.56 (say 2) million ha with an assumption of 50 tonnes per ha productivity for a felling cycle of 5 years.

Bamboo in Paper Industry

In Asia, India leads in the utilization of bamboo for paper manufacture. The estimated forest area covered by bamboo in India is nearly 8.96 million ha. The general consumption pattern of bamboo in India indicates that 8.4% of bamboo is being consumed by pulp and paper industries, while cottage, furniture and implements industries consume 65% bamboo (Tewari D.N., 1996). In earlier days, more than 70% bamboo was used for paper and paperboard production. Now there is a drastic shift in usage of bamboo as it reduced to less than 10% and even some mills have completely shifted to hardwood. Presently, 8 paper mills use 0.82 million tonnes of bamboo per annum. Few mills have given up bamboo and replaced it with hardwood, thus promoting more pulpwood plantations.

Recent Wood Availability Crisis

Though there appear to be excellent efforts of raising plantations by the paper industry, the availability of wood has dwindled in the recent past on account of uprooting of plantations by the farmers due to Eucalyptus Gall disease and fall in wood prices during 2009-2011. Cultivation expenses of pulp wood plantations have gone up considerably and farmers shifted to agricultural crops. This situation could have been averted provided certain mills have raised plantations commensurate with their consumption. Now mills are procuring wood spending huge amount on transport from far away distance. The wood prices increase from May 2012 is more than 46% (Rs. 1400 on Rs. 3000 mill delivered). Mills are exploring import of wood chips or logs to overcome the present shortage.

ITC R&D and Farm Forestry

ITC Paperboards and Specialty Papers Division, Unit: Bhadrachalam, which is one of the leading pulp and paper manufacturing companies in India, took major initiatives to launch a Tree Improvement Research (Kulkarni H.D., 2001, 2004). The division produces 300,000 tonnes of pulp with 1125,000 tonnes of wood. To meet the wood requirement, the company embarked on an ambitious farm forestry project with dual objective of achieving self-sufficiency and improving productivity, while giving agri farmers an alternative land use option. ITC devised the strategy to raise pulpwood plantations commensurate with wood consumption and also to fulfill its social responsibility towards the rural population.

Table 5.4 Wood requirement and generation by paper mills through farm forestry

S. No.	*Company*	*Wood Pulp (t)*	*Wood Requirement (t)*			*Area Planted (ha)*	*Estimated Wood Generation (t) @ 60 t/ha*
			Bamboo	*Wood*	*Total*		
1.	ITC Ltd.	300000	0	1125000	1125000	140989	8459340
2.	Tamil Nadu News Print Ltd. *	280000	0	400000	400000	34542	2072520
3.	Century P & P Ltd. *	280000	168750	958125	1126875	2340	140400
4.	JK Corp (Orissa)	220000	50000	850000	900000	59974	3598440
5.	JK Corp (Gujarat)	55000	80000	126250	206250		
6.	Orient Paper Mill Ltd.	80000	167000	173000	340000	33043	1982580
7.	Star Paper Mill Ltd.	70000	100000	180000	280000	72740	4364400
8.	Mysore Paper Mill *	60000	0	225000	225000	27500	1650000
9.	Sirpur Paper Mill Ltd.	120000	0	375000	375000	30921	1855260
10.	BILT Ballarpur/Asthi	220000	200000	175000	375000	36055	2163300
11.	BILT Sewa	100000	40000	335000	375000		
12.	BILT Yamunanagar	100000	0	375000	375000		
13.	BILT Kamalapur	100000	0	375000	375000		
14.	BILT Chowdwar (Not Operational)	0	0	0	0		
15.	Seshasai P & B Ltd.*	160000	0	400000	400000	18534	1112040
16.	Andhra Pradesh Paper Mill Ltd.	220000	18000	800000	818000	124040	7442400

17.	Circar Paper Mill *	0	0	5000	5000	0	0
18.	West Coast Paper Mill Ltd.	280000	0	900000	900000	44260	2655600
19.	Rama News Prints *	0	0	0	0	0	0
20.	HNL Kottayam	80000	0	300000	300000	32000	1920000
21.	HNL Naogaon	100000	0	375000	375000		
22.	HNL Kachar	100000	0	375000	375000		
23.	HNL Nagaland	20000	0	75000	75000		
24.	Nepa Paper Mill *	0	0	50000	50000	0	0
25.	Yash Paper Mill	5000	0	16500	16500	150	9000
26.	Delta Paper Mill *	80000	0	300000	300000	0	0
27.	Emami Paper Mill *	0	0	0	0	5	300
	Total	3030000	823750	9268875	10092625	657093	39425580

* Mills using Agro residues and waste paper for Pulp Production apart from Wood and Bamboo

PLANTATION PROGRAMME: A LEARNING EXPERIENCE

In the start of the programme, the company distributed 4.67 million seedlings to cover 2000 ha from 1982 to 1986. Later, from 1986 to 1995, it distributed 17.4 million seedlings covering 7441 ha of land under ITC-NABARD (National Bank for Agriculture and Rural Development) scheme. The resultant plantations showed poor survival (30 to 50%) and low productivity (6 to 10 t/ha/yr) due to high genetic variation, leaf blight disease, termite damage, hybrid breaking, primitive nursery practices, mismatch of species and provenances to site, close spacing, lack of follow-up of correct package of practices, etc. Further, because of Eucalyptus Controversy (Rajan B.K.C., 1987), farmers were also scared to take up Eucalyptus plantations and realized that tree-farming option is uneconomical. In case of Casuarina, its cultivation was largely limited to coastal belts in Andhra Pradesh. Narrow genetic base, site specificity, mycorrhizal association, frequent droughts and disease pest problems hindered extensive cultivation of Casuarina. For Leucaena also, the genetic base was narrow and psyllids and other diseases with primitive cultural practices reduced the productivity and cultivation. Hence, a decade ago, farm forestry plantations of Eucalyptus, Casuarina and Leucaena were becoming unpopular in spite of the incentives, subsidies and loans to the farmers. This adverse scenario changed after 1989 when the clonal technology based plantations started appearing on the horizon (Kulkarni and P. Lal, 1995).

The experimental site and Clonal Research Station is located at 17°40′ N latitude and 81°E longitude. The altitude of the place is 100 m above the mean sea level. The climate is subtropical with annual rainfall of 1033 mm, mostly from southwest monsoon. The maximum temperature recorded is 49°C and minimum 10°C. The predominant soils are red sandy and black cotton. Soils are either normal or alkaline. Saline soils are also found.

For Eucalyptus, Casuarina and Leucaena, the seeds were imported from CSIRO, Australia and Hawaii, USA to raise provenance trials. Candidate Plus Trees were selected from provenance trials, Government and Farm Forestry plantations. Selected plus trees were propagated vegetatively from coppice cuttings and cladodes under mist propagation. Root Trainer Technology was adopted for the production of plants. Successful ramets were planted in gene banks. The clones were tested under multi-locational field trials and promising clones were short-listed. Clonal seed orchards were established. Clonal demonstration plots were raised under the extension scheme. Inter- and intra-specific hybridization was carried out between selected best clones and other species. Social forestry scheme (Mission Sunehra Kal) was undertaken with the help of non-governmental organizations under poverty alleviation programme, while farm forestry was practised in marginal agricultural lands under rain fed conditions.

The Tree Improvement Research (TIR) on Eucalyptus, Casuarina and Leucaena actually begun in 1989 to improve the productivity and profitability of plantations for making farm forestry an attractive land use option. The major emphasis was on genetic improvement of planting stock and improvement of package of practices. Significant gains in productivity of Eucalyptus, Casuarina and Leucaena have been achieved through clonal selection, vegetative propagation combined with Root Trainer Technology and clonal testing. These clones, developed in India, for the first time, are known as

'ITC-Bhadrachalam' clones. Development and deployment of hybrids, locality-specific, high yielding, fast-growing, disease-resistant clones of Eucalyptus, Casuarina and Leucaena have been followed by rapid adoption and raising of large-scale clonal plantations under social and farm forestry. At the beginning of the programme, the main handicap faced was the non-availability of a wide genetic base for the improvement of Eucalyptus, Casuarina and Leucaena. Therefore, "***breed the best with the available best***" strategy was followed.

Eucalyptus

Eucalyptus tereticornis was the main species for improvement as it was most suited to the mill catchment area. More than 1000 trees were selected having characteristics such as straightness of stem, annual growth rate, disease resistance, crown structure, wood density, fibre morphology, cellulose/lignin balance, bark to solid wood, under bark relationships, etc. Clones were evaluated in 159 trial plots of various soil types spread over 36 ha area, based on comparative genetic superiority and 107 promising clones were short-listed.

The most important commercial clones are - 3, 6, 7, 10, 27, 71, 72, 99, 105, 115, 122, 128, 130, 223, 265, 266, 271, 272, 273, 274, 175, 277, 284, 285, 286, 288, 290, 292, 316, 319, 405, 411, 412, 413, 417, 439 and 470.

The most adaptable clones for alkaline soils are - 1, 10, 27, 71, 99, 105, 115, 116, 122, 128, 130, 158, 223, 266, 271, 272, 273, 274, 277, 290, 316, 318, 328, 410, 411, 412, 413 and 417.

The plastic clones are - 27, 71, 83, 99, 105, 116, 128, 130, 147, 271 and 285.

The outbreak of diseases caused by various fungi on Eucalyptus in nursery and field revealed main pathogens as *Cylindrocladium* spp. and *Alternaria* spp. The fungal disease resistant clones short-listed are 1, 3, 6, 7, 288 and 316.

Eucalyptus has a satisfactory health record in India concerning insect pests. However, *Leptocybe invasa* Fisher and La Salle, a gall inducing insect (Mendel et. al., 2004) attack on Eucalyptus was reported in 2002 is posing a great threat to Eucalyptus clonal forestry. Clones 10 and 27 are recorded as highly susceptibles while clones 1, 6, 7, 320, 411, 413, 513, 612, 2008, 2145, 2253, 2254, 2306 and 2313 are found free of gall attack amongst 107 Bhadrachalam clones of Eucalyptus (Kulkarni H.D., 2010, 2011a). Parasitoides *Quadrastichus mendeli* and *Megastigmus* spp. as biological control agents were introduced in 2010 and the gall insect attack is under control.

The survival percentage for majority of clonal plantations is more than 95 (Kulkarni H.D., 2004; Kulkarni and P. Lal, 1995). The productivity of "ITC-Bhadrachalam" clone's range from 20 to 58 t/ha/yr compared to 6 to 10 t/ha/yr from seedling origin plantations. Apart from increase in productivity by 4 to 6 times the rotation period is reduced by half. Therefore, farmers are now harvesting plantations at 4 years instead of 7 years. To quote a few cases:- Mr. D Satyanarayana, a farmer of Prakasam district in Andhra Pradesh, obtained a yield of 190 t/ha in 5 years under rain-fed condition, while Mr. Venku Reddy of Nellore District obtained 250 t/ha in 5 years under partially irrigated conditions.

One hundred plantations were felled in different districts to assess productivity for authenticating the clonal trial results. The Mean Annual Increment (MAI t/ha) figures are

- 23 in Khammam, 28 in Prakasam, 21 in Guntur, 24 in Krishna and 39 in West Godavari districts of Andhra Pradesh. The average MAI (t/ha) works out to 27. The farm forestry plantations average IRR per hectare in different districts worked out to 40 in West Godavari, 48 in Khammam, 32 in Prakasam, 26 in Guntur and 30 in Krishna (Manage, 2003). Further, equally higher yields and returns from second and subsequent coppice rotations of "ITC-Bhadrachalam" clones resulted in economic prosperity of farmers.

Clonal demonstration plantations (CDP) raised by the company resulted in large-scale adoption of genetically superior clones by the farmers and State Forest Departments/ Forest Development Corporations. As *"seeing is believing"*, farmers' meetings were regularly held in these plots which enabled them to pick and choose the clones most suited to their land.

The hybridization programme was initiated in 1994. A breeding orchard was set-up with *E. tereticornis, E. camaldulensis, E. alba, E. urophylla* and *E. grandis*. Inter-specific hybridization was attempted to combine desirable complementary attributes of promising clones and eliminate defects keeping in view the customer's (growers/mill) viewpoint viz., high yields (volumetric productivity), felling cycle of 3 to 5 years (economic rotation), adaptability to sites, superior wood quality and uniformity of raw material.

Development of inter-specific hybrids such as *E. tereticornis* Smith × *E. urophylla* Blake.; *E. tereticornis* Smith. × *E. grandis* Muell.; *E. tereticornis* Smith. × *E. camaldulensis* Dehnh.; *E. tereticornis* Smith. × *E. alba* Reinw. and *E. tereticornis* Smith. × *E. torelliana* Muelli; *E. urophylla* Blake × *E. grandis* Muell was attempted. Teretigrandis and Urograndis hybrids have adapted well to drought conditions and are producing maximum volume of wood. These hybrids are now planted on large scale. By controlled pollination between the best 32 clones of *E. tereticornis,* the derived full-sib hybrids have shown good heterosis at 2 years age. The full-sib progeny trial showed a maximum of 33% improvement over the parents for production of wood volume. Nearly 358 full-sib hybrid trees have been cloned. Heterobeltioisis studies on 18 hybrid clones showed 82% improvement in wood volume production over the best parent. A few hybrid clones from the crossing of clone 6, 10 & 27 gave hybrid clones 2011, 2014, 2045, 2050, 2052, 2053, 2120, 2121, 2149, 2155, 2156 & 2313 which are totally devoid of the defects and surpassed in growth. In addition, some of the clones showed a narrow crown which is required for closer planting at a spacing 3 × 1.5 m enabling harvesting of trees at 4 years age.

Wood, Pulp and Paper Properties

Significant variation in wood basic density was recorded in Bhadrachalam clones of Eucalyptus (Table 5.5). By and large wood density increased with age up to 9 years in all the clones studied. Clone 4 displayed densest wood (710 kg/m^3) which was significantly different from clone 6 with lightest wood. The basic density value higher than 600 kg/m^3 is not desired in Eucalyptus. For better pulp yield trees, clones such as 3, 6 and 7 are better compared to clone 4. Among 98 clones studied, clones 10 (699 kg/m^3) and 433 (676 kg/ m^3) also have high density. The fiber length varied from 0.7 to 1 µm and with age (for clone 3 - at 3 years 0.82 and 9 years 1 µm). Clones gained considerable superiority over seedling crop with optimal basic density and longer fibers. Wood quality is significantly influenced by various factors such as clone, age, site/soil type, spacing, irrigation, etc. Age seems to be the most crucial factor that determines the pulp wood quality up to the age of 9 years (Kulkarni H.D., 2011a).

Table 5.5 Estimated adjusted means of clones with regard to selected wood properties

Clone No.	*Tree Vol. m³/tree*	*Basic Density kg/m³*	*Moisture content %*	*Heart wood %*	*Fiber length μm*	*Fiber dia. μm*	*Lumen width, μm*	*2 x fiber wall thickness μm*	*Vessels per mm²*	*Vessel dia. μm*
3	0.238	575.8	31.8	20.3	0.88**	19**	8**	10**	2.2*	131
4	0.169	613.9**	18.8	34.8	0.81*	16*	7	9*	2.6	136
6	0.138*	573.1*	21.4	32.2	0.86	16*	7*	9	2.5	144
7	0.346**	580.7	30.4	17.8	0.84	18	8	10	2.7**	141

* Significantly lower value, ** significantly higher value

More than 225 clones are evaluated for pulp and paper properties. The high pulp yield of 50 to 53% is recorded for *E. tereticornis* clones 3, 6, 7, 10, 27, 158, 273, 274, 279, 286, 288, 411, 412, 2014, and its hybrids 2050, 2121, 2129, 2135, 2140, 2143, 2156, 2294, 2306, 2315 and 2401 as against 44% from the wood derived from unimproved seed source plantation (Fig. 5.4). Yield increase from 48.4 to 51% for clone 3 and 47.4 to 49.8% for clone 7 from year one to year four indicated influence of age on pulp yield. The holo-cellulose content varied from 54.3 to 70% with a mean value of 63.5%. Lignin (range 24.7 to 35.7%) and pentosans (9 to 17.2%) content showed high variation with mean values of 29.4 and 13.9% respectively. The strength properties indicated by strength index value varied considerably from 20 to 85 with a mean value of 41 (Table 5.6).

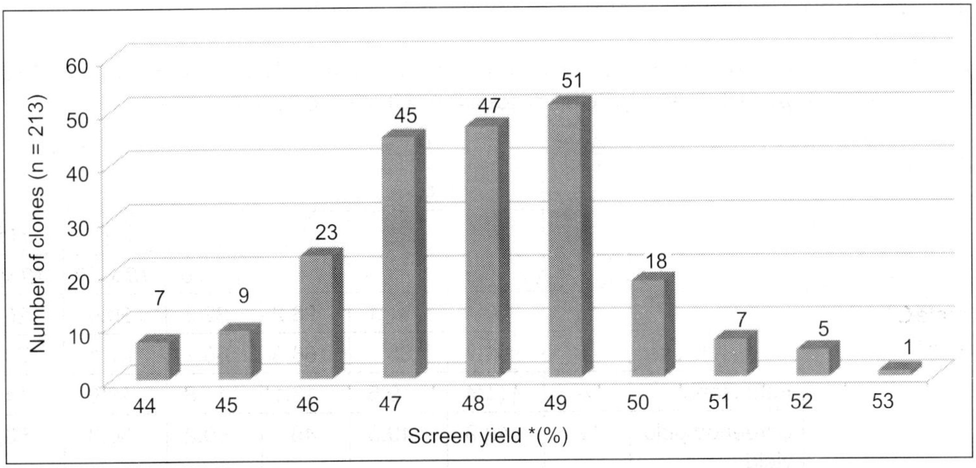

Fig. 5.4 Pulp screen yield for *E. tereticornis* clones

Table 5.6 Variation in pulp & paper properties for *E. Tereticornis* clones

Parameter	*Unit*	*Min*	*Max*	*Mean*
Ash	%	0.37	4.20	1.3
A-B extractives	%	1.24	8.40	3.3
Lignin	%	24.7	35.7	29.4
Holo-cellulose	%	54.30	70.00	63.7
Pentosans	%	9.00	17.20	13.9
Screen pulp yield	%	44.00	53.00	47.9
Rejects	%	0.23	4.10	1.2
Kappa	No.	19.00	29.70	22.0
UBV	Cps	11.90	22.80	14.8
Brightness	%	24.40	39.00	32.8
R.A.A.	Gpl	3.10	13.30	6.9
Solids	%	11.00	19.20	14.4
Organics	%	51.50	62.00	56.8
In-organics	%	38.00	48.50	43.0
Bulk	cc/gm	1.34	1.97	1.7
Strength index		20.00	85.00	41.0

Significant increase in pulp yield from 46 to 50.8% and strength index from 33 to 87.8 is shown by the hybrid progeny clones 2120, 2121 and 2156 over their parent clone 6, 10 and 27 (Table 5.7). Pulp yield and strength properties are the important characteristics which are financially rewarding and advantageous to the paper mills. Bhadrachalam clones of Eucalyptus are ranked according to the pulp yield and strength properties and top 25 clones are recommended for large-scale multiplication and planting under farm forestry programme of the company (Kulkarni H.D., 2011b).

Table 5.7 Comparison between parent and hybrid progeny for improvement of pulp and paper traits in ITC Bhadrachalam Clones of Eucalyptus

Properties		*Unit*	*Parent (at 6 yrs age)*			*Hybrid clone (at 3 yrs age)*		
			6	*10*	*27*	*2120*	*2121*	*2156*
						10 × 6	*10 × 27*	*10 × 27*
Proximate Chemical properties	Lignin	%	30.4	31.1	28.2	32.8	29.6	26.7
	Holo-cellulose	%	64.2	63	68.5	66.7	67.1	70.1
	Pentosans	%	11.4	9.5	16.1	11.8	13.9	14.1
	Screened pulp yield	%	48.9	49.5	46	50.2	50.8	50.6
Strength properties	Bulk	cc/gr	1.5	1.5	1.9	1.7	1.5	1.6
	Strength index	–	60.4	57.3	33	81	87.8	56

Clonal Nursery

For a successful clonal forestry programme, a good nursery is a prerequisite. A world class, state of the art, modern clonal nursery with an annual production capacity of 2.5 million ramets was established on 12 ha area in 2012. The old infrastructure for clonal propagation with a capacity of 1.5 million plants, started in 1989 is on 10 ha land consisting of 120 mist chambers covering an area of 12,000 m^2, hardening area of 5000 m^2 and 1,00,000 m^2 for open nursery. The clonal technology with root trainers has given considerable improvement in the production of quality planting stock. The root development is better than seedlings raised in polypots as multiple roots form in the root trainers and root coiling is totally avoided. The out planting results were quite high, thereby increasing survival and productivity.

Package of Practices

Innovative package of practices for raising and maintenance of clonal Eucalyptus plantations were developed and demonstrated the benefits of the same to the farmers. Study of soil profiles and analysis of soil samples were carried out to match adaptable clones to the planting sites. Deep ploughing of the soil with disk ploughs or mould-board ploughs in both directions is recommended for preparing the fields for transplanting of clonal saplings. Spacing of 3 × 2 or 3 × 1.5 m (with 1666 or 2222 trees per ha) is preferred for the production of poles and pulpwood, and larger spacing is desirable for production of timber. Transplanting in 30 cm^3 pits is carried out during the early parts of the monsoon rains so that plants establish and grow well benefiting from the good moisture availability throughout the monsoon rains. Soil in and around the planting pit is treated with 2 ml of Chloropyriphos in 1 litre of water to prevent damage to the young clonal saplings by termites during the critical establishment stage. Application of extract of botanical pesticides like kodesa (*Clistanthus collinus*) for controlling termites was introduced as an eco-friendly replacement to chemical pesticides. Cultural practices recommended include timely weeding and soil working, protection against damage by insect pests and cattle and raising of leguminous crops in between the 3 m wide planting rows for green manuring.

In addition, inter-cultivation with cotton, chili, tobacco, pulses, vegetables and even dry land paddy was encouraged during the first year of planting which gives additional earnings to the farmers. A new concept evolved after a series of experimental trials with different planting geometry is "agro-forestry" model (Prasad et al., 2010). Under this model, pulpwood trees are planted apart either in a single or double row leaving a wide gap of 8 m to allow maximum sunlight for growing food crops in-between the lines. Every year, the farmer can cultivate food crops and harvest the trees after 4 years interval. The income from crops ensures an annual income and trees are felled after 4 years. In case of crop losses due to drought, flood, pest and diseases, the income from tree harvest hedges the risk. The model accommodates 2000 trees per ha without impacting wood productivity. Land allocation to forestry is only 25% and remaining 75% land is available for agriculture crop (polyculture). The novelty of the agro-forestry is to improve the farm productivity and profitability, while conserving the environment and bringing diversification on farm lands. ITC is therefore popularizing this agro-forestry model for wood and food security so that more trees are planted in the farmland without sacrificing the crop production. This programme commenced in the year 2009, and by 2012 nearly 2427 ha of agro-forestry plantations have come up (Fig. 5.5 a and b).

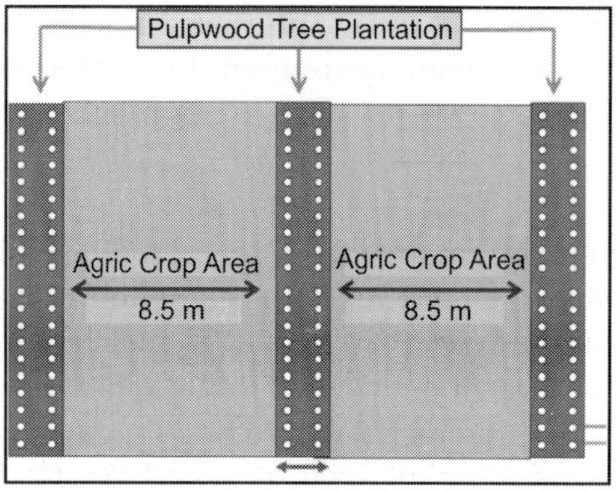

(a)

(b)

Fig. 5.5(a) Agroforestry lay out

Fig. 5.5(b) Agroforestry plantation

As most of the soils in India are deficient in nitrogen and phosphorous, fertilizers to supplement availability of these deficient plant nutrients were applied. Soil and water conservation measures like raised field boundaries and staggered trenches were taken up in well-drained planting sites for holding the rainwater. However, in low-lying areas or poorly drained heavy black cotton soils, drainage was improved to drain excess water in rainy season.

Casuarina

The company started TIR on Casuarina in the year 1993. So far, 240 CPTs from *Casuarina equisitifolia* and *C. junghuhniana* have been selected and cloned. The plus tree ramets produced through cladodes from mature trees were planted in a hedge orchard (Gurumurthi and Kumar, 1996; Santhakumar et al., 2001) and were maintained at 1 to 2 m height as hedge. Juvenile cuttings were obtained for clonal propagation. Six to seven year old clonal trials revealed that the MAI of 27 t/ha/yr was put up by the best performing Casuarina clones in comparison to 6 t/ha/yr by the seedling plantations. Casuarina clone 50 put up 162 t/ha compared to the plot mean of 77 t/ha. Promising clones CHCE 893004, CPCE 890108, ITC-Bhadrachalam 28, 32, 49, 50 were short-listed with respect to growth and apparent disease resistance for mass multiplication for raising farm forestry plantations. Blister bark wilt disease caused by *Trichosporium vesiculosum* was recorded on Casuarina trees and it was also reported on Casuarina hybrids in Tamil Nadu (Narayanan and Sasidharan, 1998). Casuarina trees were also affected by Loranthus epiphytic parasite.

Leucaena

In recent years Leucaena is becoming an increasingly popular species for production of pulp and paper, especially in southern India. The Leucaena TIR was started in the year 2001. ITC is promoting plantation of 5000 ha annually under farm forestry in Andhra Pradesh. Several Leucaena varieties were planted in India. Pure seeds of K-8 and K-28, which were planted on large scale, are not available because of seed mixture. The species (Hughes, 1998) introduced are *L. leucocephala*, cultivars, K-8, K-28, K-636; *L. diversifolia* cultivars, K-156, K-784; Leucaena hybrid cultivars, KX2 (*L. leucocephala x L. pallida*) and Leucaena hybrid cultivar, LxL (inter-specific hybrid of five subspecies of *L. leucocephala*) are reintroduced into Leucaena TIR programme from Hawaii.

The selection of CPTs was made in existing plantations of K-8, K-28 and K-636. Nearly 141 CPTs are selected and domesticated. The coppice cuttings were rooted under mist conditions. Nearly, 76% rooting is achieved. Age of coppice cuttings is very much crucial as three-month old coppice cuttings showed less (20%) rooting, while one month old juvenile cuttings rooted (76%) well. The ramets belonging to 141 CPTs were planted in one hectare gene bank in mill premises. The assessment of clonal trials revealed that clone 16 attained 33 t/ha/yr and the average productivity was 27 t/ha/yr. Whereas, the productivity of seed route plantations was 8 t/ha/yr. So far, 12 promising clones were short-listed. The best spacing was found to be 2.5 × 1.5 m.

ITC Form Forestry Plantations

The company has promoted farm forestry plantations over 140,989 ha with 607 million saplings from 1992 to 2012 (Figs. 5.6 and 5.7). The species-wise plantations promoted are - Eucalyptus 60,352 ha, Leucaena 67,007 ha and Casuarina 13,630 ha. The core area planting (200 km radius of the mill) is 69,283 ha and in non-core area 71,706 ha involving nearly 75,000 farmers. ITC provides R&D support, genetically superior high yielding disease resistant clonal planting stock, marketing and extension services to growers from selection of sites, planting and maintenance till harvest of produce along with buy-back arrangement (Kulkarni H.D., 2008). By 2020, ITC plans to cover additional 160,000 ha and with 140,989 ha, the total plantation area will be 300,000 ha.

Societal Gains

Asset Creation and Employment Generation

The plantation programme has introduced an alternative means of livelihood for the rural community. It is estimated that the plantation of 140,989 ha over a period of 33 years (1992 to 2025) gives an estimated wood asset value of US $ 2727 million (Rs. 13,637 crore) at an average farm gate price of Rs. 2500 per tonne (price range from Rs. 700 in 1992 to Rs. 3000 in 2012 and likely to be Rs. 4500 by 2025). The average net income to farmer from clonal plantation is about US $ 500/ha/yr under rain-fed condition and

4-year-old Clonal Eucalyptus Plantation

3-year-old Clonal Casuarina Plantation

3-year-old Clonal Leucaena Plantation

Fig. 5.6 Colon pulpwood plantations

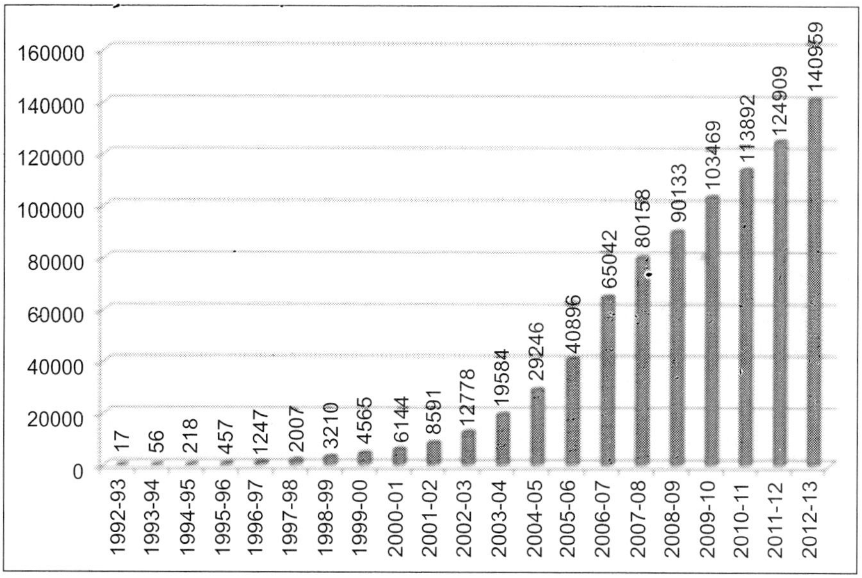

Fig. 5.7 ITC agro farm forestry plantations (1992 to 2012)

US $ 800/ha/yr with irrigation on a four-year rotation cycle. This is significantly higher compared to traditional crops grown in the operational area and at much lower risks. The impact of clonal technology can be measured from the fact that since 1992 when meager 17 ha was under cultivation, the situation has drastically changed in 2012 as 140,989 ha area is planted. Thus, plantations are playing an increasingly important role in rural economic development and poverty alleviation in company's catchment area. Since employment generation for farm labours is important for reducing poverty, the 140,989 ha of plantations provide an estimated 63 million person days of employment from tasks such as nursery, planting, logging and maintenance operations. Social and farm forestry have already contributed greatly to rural livelihood and created more jobs for local people.

Environmental gains

Clonal plantations have additionally been able to mitigate environmental degradation on a large scale greening efforts. It is estimated that 140,989 ha of plantation has an estimated potential to sequester 93.28 million tonnes of carbon by reducing 171 million tonnes of CO_2 over a period of 33 years (1992 to 2025). On an average, the annual CO_2 reduction works out to 5.18 million tonnes. ITC therefore is a carbon positive company as it sequesters more carbon through plantation than it emits from all of its operations. By promoting plantations on such a large scale, the project has increased the green cover in the country. Apart from the innumerable benefits of such large-scale afforestation, clonal plantation directly contributes to in situ moisture conservation, groundwater recharge and significant reduction in topsoil losses due to wind and water erosion and help in conservation of natural forest resources. Equally, as a result of the leaf-litter from multi-species plantations and the promotion of leguminous inter-crops between rows, depleted soils are getting constantly enriched, making these farmlands more productive. In the near future, the increase in soil fertility will lead to a decline in fertilizer and pesticide consumption, thus reducing the pollution of groundwater sources through leaching of such chemicals.

REFORESTATION CDM PROJECT

The Clean Development Mechanism (CDM) reforestation project (UNFCCC Reg. No. 2241) covering 3070.19 ha in Khammam district of Andhra Pradesh is registered on 5-6-2009. This project generates 57791 Certified Emission Reduction (CERs) annually, and for a period of 30 years 1.87 million CERs are estimated. The estimated value for these 57791 annual CERs at US $ 4 per CER works out to US $ 231,164 (Rs. 1.15 crore). There are 3398 tribal beneficiaries in the project (Kulkarni H.D. 2003, 2007; Sudha et al., 2007). The project has been verified recently for a period of 9 years (2001 to 2009) and on 14-12-2012, UNFCCC has issued 403,610 CERs. Such CDM forestry projects will help paper mills in generating more wood through plantations.

Forest Stewardship Council Certification

Forest Stewardship Council (FSC) certification is becoming a non-trade barrier for the Indian paper industry and the farmers. ITC took proactive initiatives in obtaining Chain-of-Custody (C-o-C) and Forest Management (FM) certification. The FSC FM certificate for 8028 ha is the first of its kind for the paper industry in India. Farmers are paid premium price for the FSC FM wood.

Hence, the efforts of ITCs social farm forestry lead to greening the bottom line such as:

- for the farmer it is commercially attractive land use option in taking care of climatic uncertainties;
- for the agricultural labourers there is additional on farm employment every year;
- for ITC it is cost effective procurement of pulp wood at its door step;
- for the nation it is converting waste lands into productive assets adding to the green cover; and
- saving public forests and foreign exchange.

Research Needs

As degraded and exhausted sites are to be afforestated, agronomic research becomes a necessity to grow variety of tree corps for obtaining sustained productivity. Presently, block planting method is adopted on large scale in farm forestry plantations; research on various agro-forestry models is essential for maximizing productivity and returns per unit area. Industry should intensify sustainable forest management (SFM) based plantation research to ensure that the trend in plantation productivity is non-declining and increase over successive years. The insect-borne diseases on Eucalyptus such as the little leaf disease (caused by Thrips) and gall disease (caused by Leptocybe invasa gen. & sp.n.) (Mendel et. al., 2004; Kulkarni, 2010) reported in the recent times in India are creating havoc in nurseries and plantations and Eucalyptus clonal forestry is under threat. Research on evolving control measures and long-term breeding programme to breed resistant varieties is therefore essential.

Plantation Strategy for Future

Paper consumption is going to increase to 24 million tonnes by 2025, which requires planting of 1.2 million hectares to obtain 12 million tonnes of wood. A strategy to enlarge farm forestry and raise plantations with the help of forest development corporations appears to be the solution for meeting the demand. Clonal plantations covering 1.2 million ha of the degraded forest areas, if taken up, can yield 25 million tonnes of pulpwood annually. That would be sufficient for meeting India's entire pulp and paper requirements. Likewise, high yielding short rotation clonal plantations on 20 million ha of wastelands/degraded forestlands could meet country's current firewood requirements on a sustainable basis. That would minimize biotic pressures on remaining natural forests and conserve their rich biodiversity. In addition to restoring marginal lands to high sustainable productivity, clonal plantations will generate vast employment opportunities for the rural poor, contribute to environmental amelioration and help conservation of precious soil and water resources. Such plantations will also create opportunities for significant value addition through local processing of plantation wood and save large amounts of scarce foreign exchange used to import of wood based products. Therefore, clonal forestry can bring "***Brown revolution***" in India.

Acknowledgements

The author is grateful to the management of ITC company who encouraged and supported the plantation research and development team. Thanks to the Central Lab, research, production, marketing and extension manager for active support and encouragement. Various farmers, State Forest Departments/Forest Corporations who adopted "ITC-Bhadrachalam" clones, served as the main source of inspiration and contributed to the success of the programme.

REFERENCES

1. RISI (2010). "Forecast, August 2010", Annual Review 2010. RISI. USA.
2. Anon., "Forestry – Enhancing the natural capital", Proc. Sustainability Summit Asia-(2006). Promoting excellence for sustainable development. CII-ITC centre of excellence for sustainable development, New Delhi, pp. 25-40.
3. Ahuja, S.P. (1980). "Paper Industry in India. Retrospect and prospects", The Institute of Economic and Market Research, New Delhi, pp. 39-48.
4. Anon (1996). "A panorama of paper industry in India", Joint committee of paper industry (JCPI), Indian Paper Makers Association, New Delhi, pp. 3-8.
5. GoI (2011). "Pulp and paper sector – report of the working group for 12th five year plan 2012-17". Department of Industrial Policy & Promotion, Ministry of Commerce & Industry. New Delhi, pp. 2-14.
6. Gopalratnam, N. (2007). "Paperex background". Proc. 8th Int. Tech. Conf. Pulp, Paper, Conversion and Allied Industry, Paperex, 2007, New Delhi, India, pp. 1-2.

7. Suri, R.K. (2007). "Emerging trends in environmental management in paper industry". Proc. 8th Int. Tech. conf. Pulp, Paper, Conversion and Allied Industry. Paperex 2007, New Delhi, India. pp. 64-73.
8. CPPRI. "Development of statistics of Indian paper industry", Proc. of Second Workshop on Frontier Technologies Addressed under CESS Projects and Energy norms for the Pulp and Paper Industry. CPPRI, Saharanpur/New Delhi, (2005).
9. Toland, J. (2006). "India goes upscale". Pulp and Paper International, 48(8): 12–32.
10. FAO (2005). "Global forest resources assessment 2005", Progress towards sustainable forest management, FAO forestry paper 147, FAO, Rome, Italy.
11. Pandey, M.C. (1992). "Farm forestry in context to raw material crisis in paper industry", IPPTA, 4(1): 7-11.
12. GoAP (1999). "Andhra Pradesh agricultural produce and live stock market Act, 1958/1966", GO MS No. 158, dated 10-6-1999, Hyderabad, Andhra Pradesh, India.
13. Saigal, S., and Kashyap, D. (2002). "The second green revolution: Analysis of farm forestry experience in western tarai region of Uttar Pradesh and coastal Andhra Pradesh", Instruments for Sustainable Private Sector Forestry, India Country sub study No. 6. ETS and International Institute for Environment Development, New Delhi, India, pp. 95-126.
14. FSI (1999). "Executive summary. State of forest report 1999", Forest Survey of India, Dehra Dun, India, pp. 9-10.
15. FSI (2000). "Valuation and evaluation of trees outside forests of India". Forest Survey of India, Dehra Dun, India, pp. 9-27.
16. FSI (2005). "Executive summery. State of forest report 2005", Forest Survey of India, Dehra Dun, India, pp. 4-5.
17. Tewari, D.N. (1996). "A monograph on bamboo", International Book Distributors, Dehra Dun, India, pp. 207-251.
18. Kulkarni, H.D. (2001). "Eucalyptus hybrid breeding in ITC Bhadrachalam, India", Proc. IUFRO International Symposium developing the Eucalyptus for the Future. (Unit. 2.08.03 Improvement and Culture of Eucalyptus), Valdivia, Chile.
19. Kulkarni, H.D. (2004). "Clonal forestry for industrial wood production. An ITC experience". In: Parthiban, K.T. et al., (Eds.)., Compendium on Clonal Forestry, Forest College and Research Institute, Tamil Nadu Agri. Univ., Mettupalyam, India, pp. 92-113.
20. Rajan, B.K.C. (1987). "Versatile Eucalyptus". Diana Pub., Bangalore, pp. 183-201.
21. Kulkarni, H.D., and Lal, P. (1995). "Performance of Eucalyptus clones at ITC Bhadrachalam India". In: Potts, B.M. et al., (Eds.)., Eucalyptus plantations: Improving Fiber yield and quality, Proc. CRCTHF - IUFRO Conf., Hobart, 19-24 Feb, 1995, pp. 274-275.
22. Mendel Z., Protaspv, A., Fischer, N., and Salle, J.L. (2004). "Taxonomy and biology of *Leptocybe invasa* gen. and sp.n. (Hymenoptera: Eulophidae), an invasive gall inducer on Eucalyptus, Aust. J. Entomology, 43: 101-113.

23. Kulkarni, H.D. (2010). "Screening Eucalyptus clones against *Leptocybe invasa* Fischer and La Salle (Hymenoptera: Eulophidae)", Karnataka *J. Agric. Sci.*, 23(1): 87-90.
24. Kulkarni, H.D. (2011a). "Wood quality studies of ITC Bhadrachalam Clones of Eucalyptus". In: J.L. de M. Goncalves. et al. (Eds.)., Proc. Improvement and Culture of Eucalyptus, IUFRO Working Group 2.08.03, Porto Seguro, Bhaia, Brazil, pp. 278-293.
25. Manage (2003). "Economics of ITC Clonal Eucalyptus", Manage, Hyderabad, p. 10.
26. Kulkarni, H.D., (2011b). "Variation in pulp yield and paper properties of ITC Bhadrachalam clones of Eucalyptus". In: J.L. de M. Goncalves. et al. (Eds.)., Proc. Improvement and Culture of Eucalyptus, IUFRO Working Group 2.08.03, Porto Seguro, Bhaia, Brazil, pp. 511–523.
27. Prasad, J.V.S.N., Korwar, G.R., Rao, K.V., Srinivas, K., Srinivasarao, Ch., Pedababu, B., Venkateswarlu, B., Rao, S.N., Kulkarni, H.D. (2010). "On-Farm evaluation of two fast growing trees for biomass production for industrial use in Andhra Pradesh, Southern India", New Forests., (Springer, DOI 10.1007/s 11056-010-9236-6), Germany.
28. Prasad, J.V.S.N., Korwar, G.R., Rao, K.V., Srinivas, C.A., Rama, Rao., Srinivasarao, Ch., Venkateswarlu, B., Rao, S.N., Kulkarni, H.D. (2010). "Effect of modification of tree density and geometry on intercrop yield and economic returns in Leucaena based agro forestry systems for wood production in Andhra Pradesh, Southern India", Expl. Agric., pp. 1-18.
29. Gurumurthi, K., and Kumar, A., (1996). "Establishment of a clone bank of *Casuarina equisetifolia* for tree improvement". In: Pinyopusarerk, K. et al., (Eds.)., Recent Casuarina Research and Development, Proc. Third International Casuarina Workshop, Da Nang, Vietnam, pp. 159-161.
30. Santhakumar, P., Venkatesh, K.R., and Kulkarni, H.D. (2001). "Clonal trials of Casuarina at ITC Bhadrachalam", Fifth Annual Workshop on Casuarina Network. Regional Forest Research Centre, Rajahmundry, Andhra Pradesh, IFGTB, Coimbatore.
31. Narayanan, C., and Sasidharan, K.R. (1998). "Occurrence of Blister-Bark disease in Casuarina hybrid (*C. equisetifolia x C. junghuhniana*)", Ind. For., 124 (2): 167-168.
32. Hughes, C.E. (1998). "Leucaena a genetic resources handbook", Oxford Forestry Inst., Oxford, pp. 107-119.
33. Kulkarni, H.D. (2008). "Private farmer - private industry partnerships for industrial wood production. A case study", *Int. For. Rev.*, 10(2): 147-155.
34. Kulkarni, H.D. (2003). "Forest sector mitigation projects: Carbon benefits of industrial plantations", *IIPTA*, 15(2): 47-51.
35. Kulkarni, H.D. (2007). "Carbon trading - clean development mechanism opportunities for pulp and paper sector", Proc. 8th Int. Tech. Conf. Pulp, Paper, Conversion and allied Industry, Paperex 2007, New Delhi, India, pp. 34-42.
36. Sudha P., Ramprasad, V., Nagendra, M.D.V., Kulkarni, H.D., and Ravindranath, N.H. (2007). "Development of an agroforestry sequestration project in Khammam district, India". Mitig Adapt Strat Glob Change, Springer, Bonn, Germany.

6

Sustainable Management of Natural Gums and Resins in Madhya Pradesh

Pratibha Bhatnagar and Radhika Urmalia
State Forest Research Institute, Jabalpur (M.P.)
Email: pratibhasfri1@gmail.com

ABSTRACT

NTFPs sustain rural livelihood and alleviate poverty. Millions of tribals depend upon NTFPs for food, medicine, fuelwood, fodder and cash income. Gums and resins are special products and important components of forest ecosystem. Forests support about 100 million people inhabiting near forest areas and 70 per cent employment is from NTFP collection. This paper discusses important gum yielding species (*Sterculia urens*) of Madhya Pradesh. In past decades, the area under forest and natural vegetation has declined and the unsustainable use and over exploitation for commercial trade has increased to a greater extent. This is particularly in the case of gums, which are commercially important. The management of gums and resins has not been given adequate attention to check the loss of trees and regeneration. Decline in gum production is due to increasing demand of gums and resins, leading to uncontrolled tapping practices of gum yielding species, especially *Sterculia urens*. Village level skill development programmes for gum tappers and staff of forest department in Sheopur forest division conducted by Indian Institute of Forest Management (IIFM) have suggested sustainable management practices for harvesting, processing, grading and storage purpose. Primary processing of gums may increase the income of local tribal community manifold on their collection by drying and grading. Though sustainability depends on various parameters, natural regeneration of the species should be given priority by the forest department for restoration of this species in future.

INTRODUCTION

Madhya Pradesh is endowed with rich forest wealth. The total geographical area is 3,08,245 sq. km out of which forest area covers 7700 sq. km (FSI, 2011). Several important NTFPs are found in the forests, e.g., Tendu leaves (*Diospyros melanoxylon*), Sal seeds (*Shorea robusta*), Harra (*Terminalia chebula*), Chironji (*Buchanania lanzan*), flowers and seeds of Mahua (*Madhuca indica*) and gums, etc.

NTFPs sustain rural livelihood and alleviate poverty. Millions of tribals depend upon NTFPs for food, medicine, fuelwood, fodder and cash income. Gums and resins are an important component of NTFPs. Some plants exude only gums, some others only resins and yet others both, as a consequence of the phenomenon occurring due to the attack of disease or injury to the bark or wood. These gums/resins are produced as exudation, usually from the stem and sometimes from the roots, leaves and other portions also. Most of the gums and resins can be harvested from three to nine months in a year. Forests support about 100 million persons inhabiting near forest areas, and 70 per cent employment is from NTFP collection (FAO, 1995).

This paper discusses important gum yielding species of Madhya Pradesh as given below:

1. *Acacia arabica var. nilotica* (Arabic/ babool gum)

An erect tree, with straight spines, commonly known as babul, found widely distributed throughout the state. Tree exudate gum Arabic naturally. The tapping method commonly followed consists of making incision on the trees with axe. The gum consists of tears or broken pieces of light dusky brown tint. When dry, it is permeated by cracks and is very brittle.

2. *Acacia catechu* (Khair gum)

It is a moderate-sized deciduous tree, with dark brown, much cracked bark and short-hooked spines in pairs, commonly found in dry forests throughout the state. Khair gum obtained from the tree is not collected separately and is generally mixed up with other *Acacia* gums.

3. *Anogeissus latifolia* (Gum ghatti)

It is a large tree found widely throughout the state. So far, no systematic method of tapping has been worked out. The prevalent method of tapping consists of making incision on the tree with an axe. The gum oozes out of the wounds in round or vermicular tears. There are five grades which are currently recognized in trade.

4. *Azadirachta indica* (Neem gum)

It is a tree having short, straight bole, dark brown to grey bark, with dense rounded crown of pinnate leaves. Commonly distributed throughout the warmer parts of the world. Its bark yields clear, bright and amber-coloured non-bitter gum.

5. *Butea monosperma* (Kamarkas gum)

It is also known as flame of the forest and is common throughout the country except in very arid parts. Tree exudates gum called Kamarkas found in small, brittle glistening pieces and reddish-black in colour. They are odourless with a very astringent taste and stick to the teeth when chewed and make the saliva bright red. It is also known as Bengal Kino, and is considered valuable by druggists because of its astringent qualities, and by leather workers as tans. Too many incisions are usually made all through the tree bole, and even on branches for gum.

6. *Lannea coromandelica* (Jhingan gum)

It is moderate to large-sized deciduous tree found commonly throughout the greater part of India and up to 1500 m in the sub-Himalayan tract. The mucilaginous gum is exuded from wood and cracks of bark. Tapping is done by making shallow cut into the bark. The gum collection is done mostly in the eastern part of the state.

7. *Cochlospermum religiosum* (Katira gum)

A small or medium-sized deciduous tree of 3- 6 m in height. The tree yields a gum similar to karaya in appearance and properties; also called Indian or country tragacanth.

8. *Sterculia urens* (Karaya gum)

Kullu is found in dry deciduous forests, mostly on stony or rocky soils. It occurs usually in association with *Boswellia serrata*. Almost fifty percent of gum karaya is produced from

Table 6.1 Classification of gums and resins of Madhya Pradesh

S. No.	*Category*	*Trade/ Common name*	*Source (Botanical Name)*	*Physio-chemical properties of exudates*	*Distribution in Madhya Pradesh*	*Uses*	
A.	Gums					Medicinal	Industrial
A.1.	Gums	Arabic/babul gum	*Acacia arabica Var. nilotica*	Pale straw to a dark reddish brown or almost black colour, complex, slightly acidic, polysaccharide, translucent, viscous and completely soluble in water.	Throughout the state especially in Gwalior, Bhind, Morena, Shivpuri, Sheopur, Guna, Rajgarh, Bhopal, Sehore, Indore, Dewas, Ratlam, Mandsour, Ujjain, Khandwa, Khargone, Betul, Hoshangabad, Shahdol, Umaria, Rewa, Satna, Panna, Chhatarpur, Sagar and Damoh.	Diarrhoea, dysentery, diabetes and given to recently delivered women as a tonic, combination with quinine in fever.	Cosmetic, adhesives, paints and confectionery industries as an emulsifier and as a glaze in candy products.
		Khair gum	*Acacia catechu*	Pale-yellow often occurring in tears, 2.5 cm in diameter mostly bright coloured varying to dark amber, sweet and soluble in water.	Common in dry forests throughout the state.	Diarrhoea, dysentery, diabetes, haemorrhage, cough, throat, pulmonary and catarrhal infection.	Dyeing, sizing material for silk and cotton, paper manufacturing and as adhesives in industries.
		Ghatti gum	*Anogeissus latifolia*	Soluble in water, forming colourless viscous mucilage and consists of pentose and galactose.	Common in dry forests, throughout the state especially in Sheopur.	Scorpion sting and snake bite.	Textile sizing, calico printing, pharmaceutical and paint industries.
		Neem gum	*Azadirachta indica*	Clear, bright, amber coloured tears or fragments, non-bitter in taste and is soluble in cold water, contains 35% of proteinaceous material.	Throughout the state.	Exudation used as a tonic and bulking agent to prepare special food for diabetic patients.	Used as a dye in textiles.

		Kamarkas/ butea/bengal gum	*Butea monosperma*	The juice exuded by bark hardens into brittle ruby coloured gum beads.	In Tikamgarh, Chattarpur, Damoh, Sagar, Panna and Umaria.	Used as a tonic, astringent, aphrodisiac and diuretic.	Food, fibre and raw material for household industry.
		Jhingan gum	*Lannea coromandelica*	Round tears or colourless angular fragments, soluble in water, forming thin mucilage with good adhesive properties.	Umaria, Damoh and Sagar	Applied to sprains and bruises after making it into an ointment with coconut milk also given in asthma and as a cordial to women during lactation.	Paper and cloth sizing, making inferior varnishes, reserving fishing nets and confectioneries.
A.2.	Gums	Katira gum	*Cochlospermum religiosum*	Pale and semi-transparent, transverely stirated, insoluble in water, but swells into a pasty transparent mass.	Jabalpur, Mandla, Damoh, Sagar, Chhatarpur, Datla, Panna, Rewa, Satna, Seoni, Balaghat, Chhindwara, Gwalior, Bhind, Morena, Shivpuri, Sheopur, Bhopal and Guna.	Cough, hoarse throat, scalding in urine, diarrhoea and dysentery.	Edible, calico printing, leather dressing and to give polish to tassar silk.
		Karaya gum	*Sterculia urens*	Transparent white, viscous, acidic, partially acetylated polysaccharide and absorbs water rapidly to form viscous mucilage at low concentrations.	Sheopur, Shivpuri, Seoni, Damoh, Burhanpur, Umaria, Dewas and Khandwa.	Bulk laxative, appetite suppressant. Applied to dentures to reduce bacterial adhesion, used as post-surgical core sensitive skin and colostomy bag fixing.	Used in food, pharmaceutical, paper, textile and cosmetic industries.

Contd...

Contd...

B.	Gum-oleo resin	Guggal gum	*Commiphora wightii*	Pale yellow, reddish brown or dull green in colour consists of Diterpene hydrocarbon and alcohol, Z-guggulsteron, guggul sterol-1 besides sesamin and cholestrol with bitter aromatic taste.	Morena, Bhind, Shivpuri and Sheopur	Treat dysmenorrhea, dyspepsia, bronchitis, stomatitis, malaria, hyperlidemia and hypercholesterolemia reduces cholesterol, thyroid stimulant, etc.	Pharmaceutical and perfume industries.
		Indian olibanum/ Frankincense	*Boswellia serrata*	Golden brown to dark brown or dark greenish brown in colour, on hydrolysis mainly yields pentose with high proportion of arabinose while Galactose and xylose in small quantities.	Dry deciduous forests, especially in Sheopur, Shivpuri Tikamgarh, Gwalior, Khandwa, Badwani and Jhabua.	Rheumatics, nervous disorders, diaphoretic, astringent, antipyretic, antidysentric, expectorant, diuretic and stomachic.	Paper and pulp industries, packing cases, splints, food, pharmaceuticals, textile, binding agent in tablets, creams and lotions.

Sheopur, Shivpuri, Damoh, Dewas, Mandla, Dindori, Narsinghpur, Indore and Katni districts.

Although tapping should be confined to trees above 90 cm GBH, in practice rules are not adhered where dimension of the blaze is concerned, causing damage to the trees. Sustainable method of harvesting karaya gum has been evolved earlier but is being explored for potential of further improvement.

9. Commiphora wightii (Guggal gum)

It is a perennial, medium sized tree, distributed in the arid and semi-arid regions of the world. The trees exude gum-resins obtained by tapping mature trees through an incision on the trunk.

10. Boswellia serrata Roxb. (Salai oleo resin)

Boswellia serrata, locally known as salai, occurs frequently in dry deciduous forests of Madhya Pradesh. It yields Salai oleo-resin found mainly in Sheopur and Shivpuri. Tapping and collection of salai is usually done by Sahariya tribes. First blaze is given in the month of November and first collection after 15-20 days after the initial blazing. Tapping is done from about 15 cm wide band on the trees above 90 cm girth. This interval is known as the ripening periods in which wide cracks appear in the exposed phloem and it is through them that globular drops of oleo resin appear and start producing oleo resins.

In past decades due to decline in the area under forest and natural vegetation, unsustainable use and overexploitation for commercial trade has increased to a great extent. This is particularly in the case of gums, especially those which are commercially important. The management of gums and resins has not been given adequate attention to check the loss of trees and regeneration. Although prospects in trade are promising, the production rate has gone down. Some gums and resins important in trade are classified and detailed information on gums, phytochemical properties, distribution and medicinal and industrial uses is given in Table 6.1. (Anon, 1972, Bhausar and Tiwari, 1995 and Shiva, 2006).

Table 6.1 gives detailed information on gums, phytochemical properties, distribution and medicinal and industrial uses.

The exudate gums of commercial importance in the state are gum karaya (*Sterculia urens*), dhaora gum (*Anogeissus latifolia*), salai guggal (*Boswellia serrata*), khair gum (*Acacia catechu*) and babul gum (*Acacia nilotica*). These are notified as "specified forest produce" under the M.P. *Van Upaj* (*Vyapar Viniyaman*) Act, 1969. Karaya and salai gums are grouped together as a class I gums, whereas dhaora, babul and khair as class II gums. These gums are obtained by tapping trees.The production of class I and II category gums is given in Table 6.2.

Table 6.2 Collection of class I and II gums in year 2000 to 2013

S. No.	Collection Year	Class I gums		Class II gums
		Karaya gum collection (Quntls.)	Salai gum collection (Quntls.)	Dhaora, khair and babul gum collection (Quntls.)
1.	2000-01	352.20	7036.00	–
2.	2001-02	328.91	5947.00	1196.12
3.	2002-03	529.40	5072.00	904.23
4.	2003-04	215.93	–	403.53
5.	2004-05	422.00	–	742.16
6.	2005-06	333.90	–	145.00
7.	2006-07	567.26	–	141.58
8.	2007-08	234.89	–	306.00
9.	2008-09	231.97	–	560.00
10.	2009-10	87.44	–	619.25
11.	2010-11	64.81	–	26.74
12.	2011-12	–	–	64.50
13.	2012-13	–	–	955.00

Source: MP Minor forest produce, 2012; URL: www.vesdata.com/websit.html?webinfo=MP-State-Minor-Forest

An erratic and decreasing trend in gum production is observed from the above data. Decline in gum production is due to increasing demand of gums and resins which had been traced back since 1930s leading to uncontrolled tapping practices of gum yielding species, especially *Sterculia urens*. This activity had resulted in large-scale destruction and loss of trees in central provinces; a four-year rest period was enforced, but illicit tapping of trees continued (Singh, 1935). The state government had banned the extraction of class I gums from the year 1982 to 1992 respectively, but in 1995 and 1997 the ban was lifted and extraction of gum karaya was permitted in Morena, Bastar and Khandwa districts and of salai gum in Gwalior, Shivpuri, Ujjain and Khandwa forest circles, Narmada Sarovar submerged areas of Badwani and Jhabua forest divisions.

According to a survey conducted by Indian Institute of Forest Management (IIFM, 2002), about 30% of Kullu (*Sterculia urens*) trees were reported to die or permanently damaged in the state due to pollarding, lopping and disease attack. On the other hand, 76% followed by 56, 34 and 16 percentage of high intensity tapping practices on Kullu (*Sterculia urens*) trees were reported in Sheopur, Umaria, Damoh, Khandwa and Seoni districts (Bhatnagar et al., 2012). Similar destructive tapping practices, up to 72.09% followed by 33.33% on *Butea monosperma* was reported in Umaria and Shivpuri districts, whereas 20.58% on *Boswellia serrata* in Sheopur district. *Commiphora wightii* is also facing decline in the number of trees in Chambal ravines due to intensive tapping activities by members of *Malhar* community living near the river side (Thomas et al., 2010).

Thomas et al. (2010) stated that properly managed tapping practices do not necessarily reduce vitality, growth ability and reproduction of tree. According to Bhattacharya et al. (2002), 43% trees could recover to reach class I status (healthy trees) if imparted training to gum collectors in non-destructive gum-tapping techniques. Village level skill development programmes for gum tappers and staff of forest department in Sheopur forest division conducted by Indian Institute of Forest Management (IIFM) also suggested sustainable management practices for harvesting, processing, grading and storage purpose. Gums are sold by tribal collectors without drying and grading which results in loss of income. There is considerable difference in market rates for different gum yielding species.

Above discussed facts highlight the need for sustainable management practices. Some of the sustainable management practices recommended for gums and resins are listed below:

- Assessment of distribution of gums and resins yielding species.
- Monitoring of ban on unsustainable tapping practices and imparting training to gum collectors. Awareness to local communities towards sustainable harvesting practices. Training to all gum collecting families across nation as a national programme.
- Field level harvesting workshop from time to time for monitoring and establishing demonstration plot for demonstrating sustainable harvesting regime.
- Research efforts to be focussed on value-addition and processing of different gums.
- Detailed information should be collected on market network, price and trader-collector continuum. Bridging the gap between gum collectors and buyers by avoiding middle man traders.
- Motivation to farmers to plant declining tree species specially *Sterculia urens, Butea monosperma, Boswellia serrata, Commiphora wightii, Anogeissus latifolia, Buchanania lanzan*, etc.

REFERENCES

1. Anonymous (1972). Indian Forest utilization. Vol. II, Manager of Publication Delhi.
2. Bhatnagar, Pratibha., Goswami, M., and Jain, S., (2012). Conservation and harvesting of *Sterculia urens* (Kullu) gum in Madhya Pradesh. *Vaniki Sandesh*, 3(3): 1-5.
3. Bhattacharya, P., Joshi, B., Bhagat, N.K., and Hayat, F., (2002). Sustainable harvesting of Kullu (*Sterculia urens*) gum. Indian Institute of Forest Management. *Newsletter*, V (1 and 2): 3-5.
4. Bhausar, K.N., and Tiwari, A.K., (1995). State of art of availability, collection of conservation of gum yielding plants of M.P. *Vaniki Sandesh*, XIX (1): 8-28.
5. FAO (1995). Gums, resins and latexes of plant origin. *Non Wood Forest Products 6*, Rome.
6. FSI (2011). *Indian State of Forest Report*. Ministry of Environment and Forests, Govt. of India, Dehradun.
7. Shiva, M.P., (2006). Types of gums and resins and their industrial uses. *MFP News*, Vol. XVI (3): 4-6.
8. Singh, K., (1935). Kullu (*Sterculia urens* Roxb.) wailing from Damoh C.P. *Indian Forester*, LXI (3): 144-149.
9. Thomas, M., Shrivastava, A., and Tomar, S.S., (2010). Guggul: Production and Conservation. Directorate of Research Services, JNKV, Jabalpur, M.P. (India).

7

Livelihoods Interventions Under Joint Forest Management (JFM) Programme

Lolita Pradhan*, A.K. Kandya and Prodyut Bhattacharya†**

**Indian Institute of Forest Management, Nehru Nagar, Bhopal–462003*

***Department of Botany, Dr. H.S. Gour Central University, Sagar (M.P.) – 470003*

†University School of Environment Management, GGS Indraprastha University, Block-A, Sector-16-C, New Delhi – 110075

E-mail: pradhanlolita2013@gmail.com & prodyutbhattacharya@yahoo.com

ABSTRACT

The concept of Joint Forest Management (JFM) has come into sharp focus due to multiple uses of forest resources like the increasing presence of forest products in the market economy and greater understanding of the non-tangible benefits from forests. Community-based programme in forest management (which is popularly known as joint forest management in India) is fundamentally a decentralized grass roots/ bottom up movement under localized natural resource management programme initiated by forest fringe communities and government to strengthen communities' livelihood base and to protect natural forests from further degradation. Scientists and researchers have advocated this programme for sustainable forest management as well as livelihood generation. There is an intricate relationship between livelihood pursuits of tribal communities and surrounding ecosysytem. The critical balance between the two is essential for sustainable livelihoods of forest dwellers in the world, in general. In this perspective, the main objective of this article is to assess the impact of participatory initiative of JFM Programme on Rural Livelihoods of the across forest fringe communities in JFM areas of Madhya Pradesh.

INTRODUCTION

In most of the developing countries, open-access forest, inter alia common pool resources, serves as an important source of livelihood for a large population of the forest fringe dwellers. WCFSD (1999) states that around 350 million of the world's poorest people depend almost entirely for their subsistence and survival needs on forests, and in India, around 275 million landless people and small farmers benefit from gathering resources they find within adjacent forests (Das, 2007) (Pandey, 1996).

Of the 68 million tribal population of India, 50% still depend on the forest resources for various livelihood requirements, viz., food, fodder, medicine, small timber and a variety of other NTFPs (Bhattacharya and Hayat, 2004). Though forest products are normally not the principal cash earners of poor households, they can have important 'gap filling' functions. Unlike regular income contributions, these benefits accrue unevenly over time. The importance of forest product income is usually more in the way it fills gaps and complements other incomes, than in its absolute magnitude or share of overall household income (Byron and Arnold, 1999).

The term 'livelihood' connotes the activities, entitlements and assets by which people make a living. Assets in this particular context are defined as not only natural/biological (land, water, common property resources, flora and fauna) but also social (i.e., community, family, social networks), economic (i.e., job, savings, credit), political (i.e., participation, empowerment), human (i.e., education, labour, health, nutrition), and physical (i.e., roads, markets, clinics, schools, bridges). Livelihood interventions are conscious efforts by an agency or an organization to promote and support livelihood opportunities for a large number of people (other than those directly or indirectly employed by them) (Dutta *et al.*, 2004). Livelihood can be sustainable if it can cope with and recover from stresses and shocks and maintain or enhance its capabilities and assets, both now and in the future, while not undermining the natural resource base (Carney, 1998).

In 1987 World Commission Environment and Development (WCED) recognized the concept of sustainable livelihoods (Chambers and Conway, 1991; (Chambers, 2005), which was further acknowledged by Brundtland Commission on Environment and Development and UNED by recognizing it towards eradication of poverty of rural poor (Krantz, 2001). This concept of sustainable livelihood is based on people centered approach, which contains security to basic human needs, food security, sustainable agriculture, practices and poverty. After introduction of sustainable livelihood, this concept has been adopted by a number of government, non-governmental organizations and multilateral organizations like DFID, INDP, OXFAM, CARE.

Major Livelihood Generation Interventions in India

India's strategy for poverty reduction is accelerated and sustained labour, intensive growth and arrangement of debt-providing institutions. The Government of India is providing various kinds of social, economic development and protection programmes as safety nets for the rural poor; these programmes aimed to uplift of these people. Some of the important ones are Integrated Rural Development Programmes (IRDP), Employment Generation Schemes (EAS), Sampoorna Gramin Rojgar Yojna (SGRY), Swarnajayanti Gramin Swarojgar Yojna (SGSY), Integrated Wasteland Development Programme (IWDP), Drought-Prone Desert Programme (DPAP), and Desert Development Programme (DDP)(GOI,2009). Some other livelihood based interventions are also undertaken under the following programmes run by the state and the central government: Joint Forest Management, National Afforestation and Eco-developmental Supported Programmes (NAP), Grant in aid for Greening India, Eco-development Programmes, etc., Besides international banks like World Bank, JBIC, etc., have also initiated livelihood projects in different parts of the country like Odisha, Madhya Pradesh, Karnataka, Tamilnadu, etc. Livelihood projects have also been initiated by DFID, Rural Livelihoods Programmes on PRI's or Administrative Units, Watershed Programmes, etc. (Malhotra and Bhattacharya, 2010) (Reddy et al., 2004).

ROLE OF FOREST IN LIVELIHOOD GENERATION

Livelihood security is dependent on two related factors – one, the access to resources to meet the basic needs of a community and, second the state policies in this regard and the attitude of the civil society as reflected in the state policies (Sudarsen and Sumathi,

2003). There is an intricate relationship between livelihood pursuits of tribal communities and surrounding natural resources like forest, land, water-bodies and other flora and fauna. The critical balance between the two is essential for sustainable livelihoods of forest dwellers in the world in general. The coping mechanisms developed by them are cultural responses to combat the scarcity and poverty conditions that threaten them periodically (Prasad and Eswarappa, 2005). Sustainable Livelihood Approach aims to promote a holistic vision of development that includes income generation, natural resource management, people's empowerment, use of appropriate technology, financial services and good governance.

At global level, more than two billion people are dwelling in forest, depending on NTFPs for subsistence, income and livelihood security (Vantomme, 2003). NTFPs are considered to be important for sustaining rural livelihoods, reducing rural poverty, biodiversity conservation, and facilitating rural economic growth (Global NTFP partnership, 2005). An estimated 80% of the population of the developing world uses NWFP to meet some of their health and nutritional needs (FAO, 2008).

The importance of forest resources to the livelihoods of people living in and around forests has increasingly been recognised during the last three decades. It is indicated that about one billion of the world's poor depend on forest resources to sustain their livelihoods (Scherr et al., 2003). Studies in sub-Saharan Africa (Babulo et al., 2008; Campbell et al., 2002; Cavendish 2000; Fisher 2004; Kaimowitz 2003; Mamo et al., 2007; Paumgarten 2005; Shackleton et al., 2007); Asia (Adhikari et al., 2004; Reddy and Chakravarty; 1999), and Latin America (Gavin and Anderson, 2007; Pattanayak and Sills 2001; Takasaki et al., 2001) have shown that rural households regularly supplement their income from forest resources.

Forest, trees and CPRs support livelihoods of local communities/groups in three ways (Chambers, Saxena & Shah 1991):

- by providing for subsistence needs of fodder, fuel etc.;
- as a source of income; and
- as capital goods or savings to be cut and encashed to meet contingencies.

In India over 50 million people depend on NTFPs for their subsistence and cash income (Hegde et al., 1996). This provides 50% of household income for 20 to 30% of rural population, particularly for tribals. Potentially, around 3000 species of forest products are found to be useful, but only 126 have developed marketability (Maithani 1994). Around 50% of forest revenues and 70% of forest based export income of the country comes from NTFPs. Thus, it can be inferred that NTFPs form one of the mainstays of income and sustenance for many tribal communities (Rao, 1987; Gauraha, 1992; Chopra, 1993; Mallik, 2000). Concepts such as forestry, community forestry, joint forest management and conservation and development projects were meant to reflect such recognitions (Carter and Gronow 2005, Shepherd 2004).

Joint Forest Management (JFM), which is now a decade old, is operating on the "Care and Share" principal and aims at empowering local people as partners in the management of forest resources besides meeting their livelihood needs. About 22.02

million ha of largely natural degraded forest are being managed under JFM in India involving 170,000 forest fringe villages (Bahuguna and Upadhya, 2004; Pai and Datta, 2006). In India, JFM programmes currently span in 29 states, comprising 106,482 JFM committees encompassing an estimated 8.3 million families, half of which are scheduled castes and scheduled tribes, and cover more than 22 million hectares of forest land (Pai and Datta, 2006). Besides, regeneration of degraded forest, it was thought that JFM would also improve the employment and livelihood opportunities of the forest fringe dwellers. Though JFM guidelines are more or less similar, but evolution varies with a complex networks on interrelated parameters affects the implementation of JFM programmes. Livelihood generation is an important parameter for the state forest department for the promotion of JFM in any state. This method had been successfully implemented in the JFM areas of Madhya Pradesh, Tamil Nadu, West Bengal, Odisha, etc. (Khare et al., 2000). JFM has been (fore) seen as a means of poverty alleviation in India and as a part of integrated land use, in which agriculture, pasture, plantation and NTFP processing, all can be fitted in as complementary activities in an area (Rangachari and Mukherji, 2000).

Although the mandate of the department was not rural development, a number of its entry-point activities, when JFM was introduced, had a significant impact on livelihoods. Some cases are been given below.

Livelihood Opportunities Generated through JFM

Livelihood interventions are conscious efforts by an agency or organization to promote and support livelihood opportunities for a large number of people (Datta, Mahajan and Thakur, 2004). The interventions are spatial, segmental and sectoral. These promote activities intended to help the economically disadvantaged.

Madhya Pradesh

JFM programme initiated under the MP Forest Project created income generating activities (IGA) in many villages. These income generating activities and resource development work carried out in the villages lead to subsequent betterment in the socio-economic condition of the people and poverty reduction (Shukla, 2004). Thus, the programme has benefited the landless and poor through wage employment. In the year 1999-2000 JFM programme was able to generate 11.69 million mandays of employment, giving an income of 14.61 million USD. Within the organizational development and institutional mechanism the project took various steps in asset creation; including natural, social, human, political and physical assets to generate livelihood options (Debnath and Dasgupta, 2006). Prasad and Kant (2003) have reported that before JFM was initiated, in most cases collection and sale of NTFP was highly unorganized in Madhya Pradesh. This resulted into non-remunerative returns to the NTFP gatherers. Middlemen and contractors apportioned the lion share of profit. The bargaining power of the NTFPs collectors was low which forced them to sell the products at nominal rates (Prasad, 1999). Realizing the gravity of the problem, the Forest Department of Madhya Pradesh decided to facilitate tribals through village FPC in collection of NTFPs, their semi-processing activities and organized marketing in some predominant tribal areas of northern

Madhya Pradesh. In 2000-01, Rs. 9.13 crore from the sale of timber and Rs. 32.19 lakh from bamboo was distributed amongst the JFMCs. This amount increased to Rs. 3.92 crore, Rs. 10.95 crore and Rs. 8.21 crore in 2002, 2003 and 2005 respectively. Various forestry operations under JFM have generated wage employment worth Rs. 359.88 crore between 2001-02 and 2005-06.

In Dhamtari Forest Division of Chhattisgarh Village Resource Development, activities have not only altered the cropping pattern but have also contributed to enhanced productivity. On an average, the irrigation facilities created have been responsible for increasing the paddy yield up to 10 times in many households, resulting in increased income in Jabarra Forest Village in Dhamtari. The per capita household income on an average ranges between Rs. 13,000-Rs. 22,000 per year as a result of which many Below Poverty Line (BPL) families have now come under the above poverty line category (Bhattacharya et al., 2008).

In different JFMCs of Hoshangabad Division, forest department along with NGO PRADAN initiated poultry raising activity with the women SHG members of the JFMC. Besides giving them training and chicken at subsized rates, they also give them training for maintenance. They have also made a buyback arrangement. Through this activity the women members earn about 15,000-20,000 per annum. This activity has completely reduced dependency of women on forests of fuelwood which was a initially a major source of earning. But now women spend their maximum time in poultry raising and rarely get time for fuelwood collection. Besides, women SHG members have also started paper-mesh craft making activity in EDC of Bamnai, through which they earn Rs. 3000-6000/- monthly based on the market demand (Pradhan, 2012).

Tamil Nadu

The VFCs constituted under Tamil Nadu afforestaion project were given the authority over regulating the access to forest, resolving intra-village conflicts and in ensuring equitable distribution of benefits. The VFCs were given legal status and certainty in their functioning under the Tamil Nadu Societies Registration Act, 1975. All the forest produce, viz., fuelwood, fodder, green manure and NTFPs that could be harvested from the restored forest on sustainability yield basis went to the VFC free of cost. The proceeds of any surplus produce sold is equally distributed among VFC members after remitting 25% of it to a specially constituted fund, Village Development Fund (GoTN, 1997). First, second and the third years of the project provided Rs. 3,00,000, Rs. 2,00,000 and Rs. 1,00,000 respectively as seed money to the VDF. Out of which 70% was for providing alternative livelihood and 30% for general development activities. After the implementation of the project, 66% of the households living below the poverty line came down to 34%. Similarly, the proportion of household living in thatched houses came down to 33% in repeat survey when compared to 44% noted at the initiation of project. With the initiation of the TAP project, the dependence on the forest resources reduced considerably. Per captita fuelwood consumption declined from 165 tonnes/household to 149 tonnes/household with considerably increased the use of other energy sources (increased from 3% to 13%) (Sreedharan and Matta, 2010).

Gujarat

In Gujarat the entry–point activities and the gap-planting activities have generated marginal employment, but the communities were not involved in forestry operations and silvicultural interventions. In Karnataka in 72% of the JFMCs the forest department utilized local labour for plantation activities in 16% of the JFMCs and in other employment generation activities were also implemented. All the JFMCs in West Bengal reported an increase in the income generation activities (Ravindranath and Sudha, 2004).

Andhra Pradesh

In the state of Andhra Pradesh an investment of Rs. 2,114.1 million was made during the year 1994-2000, out of which 85% was wage component and about Rs. 1797 million was paid as wage labour. Impact studies conducted by independent agencies have confirmed that there was a sharp decline in migration. The study also shows that there is an increase in the production as well as income from NTFPs. With 50% sharing of the revenue from beedi leaves, the VSS have received an amount of Rs. 49.671 million (Mukherji, 2007). JFM committees (VSS) such as Koyagudem, Algaon, Kethanpally, Cherupalli and Shankarapur (S) have shown significant performance in promoting livelihood opportunities among the poor forest dwellers and have generated significant revenue from forests. The recorded revenues are Rs. 2,11,75,877 (Koyagudem), Rs. 32,20,248 (Algaon) and Rs. 28,89,757 (Shankarapur) (Pai and Dutta, 2006).

JFM activities resulted in an increase in the availability of wage employment, savings and agricultural productivity, along with a reduction in migration, particularly distress migration. The ST households with an annual income below Rs. 20,000 reduced from 97.34% to 50.42% between 2002 and 2005. During 2003-2006, forestry activities such as plantation, thinning, bamboo, NTFP collection and others generated a total income Rs. 6,779 lakh to communities.

Chhattisgarh

There are micro-level experiences throughout the country on this aspect. In Chhattisgarh, one of the main incentives to people from JFM was employment generation through wage employment. An estimated 7820 JFM committees received around Rs. 7,682.20 lakh through wage employment between 2003-04 and 2005-06. Another Rs. 1,431.76 lakh were spent on infrastructural development. In Chhattisgarh, under JFM, income-generating activities such as processing and value addition of NTFPs are also undertaken on a large scale. Foresters have identified several innovative activities under JFM provision to increase wage employment opportunities along with FPCs in local meetings. A cold storage has been set up in Dantevada, where JFM committees store packaged tamarind before it goes for auction. There are also wholesale and retail outlets for herbal medicines, which are linked with Sanjeevani and other government departments. Chhattisgarh is one of the few states that has distributed benefits from final felling among the JFM committees. Around Rs. 34 crore, which is about 15% of the net profit, was disbursed among committees between 2003 and 2006, the details of which are summarized in Table 7.1.

Table 7.1 Benefits distributed to the JFM committees in Chhattisgarh

Production Year	*Number of Committees*	*Amount (in crores)*	*Payment Year*
2000 & 01	219	3.08	2003&04
2001 & 02	358	6.13	2003&04
2002 & 03	446	8.29	2003&04
2003 & 04	313	8.20	2004&05
2004 & 05	415	9.01	2005&06
Total	1751	34.71	

Source: R.C. Sharma, 2005

Tripura

In Tripura 2,185,108 man days of the employment was generated under various schemes. Usufructuary benefits under JFM and entry-point activities have also enhanced livelihood opportunities and improved the economic condition of JFM committee members.

Karnataka

In Karnataka, initiation of forest-development activities such as clearing, singling, planting and soil and moisture conservation works generated additional employment of 2410 man days, fetching Rs. 94,000 as wage amount per year per village by engaging 57 families. This directly resulted in 60% reduction in migration (Sudha et al., 2003). In Madhya Pradesh, NTFPs have been providing an average of 20-54% of household income. In West Bengal, NTFP-related income ranged from Rs. 1,000-Rs. 10,000 per household.

Jharkhand

Dubrajpur in Dhanbad has set an example of forest-based enterprise development. With the abundance of Arjun Trees in the area, the JFM Committee approached the Department of Industries for organizing a training programme for them on silk rearing. Following the training, 200 trees were inoculated at a cost of Rs. 2000. Consequently, within three months, about 30,000 cocoons were sold at Rs. 20,000 (Sharma, 2005).

Domunda VFMPC, on the other hand, successfully tied up with the Hariyali Yojana for developing infrastructure such as solar lights and pond construction in the village. The community in Tilaiya Village is linked with the Annapurna Beej Gram Yojana of the State Department, which helps villagers get better production of potato and paddy.

West Bengal

The average annual income per household in FPCs from share of the final harvest, employment, collection, use and sale of NTFPs was Rs. 5,090, which was over 11% of the total average household income. In case of the poorest sections, this proportion was 32%; JFM committee members received maximum benefit from wage employment (Pai and Dutta, 2006). The share of benefit realised under JFM is given in Fig. 7.1.

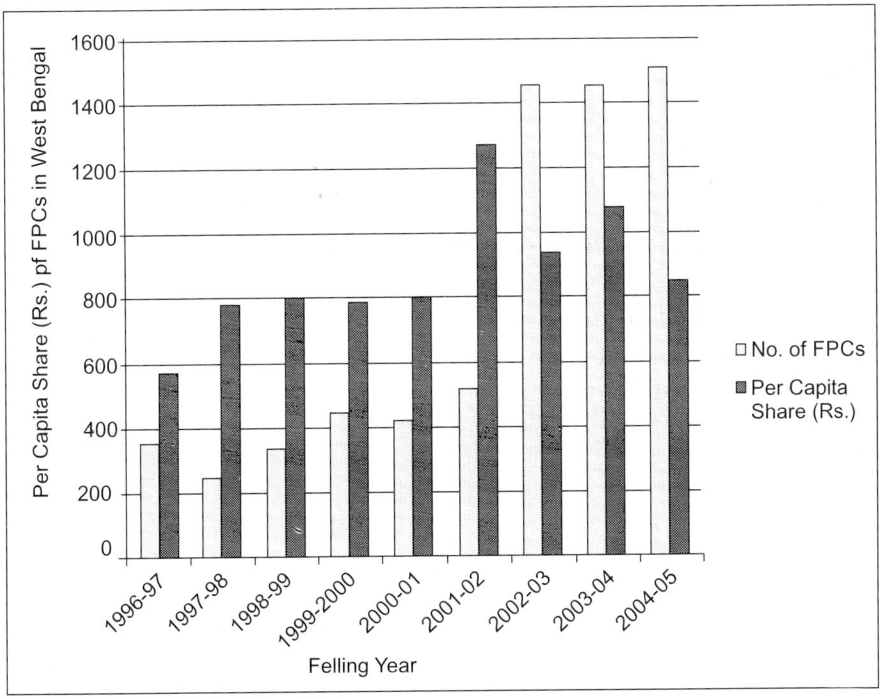

Fig. 7.1 FPC share realized from the final harvest in West Bengal

Rajasthan

In Rajasthan, the VFPMCs have created a maintenance fund by contributing a part of their daily wages, and a corpus fund of Rs. 1,00,000 was added to it under the Japan Bank for International Cooperation JBIC supported project. Now every village has between Rs. 5,00,000 and Rs. 6,00,000 to maintain the assets created. Cultivation of medicinal and oil-yielding plants through inter-planting of species such as Jatropha have provided economic benefits to the communities.

The committee has also received benefits through bamboo harvesting, the details of which are summarized in Table 7.2.

Table 7.2 Revenue generated through bamboo harvesting in JFMC of Rajasthan

Name of the VFPMC	*Total Revenue (lakhs)*	*Benefit to the VFMC (lakhs)*
Dungri Pada	2.24	0.83
Kuwania	2.46	0.86
Akyawad	2.41	0.85
Amaltha	0.65	0.26
Plasma	1.54	0.68
Rama	0.44	0.18

Contd...

Contd...

Lakhwali	1.70	0.66
Khanwal	0.94	0.39
Dhar	3.51	1.28

Source: Pai and Dutta, 2006

Developing Governance Strategies for Improving the Livelihood Opportunities of the Rural People

Providing sustainable livelihood to the rural people

The above livelihood interventions reveal that the success of the JFM programme depends on ensuring sustained economic returns to the beneficiaries. The initial financial support through entry-point activities should be a prerequisite for starting JFM in a village, and these activities need to be linked to rural development activities in the village through the panchayats and other line departments. The role of the panchayats in most of the studied JFMC villages has been negligible and this should be strengthened.

In order to avoid unsustainable harvesting clear-cut guidelines for sustainable harvesting needs to be developed. There is need to develop NTFP based forestry which would provide a sustained and regular means of employment. The focus of JFM should be on sustainable production and management of the NTFPs, which will be the main source of earning from the forests for the beneficiaries.

Providing Greater Access to Efficient Markets for Goods and Services from the Forest

New approaches to market access by the communities and market power of communities needs to be developed. State level marketing federations of the forest communities should be encouraged to strengthen their market position, facilitate the establishment of storage areas, train for value addition and allow consolidated consignments of NTFPs to be sold directly by the communities to large processing or marketing firms through auctions or contract agreements (Bhattacharya et al., 2010). Creating value in the existing value chain by scientific and technical intervention can benefit NTFP collectors.

Capacity Building in JFMC Members

There is a need for continuous training and capacity building of communities, foresters and other stakeholders. Capacity building programmes are necessary to enhance the ability of the community to view JFM as their own programme and provide the villagers with skills to plan, implement and manage JFM activities. The VFCs/VSSs should be activated and reoriented towards the task of value addition (Bhattacharya A.K., 2001).

The JFMC members should also be trained on enhancing the yield from JFM areas to sustain and improve the forest under JFM. For this purpose, VSS members need to be trained appropriately for undertaking cleaning, pruning, thinning and NTFP collection in silviculturally correct manner. This activity can be undertaken till the canopy closes.

In order to empower the JFMCs and the local communities, an integrated strategy needs to be devised at policy level to exploit the opportunities available in recent national agenda like Greening India Programme, MNREGA and Forest Rights Act. A more efficient, flexible, and transparent monitoring and control system is required in such a big national programme, where the roles and responsibilities among the forest departments and the communities are complementary for betterment and efficiency in the system (Bhattacharya et al., 2010).

CONCLUSION

Improving forest livelihoods in most communities in India must be integrated into broader rural livelihood initiatives, such as agriculture intensification and marketing. There is need for development of livelihood opportunities and entrepreneurship at the grass roots that can ensure sustainability of JFM committees. There is a need for flow of sustainable income to villagers through enterprise development at community as well as individual level. It will require skill development of the individuals, supplementing them with micro-finance and utilizing the opportunity for mass production.

REFERENCES

1. Adhikari B., Di Falco, S., and Lovett, J.C. (2004). Household characteristics and forest dependency: Evidence from common property forest management in Nepal. *Ecol. Econ.* 48: 245-257.
2. Babulo B., Muys, B., Nega, F., Tollens, E., Nyssen, J., Deckers, J., and Mathijs, E. (2008). Household livelihood strategies and forest dependence in the highlands of Tigray, Northern Ethiopia. *Agricultural Systems* 98: 147-155.
3. Bahuguna, V.K., and Upadhyay, A. (2004). Monitoring needs for JFM: The perspective of policy makers. In: Root to Canopy: Regenerating Forests through Community-State Partnerships. (Ed.) V.K. Bahuguna, K. Mitra, D. Capitrano, and S. Saigal. Winrock International and Commonwealth Forestry Association, Delhi. pp. 12-16.
4. Bhattacharya, A.K. (2001). Community Participation and Sustainable Forest Development - Global perspective, "Published by Concept Publishing Co. Pvt. Ltd. New Delhi.
5. Bhattacharya, P., and Mittra, B. (2001). Self-Initiated Forest Protection Initiative in India, Contemporary studies in Natural Resource Management in India, Inter India Publication Delhi.
6. Bhattacharya, P., and Hayat Sayed, F. (2004). Sustainable NTFP management for rural development: A case from Madhya Pradesh, *India. International Forestry Review.* Vol. 6, No. 2, pp. 161-168.
7. Bhattacharya, P., Kandya, A.K., and Krishnakumar, K.N. (Ed.) (2008). Joint Forest Management in India. 2 Volumes, Aavishkar Publishers, SMS Highway, Jaipur, p. 657.
8. Bhattacharya, P., Pradhan, L., and Yadav, G. (2010). Joint forest Management in India: Experiences of two decades. *Resource Conservation and Recycling.* 58(8): 469-480.
9. Byron, R.N., and Arnold, J.E.M. (1999). What futures for the people of the tropical forests? *World Development*, 27(5): 789-805.

10. Campbell, B.M., Jeffrey, S., Kozanayi, W., Luckert, M., Mutamba, M., and Zindi, C. (2002). Household livelihoods in semi-arid regions: Options and constraints. Centre for International Forestry Research.
11. Carney, D. (1998). Implementing the Sustainable Rural Livelihoods Approach. In Carney, D. (Ed.) Sustainable Rural Livelihoods: What contribution can we make? Papers presented at the DFID's Naturals Resources Advisers Conference, July, (1998).
12. Carter, J., and Gronow, J. (2005). Recent experience in collaborative forest management: A review paper.
13. Cavendish, W. (2000). Empirical regularities in the poverty-environment relationship in rural households: Evidence from Zimbabwe. *World Development* 28(11): 1979-2003.
14. Robert, C., Saxena, N.C., and Shah, T. (1991). To the Hands of the Poor-Water and Trees, Oxford & IBH Publishing Co., Pvt. Ltd, Delhi.
15. Chopra, K. (1993). The Value of Non-Timber Forest Products: An estimation for tropical deciduous forests in India, *Economic Botany*, Vol. 47, pp. 251-257. In: Ravi et al (ed.) Poster paper prepared for presentation at the international association of agricultural economists conference, Gold Coast, Australia, August 12-18, 2006.
16. Mahajan, D., and Thakur (2004). A resource book for livelihood promotion. pp. 1-6.
17. Fisher, M. (2004). Household welfare and forest dependence in Southern Malawi. *Environ. Dev. Econ.* 9: 135-154.
18. Das, A. (2007). Poverty induced forest degradation in JFM regime: Evidence from India. Centre for Development Studies, Thiruvananthapuram, Kerala, India. Online at http://mpra.ub.uni muenchen.de/18639/. MPRA Paper No. 18639, posted 16. November 2009 00:07 UTC.
19. Debnath, D., and Dasgupta, S. (2006). Livelihood generation and poverty reduction attempts in joint forest management Activities. *International Forestry Review*. 8: 241-250.
20. Dutta, M., Roy, S., Saha, S., and Maity, D.S. (2004). Forest protection policies and local benefits from NTFP: Lessons from West Bengal. *Economic and Political Weekly*. 23(2): 58-66.
21. FAO (2008). Non Wood Forest Products, Rome, Italy.
22. Gauraha, A.K. (1992). Micro-economic analysis of a tribal village, *Ind. J. Agricultural Econ.*, 47(3): 446-447. In: Ravi et al. (Ed.) Poster paper prepared for presentation at the international association of agricultural economists conference, Gold Coast, Australia, August 12-18, 2006.
23. Gavin, M.C., and Anderson, G.J. (2007). Socio-economic predictors of forest use values in the Peruvian Amazon: A potential tool for biodiversity conservation. *Ecol. Econ.* 60:752-762.
24. Government of India (2009). A report of the planning commission on 10th Five year Plan.

25. Hegde, R., Suryaprakash, S., Achoth, L., and Bawa, K.S. (1996). Extraction of NTFPs in the Forests of B.R. Hills. Contribution to Rural Income. *Econ. Bot.*, 50, 243p. In: Uma Shankar R, et al. (Ed.) 2004.
26. Kaimowitz, D. (2003). Not by bread alone. Forests and Rural Livelihoods in Sub-Saharan Africa. In: OKSANEN, T., PAJARI, B., and TUOMASJUKKA, T. (Eds.) Forests in Poverty Reduction Strategies: capturing the potential. EFI Proceedings No. 47. p. 45-63.
27. Khare, A., Sarin, M., Saxena, N.C., Palit, S., Bathla, S., Vania, F., and Satayaranana, M. (2000). Joint Forest Manangement: Policy, Practice and Prospects. India Country Study. Policy that works for Forests and People series No. 3. A collaboration between WWF for Nature, New Delhi, India and IIED, London.
28. Krantz, L. (2001). The sustainable livelihood approach to poverty reduction. SIDO, Division of Policy and Socio-economic Analysis.
29. Maithani, G.P. (1994). Management perspectives of Minor Forest Produce. MFP News, October-December, (1994). Dehradun.
30. Mallik, R.H. (2000). Sustainable management of non-timber forest products in Orissa: Some issues and options, *Ind. J. Agri. Econ.*, 55(3): 384-397. In: Ravi et al. Poster paper prepared for presentation at the international association of agricultural economists conference, Gold Coast, Australia, August 12-18, (2006).
31. Malhotra, K.C., and Bhattacharya, P. (2010). Forest and Livelihood. CESS, Hyderabad, p. 246.
32. Mamo, G., Sjaastad, E., and Vedeld, P. (2007). Economic dependence on forest resources: A case from Dendi District, Ethiopia. *Forest Policy and Economics.* 9: 916-927.
33. Mukerjee, A. (2007). Joint Forest Management in West Bengal. In: Forests People and Power: The Political Ecology of Reform in South Asia. (Ed.) Springate-Baginski O. and P. Blaikie. Earthscan, UK. pp. 221-260.
34. Pai, R., and Dutta, S. (Ed.) (2006). Measuring milestones: Proceedings of national workshop on joint forest management. Winrock International India and the Ministry of Environment and Forests, New Delhi, India p. 48.
35. Pandey, D.N. (1996). Beyond vanishing woods: Participatory survival options for wildlife, forests and people, Udaipur, India: Himanshu Publications.
36. Pattanayak, S.K., and Sills, E.O. (2001). Do tropical forests provide natural insurance? The microeconomics of non timber forest product collection in the Brazilian Amazon. *Land Econ.* 77: 595-613.
37. Paumgarten, F. (2005). The role of non-timber forest products as safety-nets: A review of evidence with a focus on South Africa. *GeoJournal* 64: 189-197. Perspective, Concept Publishing Company, New Delhi.
38. Pradhan, L. (2012). Ecological and Institutional assessment of the impact of JFM in Madhya Pradesh. Ph.D. thesis submitted in Hari Singh Gour University Sagour under the supervision of Prof. A.K Kandya and Prof. P. Bhattacahrya.

39. Prasad, R. (1999). Joint Forest management in India and impact of state control over non-wood forest products. Unasylva, 50: 58-62.
40. Prasad, R., and Kant, S. (2003). Institutions, forest management, and sustainable human development - experiences from India. Environment, Development and Sustainability. 5: 353-367.
41. Prasad, S., and Eswarappa (2005). Tribal livelihoods in a Limbo: Changing tribenature relationships in South Asia presented in eight sustainable development conference, Best western, Islamabad.
42. Rangachari, C.S., and Mukherji, S.D. (2000). Old roots, new shoots: A study of joint forest management in Andhra Pradesh. Winrock International/Ford Foundation, India, p. 189.
43. Rao Gopala, N. (1987). Significance of minor forest produce in tribal economy: A case study, Kurukshetra, 7: 23-28. In: Ravi et al. (Ed.) Poster paper prepared for presentation at the international association of agricultural economists conference, Gold Coast, Australia, August 12-18, 2006.
44. Ravindranath, N.H., and Sudha, P. (2004). Joint Forest Management in India. Spread, Performance and Impact. Universities Press (India) Private Limited, Hyderabad, p. 349.
45. Reddy, S.R.C., and Chakravarty, S.P. (1999). Forest dependence and income distribution in a subsistence economy: Evidence from India. *World Dev.* 27: 1141-1149.
46. Reddy, V.R., and Soussan, J., (2004). Assessing the impact of watershed development programmes: A sustainable rural livelihoods framework. *Ind. J. Agri. Econ.*, 59(3).
47. Sarker, D. (2007). How to Execute the Joint Forest Management Programme with Sustainability? A Study of Joint FPC and Female FPC in four divisions under Mednapore and Bankura districts in West Bengal, Report of UGC Minor Research Project(F PHW 074 04-05(ERO).
48. Scherr, C.J., White, A., and Kaimowitz, D. (2003). Making markets work for forest communities. *International Forestry Review* 5: 67-73.
49. Shackleton, S., Campbell, B., Lotz-sisitka, H., and Shackleton, C. (2008). Links between the local trade in natural products, livelihoods and poverty alleviation in a semi-arid region of South Africa. *World Dev.* 36: 505-526.
50. Sharma, R.C. (2005). Forest Poverty Alleviation: Chhattisgarh Experience. In Sim, H.C., Appanah, S. and Hooda, N (Eds.). Forest for Poverty Reduction: Changing Role for Research, Development and Training Institutions. Proceedings of the Workshop Organized on 17-18th June, 2003 Dehradun, India. Published by FAO.
51. Shepherd, G. (2004). Poverty and forests: Sustaining livelihoods in integrated conservation and development. In: T.O. Mcshane, and M.P. Wells (Eds.) Getting Biodiversity Projects to Work: Towards More Effective Conservation and Development, pp. 340-371. Columbia University Press.
52. Shukla, A. (2004). State Experience: Madhya Pradesh. In Root to canopy: Regenerating Forest through Community State Partnership, Bahuguna V.K., Mitra K., Capistrano D. and Saigal S, (Ed.) Winrock International, New Delhi, pp. 111-118.

53. Sreedharan, C.K., and Matta, J.R. (2010). Poverty alleviation as a pathway to sustainable forest management. *Environment, Development and Sustainability* December 2010, Vol. 12, Issue 6, pp. 877-888.
54. Sudha, P., Bhat, P.R., Jagannatha Rao, R., Nagaraja, B.C., Hegde, G.T., Shastri, C.M., Hegde, G.N., Shetty, D.M., Murali, K.S., Bhat, D.M., and Ravindranath, N.H. (2004). Joint Forest Planning and Management in Karnataka: Its spread, performance and Impact. In: Joint Forest Management in India. Spread Performance and Impact. (Ed.) N.H Ravindranath and P. Sudha, Universities Press, Hyderabad, pp. 86-121.
55. Sudarsen and Sumathi (2003). State and the forest-related communities: Issues andparadigms of development, Paper to be presented at the conference on Livelihood strategies among forest-related tribal groups of South India to be held at the centre for Indian Studies, Mysore, during October, 17-19, 2003.
56. Takasaki Y., Barham, B.L., and Coomes, O.T. (2001). Amazonian Peasants, Rain Forest Use, and Income Generation: The Role of Wealth and Geographical Factors. *Society and Natural Resources* 14: 291-308.
57. Vantomme (2003). Forest Products Division, Forestry Department, FAO, Rome varadarajan, D. (1980). A Demographic survey of kota tribals in Nilgiris District. Ph.D. Thesis, TNAU, Coimbatore.

8

Community Oriented Technology Development for Sustainable Forest Resource Management Under Joint Forest Management

R.K. Pandey

State Forest Research Institute, Polipathar, Jabalpur – 482008 (M.P.)

E-mail: pandeyrk1@yahoo.com

ABSTRACT

Tropical forest of India is repository of several commercially important non-timber forest products (NTFPs). These resources not only secure the livelihood of forest dependent rural tribal people but also provide potential source to their annual income. Nearly, 500 million people rely on NTFPs for their livelihoods.

In recent past, the fast depletion of these resources from natural forests has been a major concern of managers, policy-makers and forest dependent people at large, being direct impact on conservation, ecological perpetuation and socio-economic security.

Increased demand of NTFPs from natural forests led the users' communities to disregard the traditional or sustainable harvesting practices. Major NTFPs of high demand are under threat in in situ conditions. The collaborative integrated approach is inevitable to conserve and sustainably manage these resources through active participation of forest dependent user communities with improved skill and capabilities not only to secure livelihood but also to conserve vulnerable/overexploited wild genetic resources.

The forest resources need immediate management interventions through innovative, effective and accepted technologies with users' communities participation for sustainable management under Joint Forest Management system.

The present paper embodied the innovative techniques developed for sustainable harvesting of commercially important NTFPs, viz., *Chlorophytum tuberosum, Curculigo orchioides, Curcuma angustifolia, Dioscorea daemona, Asparagus racemosus, Plumbago zeylanica, Bauhinia vahlii and Embelia robusta* with active participation of Baiga and Gond aboriginal forest dependent tribes in natural forest ecosystem in eastern Madhya Pradesh of Central India.

Integrated approaches through skill and capacity development and self-assessment with local communities at grass roots level found to be accepted for sustainable harvesting as well as management of utilizable forest resources in natural tropical forest ecosystem. Detailed findings of case studies envisaged an integrated approach is the effective tool for conservation and sustainable management of NTFPs in state government owned natural forests in Joint Forest Management System.

INTRODUCTION

Forests are the home of dependent people who live in and around and act as an important component of forest ecosystem and play a key role in the sustenance. The age-old traditional values attached with the various forest types and varieties of forest products have gained tremendous importance in the present century. However, instant rising demand of plant based drugs of "AYUS" system is unfortunately creating pressure on economically wild medicinal plants. An ever increasing demand of these forest resources has caused adverse effects on the existence of natural forests. There are several native species of high medicinal values, which have either become extinct or are prone to extinction.

India has a recorded forest area of 76.84 m ha or 23.38 of the total geographical area of the country (FSI, 2003). It has been in the forefront of conservation movement of biodiversity through safeguarding rich and diverse biological resources.

The forests of the country are the second largest land use after the agriculture. Almost all the forest areas are state owned and managed by state governments on the basis of scientific working plans, which are approved by central government. However, forest management in India remains concentrated on production of commercial timber to meet national demand. Non-timber forest produces (NTFPs) could not get proper attention as far as the sustainable management is concerned.

Over 1 billion population of the country, about 300 million of the rural poor and tribal communities depend on forest resources particularly, on NTFPs for their livelihoods.

More than 95% of the raw material of medicinal plants used for preparation of ayurvedic medicine is still harvested destructively from natural forest. Destructive harvesting of wild medicinal and other NTFPs of commercial use, compiled with the depleting scenario of habitat conditions have challenged the sustainable management of these resources in natural forest of the country. Virtually, these native forest resources are also representing diverse floral diversity growing in the wild. In recent past, the fast depletion of these resources from natural forests has been a major concern of managers, policy-makers and forest dependent people at large, as it has a direct impact on conservation, ecological perpetuation and socio-economic security.

Increased demand of NTFPs from natural forests led the users' communities to disregard the traditional or sustainable harvesting practices. Major NTFPs of high demand are under threat in in situ conditions. The collaborative integrated approach is inevitable to conserve and sustainably manage these resources through active participation of forest dependent users' communities with improved skill and capabilities not only to secure livelihood but also to conserve vulnerable/overexploited wild genetic resources.

The forest resources need immediate management interventions through innovative, effective and accepted technologies with users' communities' participation for sustainable management under Joint Forest Management System.

Integrated approaches through skill and capacity development and self-assessment with local communities at grass roots level were found to be accepted for sustainable harvesting as well as management of utilizable forest resources in natural tropical forest ecosystem. Research findings of various case studies envisaged an integrated approach is the effective tool for conservation and sustainable management of NTFPs in state government owned natural forests in Joint Forest Management System.

STATUS OF NTFPs

National Forest Policy (NFP) 1988 of India defines the primary goals of forest management as first to conserve the natural environment, second to meet the requirement of local people (particularly, tribal and poor population) from the forest produce and third as a source of wood and other products for industries and other non-local uses (FSI and FAO, 2004). The Policy envisages participation of communities in the management of forest resources under JFM and recognizes the symbiotic relationship between people and the forests. It is now widely accepted that the rural people living in and around protected forests are the key stakeholders in management process. According to NFP the implication of JFM is "the sharing of, responsibilities, protection, produces and decision making over forest lands between forest department (owner stakeholder)" through judicious use of limited resources on sustainable manner. Nearly, all states of Indian Union have passed the JFM resolution, 1990, and more than 65,000 forest protection committees (FPCs) have been constituted so far (FSI, 2003).

India is the centre of South Asia's export trade in medicinal plants, dyes, spices, etc. More than 960 species are traded with total consumption of 3,12,516 metric tonnes. Moreover, 123 species have been identified which have demand of more than 1000 mt/ year (GOI, 2001).

Almost 3000 plant species occurring in the various forest types are acknowledged to be the source of NTFPs in the country (Tewari, 1994). Madhya Pradesh is also one of the richest states of the country with variation in topography which supports various types of forest flora. This state alone has 1985 plant species of various categories, which belong to 785 genera and 147 families (Pandey and Shrivastav, 1996). Moreover, about 676 plant species have been recognized as important medicinal plants belonging to 125 families (Shrivastava, 1989).

Traditionally, these resources were used by forest dwellers for their daily needs and only surplus produces were traded in the local market. However, a rising interest in "green consumerism" among urban population all over the globe in the recent years has been instrumental in trading of many of NTFPs, mostly in unprocessed and semi-processed forms. NTFPs have potential to earn foreign exchange apart from providing better economy to the collectors/producers and employment opportunities for rural and tribal communities. Nevertheless, several intellectual discussions have been made through seminars and conferences across the country and globe in the recent years, emphasizing the role of NTFPs in the poverty elevation and livelihood security of forest dependent poor people as well as regional and national economy. But it is surprising that in spite of all the importance, least attention has been paid to attain several questions of conservation, in situ production (particularly, those related to ecological threat due to harmful removal from the natural forests) and sustainability.

It is an accepted fact that 60% of NTFPs production of the country goes as unrecorded. Further, the data on recorded and non-recorded removal or extraction of NTFPs are neither systematic nor structured. No systematic information is available on sustainable management with an integrated ecological approach to achieve sustainability of utilizable NTFPs in natural forest ecosystem. Surprisingly, threat to ecological sustainability of several utilizable NTFPs, and their in situ conservation on existing community equilibrium and dynamics has neither been adequately acknowledged by managers nor by researchers and user communities.

About 95 percent of raw material for pharmaceutical and cosmetic industries is still being collected from natural forests. The dominance of wild collection is continued over the long term due to numerous reasons: (i) little is known about ecology and harvesting limit of the NTFPs derived from natural forests; (ii) the time, research and experience leading to domestication and cultivation are costly; (iii) no stable and secure market system for NTFPs; and (iv) awareness on loss of regional biodiversity, etc. NTFPs have long been invisible or undervalued by government policy-makers, managers and researchers. There is no accepted procedure and reliable information/data on inventory and sustainable potential of NTFPs, particularly, in a social context of poor, rural and forest dependent communities.

NTFPs: Livelihood Security and Conservation

NTFPs based socio-economic development and conservation of natural resources in situ oppose one another because economic well-being of forest dependents is driven by unregulated/unsustainable exploitation which force to ignore the conservation ethics. There is no match to have a choice among poverty reduction, livelihood security and resource conservation in prevailing system.

A livelihood comprises the capabilities, assets (stores, resources, claim and access) and activities required for means of living; a livelihood is sustainable which can cope with and recover from stress and shocks, maintain or enhance its capabilities and assets to provide sustainable livelihood security for the next generation. However, poverty is a complex phenomenon. In the context of conservation, it is assumed that poverty, in turn, leads to further degradation of natural assets. This constitutes the vicious circle of the poverty, particularly in the resource rich but underdeveloped areas with undefined and inequitable access of the poor to the common property resources (Sharma, 2003). Moving away from income or consumption criteria, many a times poverty has been equated with the lack of livelihood security.

Goods and services from forests are more importantly NTFPs constitute a critical life line and livelihood for poor forest dwellers. Due to recurrent availability of NTFPs in terms of socio-cultural, economic, environmental and industrial development, it holds a promise to develop a mechanism for sustainable livelihood. But, in situ conservation of overexploited NTFPs is a challenging task as observed in several case studies exercised in tropical forests of Central India (Pandey et al., 2003; Pandey and Saini, 2004).

The biodiversity of tropical forests of the country is under pressure of unsustainable logging practices and over-harvesting of utilizable forest resources. This, in turn, led to degradation of forest resources by changing ecology of the area. The high demand for wide range of NTFP species has led to a substantial and rapid decline in the diversity, abundance and quality particularly, commercially important NTFPs in natural forest ecosystem. Rapid deterioration of natural forest areas coupled with increase in both the demand and value of forest resources has changed the pattern of harvesting from last three decades to meet the market demand by ignoring traditional harvesting practices and non-sustainable overexploitation.

The forests nearer to habitation have suffered more with a loss and replacement of native species by exotic and less valued species. Most of the commercially important NTFPs species have been wiped out due to indiscriminate harvesting without any care for regeneration. Loss of congenial habitats, regeneration potential, growth and development are the serious ecological threats to perpetuation of indigenous species. Adverse conditions have favoured the exotics to overcome by replacing the native species as a common phenomenon all over the forest localities nearer to habitations (Pandey and Shrivastava, 1989; Pandey, 2000).

SUSTAINABLE MANAGEMENT OF NTFPs

Accelerated deforestation, unregulated harvesting, increased demand and market access in absence of tenure security and unsustainable management regimes are leading to greater rate of resource depletion in natural forests. Inconsistent knowledge of regulation, lack of documentation, unrecorded and unregulated extraction, and inadequacy in implementation of regulatory mechanism and ignorance in sustainable management practices of NTFPs are not only depleting the NTFPs resource base but are also creating a threat on biodiversity conservation even in potentially rich forests of the country. Nevertheless, there are no systematic approaches for development of mechanisms for sustainable management of these valuable resources in situ. Surprisingly, little is known about reproduction, stock, yields, habitat requirement, ecology, and regeneration potential, etc., of almost all NTFPs which are being harvested from the forests.

Development of Sustainable Harvesting Techniques

Every NTFP species has a specific regeneration potential which provide optimum harvesting limit. If harvesting level is exceeded, the plant population as well as the community that depend on them is adversely affected. To maintain the ecological sustainability, it is important to determine the species specific harvesting limit based on the regeneration potential of the species. There may be a site-specific variation on regeneration potential of the same species. Therefore, site-specific consideration for development of sustainable harvesting technique seems to be inevitable. Strategic planning for making adequate model for sustainable management of utilizable NTFPs will require detailed information on available resources, ecology/biology including regeneration potential, threshold limit, harvesting time, etc. However, overall active participation of forest dwellers/users communities in all these steps is one of the essential parts of the strategy.

Determination of Sustainable Harvesting Limit (SHL)

Sustainable harvesting limit of commercially important harvested NTFPs is being experimentally determined by Madhya Pradesh State Forest Research Institute, Jabalpur in these PPAs by taking the local people in confidence. This seems to be an accepted model

Plate 8.1 *Andrographis paniculta* Burn.f. Wallich. (Kalmegh)

which provides species as well as the site-specific innovative sustainable management technology of utilizable forest resources through development of skill and capabilities of forest dependent communities. Within the conceptual landscape, conservation promoting and NTFP based enterprises in a multiple use systems has been presumed to be managed profitably and sustainably by local communities and forest dependent households with a hospitable niche.

Plate 8.2 Costus speciosus (Keokand)

Determination of sustainable harvesting limit (SHL) and self-assessment techniques with active participation of users communities was found to be an effective tool for conservation and sustainable management of forest genetic resources (FGR) in natural tropical forests of Central India (Pandey & Saini, 2001).

Plate 8.3 *Curcuma angustifolio* (Tikhur)

Results of species specific harvesting limits determined by the State Forest Research Institute, Jabalpur, (M.P.) were found to be quite alarming, particularly for annuals and biennials whose roots are harvested (Table 8.1). In this category, *Chlorophytum tuberosum* (Safed Mushli), *Dioscorea daemona* (Baichandi), *Andrographis paniculata* (Kalmegh), *Asparagus racemosus* (Satawar), and *Costus speciosus* (Keokand) allowed only 30%, 44%, 42%, 52% and 58% of SHL respectively to maintain sustainability *in situ*. However, other species, i.e., *Curculigo orchioides*(Kali mushli) (68%), *Curcuma angustifolia (Tikhur)* (64%), *Plumbago zeylanica* (64%), *Phyllanthus amarus* (Bhuinaonla) (71%) and *Embelia robusta* (Baibidang) (90%) have shown comparatively higher sustainable harvesting limits.

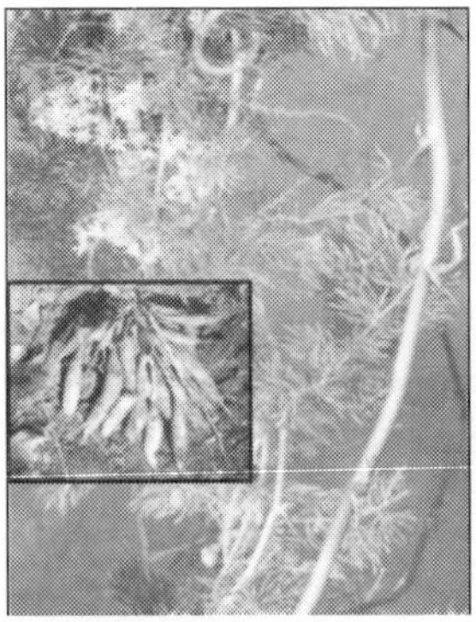

Plate 8.4 *Asparagas racemosus* (Shatavar)

Plate 8.5 *Chlorophytum tuberosum* (safed musali)

Table 8.1 Sustainable harvesting limit (SHL) of overexploited NTFPs in tribal dominated natural forest of Central India.

Name of species	*Utilizable plant part*	*SHL*
Chlorophytum tuberosum (Safed musli)	Tuberous roots	30%
Dioscorea daemona (Baichandi)	Tubers	44 %
Asparagas racemosus (Shatawari)	Tuberous roots	52%
Curculigo orchioides (Kali musli)	Roots	68%
Curcuma angustifolia (Tikhur)	Rhizomatous roots	64 %
Plumbago zeylanica (Chitrak)	Roots	64 %
Bauhinia vahlii (Mahul bel)	Leaves	58 %
Embelia robusta (Baibidang)	Seeds	90 %
Andrographis paniculata (Kalmegh)	Whole plant	42%
Phyllanthus amarus	Whole plant	71%
Costus speciosus (Keokand)	Rhizomatous roots	58%

Bauhinia vahlii, a woody climber which contributes substantial annual income (Rs. 5000-7000/family) to forest dependent communities was found to be in precarious condition. Due to repeated overexploitation of its leaves as utilizable parts, almost all plants were found to have lost their vigour in flowering, fruiting, leaf growth, etc. Leaves of this species are in high demand for making plates and cups, and harvested twice a year in summer as well as in winter seasons. Bark of the stem is also used for making ropes. Findings of this experiment have suggested that 58% harvesting of leaves irrespective of seasons should be permitted to maintain sustainability in qualitative and quantitative terms in natural forests.

Unregulated harvesting, increased demand and market access in the absence of security and non-sustainable management are leading to greater rate of resource depletion year by year. Inconsistent knowledge of regulation, lack of documentation/record, inadequacy in implementation of regulatory mechanism and ignorance of traditional practices are not only depleting the NTFP resource base but are also creating a threat on regional biodiversity. Nevertheless, there is lack of systematic/scientific approaches for development of mechanisms to halt non-sustainable harvesting of these valuable resources in situ. Surprisingly, no consideration persists on reproduction, stock, yields, habitat requirement, ecology, and regeneration potential, etc., of almost all NTFPs which are being over-harvested from natural tropical forests. In this context, development of skill and capability of local users communities was found to be a viable tool for conservation and sustainable management of forest resources in natural forest ecosystem in prevailing conditions.

Observation on phenological behavior of the NTFPs, medicinal plant species viz; germination, growth behavior, flowering, fruiting pattern, seeding, maturity period of utilizable plant parts, etc., are essential parameters for determination of SHL. These observations should be made at specific intervals round the year. (Fig. 8.1)

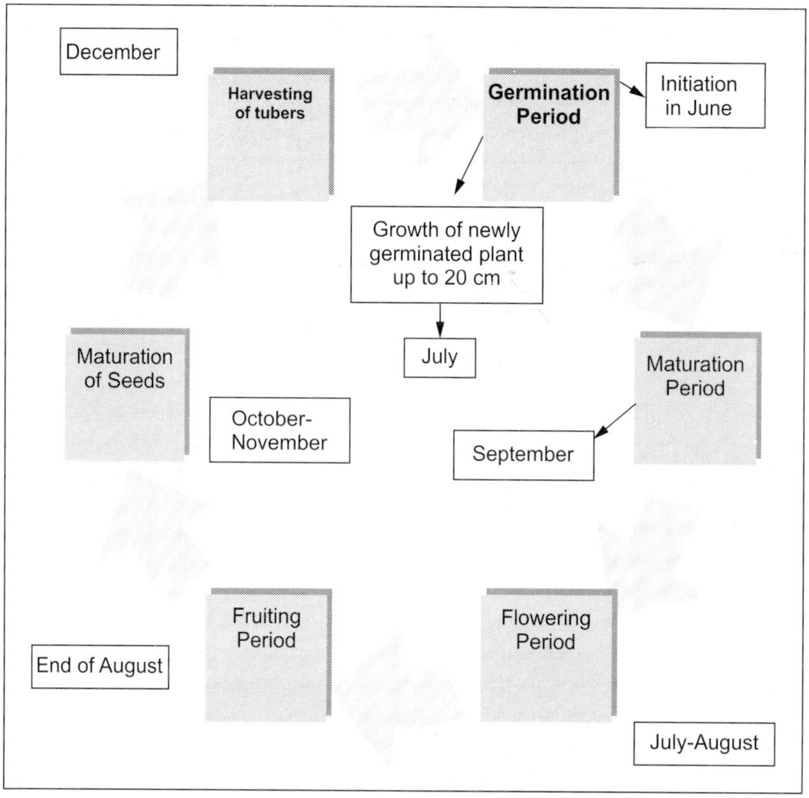

Fig. 8.1 Steps for phenological observations for determination of SHL of *curculigo orchioides* germination

Community Based Management

Strong self-regulatory institutions and community organizations with effective and transparent mechanisms for equitable sharing and conflict resolution are essential. Investments in training and organizational capacity building for the rural/tribal forest dwellers/communities should be recognized in the technology development. In this context, there is a need to develop an innovative "self-assessment technique" which could improve the capabilities of users communities for sustainable use of forest resources (Pandey *et al.* 2003; Pandey and Saini, 2004).

Case studies on forest resource assessment in the natural tropical forest obviously suggest that the prevailing harvesting trend of all the selected species is highly unsustainable and does not allow to harvest. However, in the absence of strict management interventions, due to over-harvesting, these resources are under threat for their existence insitu. Consequently, most of the species are prone to extinction. Nevertheless, there is no institutional package for conservation as well as sustainable management of these vulnerable resources from owner and/or user stakeholders. Consequences of over-exploitation of these resources have neither been realized by the local communities nor by the institutional stakeholders, and integrated strategic plans are still lacking for sustainable management of forest resources.

A community oriented sustainable resource management: Self-assessment Technique

There must be a comprehensive planning system for assessment and management of utilizable forest resources at community level. A **self-assessment system** (Pandey and Saini, 2001) seems to be an effective tool for assessing status and sustainable use of the forest resources in in situ conditions by users communities/forest dependent people at grass roots level in developing countries like India.

In one of the case studies, an innovative "**self-assessment**" method was evolved for assessment of status and use of utilizable resources through users community participation. It has shown encouraging and positive impacts in sustainable harvesting and management of NTFPs in the natural forest ecosystem in the study sites of Central India. Assessment of NTFPs through people's participation in JFM areas has two components, viz., (i) assessment of productivity and (ii) assessment of current level of harvesting. By this means species-specific information of utilizable NTFPs could be assessed and monitored in a regular way through direct involvement of local people.

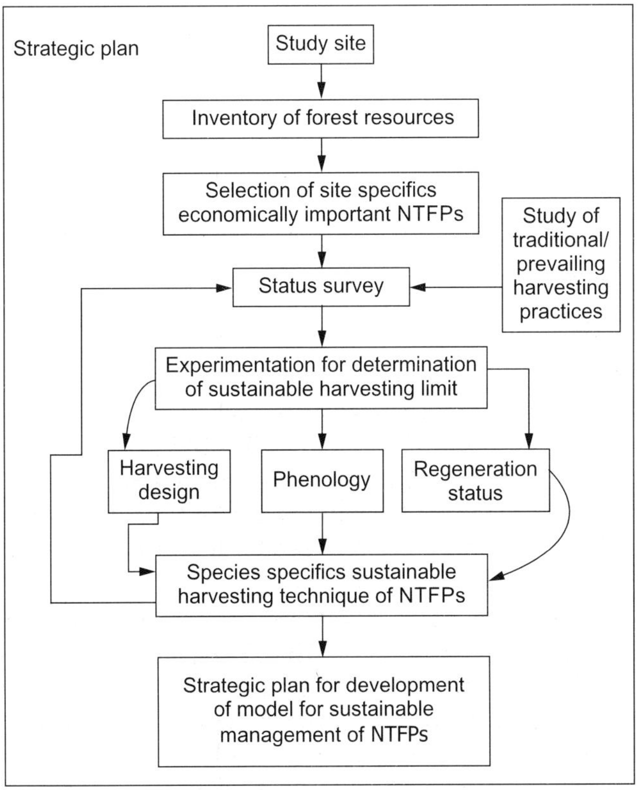

Plate 8.6 Development of model for sustainable management of NTEPs through community participation

Competent and properly motivated workforce has to be developed which should not only enlist the recommended practices, but also explain the conditions under which they apply. For development of strategies in forest resource management through people's participation, a comprehensive planning system is needed with the consideration of (i) identification of potentially rich NTFPs areas in natural forests, (ii) resource inventory, (iii) present status/availability of NTFPs (quantitative assessment), (iv) harvesting regimes, and, (v) consequences of overuse of resources on sustainability. At the community level, technology should be simple, energy efficient (economically viable) and environment-friendly. A model of forest resource assessment in natural forest ecosystem with integrated community approach seems to be a vital tool not only for sustainable management of NTFPs, but also to conserve regional biodiversity of the region.

National Forest Policy (1988), JFM resolution, Panchayati Raj Act (community governance), etc., however, provides legal framework for conferring the endowment of ownership rights of NTFPs to the local people, with the principles of: (i) harvesting of NTFPs would be on non-destructive basis; (ii) the village committee members will be free to collect NTFPs for their own consumption; and (iii) the manner, frequency and intensity of collection for any use other than bona fide domestic use by the members will be in accordance with the prescription of a management plan prepared by zila panchayat (district level committee) in conformity with the guidelines, which may be notified from time to time. This modality aims at developing positive stakes of all concerned so that rather than working at cross purposes, each one will be motivated to contribute towards the common goal (Sharma, 2003).

PEOPLE'S PROTECTED FOREST AREAS (PPAs)

In order to formulate a people-friendly framework for poverty alleviation, sustainable forest rersource management and biodiversity conservation through integrated ecosystem approach, a model of People's Protected Areas (PPAs) has been evolved considering the theme of IUCN Catergory - VI. which dictates that the site contains predominantly unmodified natural systems under management to ensure the long term protection and maintenance of biological diversity, whilst providing at the same time a sustainable flow of natural products and services to meet local community need.

Experimentally, several **People's Protected Areas** have been identified in potentially rich forest localities in Madhya Pradesh with the concept of proactive and people- friendly approach for management of floral as well as faunal biodiversity by the people and for the people. Presently, PPAs are linked to rehabilitation of degraded forest areas which are in the control of constituted forest protection committees under JFM.

Forest Department of Government of Madhya Pradesh and State Minor Forest produce T&D Co-op. Fed. Ltd. have jointly started an innovative initiative by establishing more than 25 PPAs/MPCA sites in different forest localities of various districts as a part of strategy to translate the philosophy of PPAs into implementable extension programmes. The PPAs aim at assessing and preparing a data bank on quantity, quality and value of NTFPs existing in the area by active involvement of local people along with local technology for sustainable management of forest resources in situ. Design for survey and collection of information and management of resources is based on sound statistical principles with integrated approach of community participation.

Since its inception as an experimental programme in many JFM areas, JFM has spawned institutional changes. But, later on keeping the momentum of JFM, there has been a major challenge for national and state governments when donor fund run dries. To sustain interest in the programme, JFM would have to deliver significant, continuing incentives for community participation.

Forest resource based enterprises have the potential to provide such incentives especially when linked with ex situ cultivation and processing (Capistrano, 2000). However, long-term viability of these enterprises will crucially depend on the policy environment and complimentary programmes to ensure availability of key inputs like access to productivity enhancing technologies, adequate market infrastructure, equitable benefit distribution, sustainable use of forest resources, skill and capabilities development programmes.

However, documented evidence suggests caution. A number of concerns have been raised about the possible adverse implications of commercializing NTFPs, especially those that are key inputs for livelihood security (Dove, 1994; Perez and Arnold, 1995; Arnold and Townson, 1999). The bulk of NTFPs traded in India are still collected from the wild. Questions have been raised about the validity of the implicit assumption that NTFP stocks are sufficiently abundant to withstand increased harvesting.

The ecological and genetic diversity impacts of long-term harvesting, especially when the enterprise involve collection of roots, fruits, seeds or whole plants are issues of concern (Godoy and Bawa, 1993; Peters, 1997). It has also been reported that even in the absence of destructive harvesting, regular collection of commercial quantities of fruits and seeds can cause notable changes in the structure and dynamics of a plant population of the forests (Hall and Bawa, 1993; Peters, 1994; Murali et al., 1996; Muralidharan et al., 1997 and 1999). In some species, regeneration rate is affected even by small scale harvesting of foliage and twigs over a long period of time (Singh, 1998).

Many factors determine the economic viability and resource conservation outcomes for encouraging NTFP based enterprises. Successful outcomes tend to be associated with positive state sponsored regulations (state policy) that offer clear and well-known rights to people; a harvesting time, procedure and intensity and/or technique pertaining to species-specific limit of harvesting (that does not put excessive pressure on the resource), a transparent market; well organized gatherers; and the existence of external support groups (Perez and Byron, 1999; Saini and Pandey, 2006).

Ecological Sustainability with Institutional Policy Imperative

The common property nature for use of forest resources requires secure and enforceable regulated use rights as preconditions for collective management and conservation of native forest resources. Poverty and attraction of market options, local politics, ineffectual leadership and break down of rule enforcement have undermined community institutions and left natural resources plundered and exploited beyond the limits of their regenerative capacities. Ecologically, prevailing harvesting techniques at field level were found destructive and wasteful. As a general practice, forest dwellers collect almost all available resources without due consideration of sustainability to satisfy their immediate needs due to ever increasing market demand.

Skill and Capacity Building Programme for Users Communities

Critical examination of JFM areas envisaged that success not only relates to forest protection but involves several other activities. A close scrutiny of approaches applied for monitoring, assessing and sustainably harvesting of forest resources reflects that the efforts are partly one sided either from forest department or from outside professionals. There is no sign of community initiation for inventory, data collection, analysis and production towards sustainable management of utilizable NTFPs in community managed forests.

Prevailing harvesting techniques at field level were found to be destructive and not supporting to sustainability of these native valuable genetic resources/diversity of the region. As a general practice, forest dwellers collect almost available resources without due consideration of sustainability to satisfy their immediate needs. Consequently, these resources are not able to recover the population for maintaining their sustainability in natural forests. There is a need of species-specific information on forest resources and regular status assessment and monitoring through direct involvement of local people as a key stakeholder.

Competent and properly motivated workforce is the need to develop not only enlists the recommended practices, but also to explain the conditions under which they apply. Comprehensive planning system with strict legal provisions is needed for development of strategies in forest resource management through peoples' participation. This system requires: (i) identification of potentially rich NTFPs areas in natural forests; (ii) resource inventory; (iii) present status/availability of NTFPs (quantitative assessment); (iv) harvesting regimes; (v) consequences of overuse of resources on sustainability; and (vi) value addition, market networking and trade of collected resources. At community level, imparted technology should be simple, energy efficient (economically viable) and environment-friendly.

Thus, strategic community oriented skill and capability development programmes with strict institutional management interventions have been found to be effective tools for conservation and sustainable management of forest resources in tropical forests in situ. JFMCs may play vital role in developing integrated collaborative approach with institutional stakeholders for sustainable management and conservation of vulnerable forest genetic resources in natural forests of the country.

Acknowledgements

Financial assistance from CSIR, New Delhi, Govt. of India is highly acknowledged for successful completion of this project.

REFERENCES

1. Arnold, M., and Townson, I. (1999). Assessing the potential of forest product activities to contribute to rural incomes in Africa. *Nat. Res. Pers.*, No. 37, Overseas development Institute, London. 4 pp.
2. Bangkok Thailand. 2-4 Feb. 2003, pp. No. 23-40.
3. Capistrano, D. (2000). Non-timber forest product-based enterprise in forest conservation and community development: India's evolving institutional context. Paper for presentation at the XXI IUFRO World Congress, 7-12 August, Kuala Lumpur, Malaysia.
4. DOVE, M.R. (1994). Marketing the Rainforest: 'Green' Panacea or Red Herring? Analysis from the East-West Centre 13: 1-7
5. Forest Research Institute, Trichur, Kerala. p. 36.
6. FSI (2003). Annual Report of Forest Survey of India, Dehradun.
7. FSI and FAO (2004). Pilot study on assessment of status of sustainability of forest resources in India. Forest survey of India, Dehradun with collaboration of Food and Agriculture Organization (FAO), Rome.
8. Godoy, R.A., and Bawa, K.S. (1993). The economic value and sustainable harvest of plants and animals from the tropical forest: assumptions, hypotheses, and methods, *Econ. Bot.* 47(3): 215-219.
9. GOI (2001). Report of the task force on greening India for livelihood security on sustainable development. Government of India, Planning Commission.
10. Hall, P., and Bawa, K. (1993). Methods to assess the impact of extraction on nontimber tropical forest products on plant populations. *Econ. Bot.* 47(3): 234-247.
11. Innovation in Participatory Management Technique through Self-assessment for Biodiversity Conservation. Proceedings of National Workshop on Regional Strategy for Plant Conservation" held at TFRI, Jabalpur on 26-27 Feb. 2004.
12. Murali, K.S., Shankar, U., Shaanker, R.U., Ganeshaiah, K.N., and Bawa, K.S. (1996). Extraction of non-timber forest products in the forests of Biligiri rangan Hills, India 2.Impact of NTFP extraction on regeneration, population, structure, and species composition. *Econ. Bot.* 50 (3): 252-269.
13. Muralidharan, P.K., Sasidharan, N., and Seethalakshmi, K.K. (1997). Biodiversity in tropical moist forests: A study of sustainable use of non-wood forest products in the Western Ghats, Kerala. KFRI Research Report No. 133, Kerala Forest Research Institute, Trichur, Kerala. p. 128.
14. Muralidharan, P.K., Chandrashekhara, U.M., Seethalakshmi, K.K., and Sasidharan, N. (1999). Biodiversity in tropical moist forests: A study of sustainable use of non-wood forest products in the Western Ghats, Kerala. Monitoring and evaluation of ecological and socio-economic variables. KFRI Research Report No. 162, Kerala. p. 36.

15. Pandey, R.K. (2000). Challenges for *in-situ* conservation and sustainable development of tropical forest in Madhya Pradesh. Proceeding of International Symposium on Tropical Forestry Research: Challenges in the New Millenium 2-4 Aug. 2000, p. 5, Kerala Forest Research Institute, Peechi.
16. Pandey, R.K., and Shrivastava, J.L. (1989). Status survey of medicinal wealth in natural forests of Seoni district. *J. Trop. For.* 5(3) 206-212.
17. Pandey, R.K., and Shrivastava, J.L. (1996). A material for forest flora of Madhya Pradesh State Forest Research Institute, Jabalpur, M.P. S.F.R.I. Bulletin No. 28.
18. Pandey, R.K., and Saini, S.K. (2001). Pachmarhi Biosphere Reserve ke buffer zone mein satat van-prabandhan ke adhar evum suchkank (Criteria and Indicators) ka vikas I: Ikai star per jan samuday kee bhumika (In Hindi). *Vaniki Sandesh* Vol. 25 No. (1).
19. Pandey, R.K., Saini, S.K., and Ansari, A.A. (2003). Sustainable management of degraded forest through community participation: A case study of Parraspani village in buffer zone of Pachmarhi Biosphere Reserve.
20. Perez, M.R., and Arnold, J.E.M. (1995). *Current Issues in Non Timber Forest Products Research.* Center for international Forestry Research, Bogor, Indonesia. 264 pp.
21. Perez, M.R., and Byron, N. (1999). A methodology to analyze divergent case studies of non-timber forest products and their developmental potential. *For. Sci.,* 45(1): 1-14.
22. Peters, C. (1997). Observations on the sustainable exploitation of non-timber forest products. In: Perez, M.R. and J.M. Arnold (Eds.) *Current Issues in Non Timber Forest Products Research.* Proceedings of a workshop on research on NTFP, Hot Springs, Zimbabwe. CIFOR, Bogor and ODA.
23. Peters, C.M. (1994). Sustainable harvest of non-timber plant resources in tropical moist forest: An ecological primer. The Biodiversity support Program, Washington, D.C. p. 45.
24. Pandey, R.K., and Saini, S.K. (2004). Creating awareness among different stakeholders and policy makers regarding the recent technological innovations and research advancements for application in JFM Chhattisgarh perspective. Proceedings of National workshop on technical innovations and advancements for applications in joint forest management, organized by ICFRE, Dehradun and FAO,
25. Pandey, R.K., Saini, S.K. (2007). Edible plants of tropical forests among tribal communities of Madhya Pradesh. Indian J. Traditional Knowl., 6(1): 185-190.

9

Domestication Status of Six Non-Timber Forest Produces in Madhya Pradesh and Odisha

Anshu Singh*, Sarvashish Roy, A.K. Kandya† and Prodyut Bhattacharya‡**

**PRIA, Jaipur*

***ITC, Bhopal*

† Department of Botany, Dr. H.S. Gour University, Sagar (M.P.)

‡University School of Environment Management

GGS Indraprastha University, New Delhi - 110078

E-mail: prodyutbhattacharya@yahoo.com & sbroy_chairman@yahoo.com

ABSTRACT

Non-Timber Forest Products (NTFPs) species are widely accepted as major source of earning from forests, but are getting eroded from natural forests fast due to heavy pressure for collection. In the recent past it has acquired due commercial importance with the growth of herbal industries and consumerism. Domestication of NTFPs is a scientific conservation process which enhances the capacity of species to fulfil either or both services and production functions, reveals from our study in Madhya Pradesh and Odisha. Domestication is primarily a strategy and its techniques are selected on the basis of economic choices and by the initial preference for a given cropping system within a given farming system. It is a way of responsible management wherein social equity and environmental concern are interwoven with economic prosperity. The present study provides justification and potential of domestication process under agroforestry system of NTFP tree species in the outside natural forest with ecological, social and cultural characteristics as well as the economic needs of local inhabitants for the future sustainability.

INTRODUCTION

India's forests are under severe pressure for meeting growing demands for fuel, fodder, grazing, timber and non-timber forest products for an ever increasing human and livestock population and industrial demands. Approximately one-fourth of the world's poor and 90% of the poorest rely significantly on forests for their livelihoods (WHO, 2003; Mishra, 2011; Malhotra et al., 1991). NTFP species are widely accepted as major source of earning from forests. In the recent past it has acquired due commercial importance with the growth of herbal industries and consumerism (Rawat, 2002). While the demand has been steadily increasing, the supply is heavily affected because majority of the raw material (70-90%) comes from wild sources (Bhattacharya and Hayat, 2004; Bhattarai and Croucher, 1996; Lange, 2006; Ved and Goraya, 2007). In such a scenario domestication and commercialisation of NTFPs seems to be the only viable option for maintaining the resource base in the wild and at the same time meeting the commercial demand of the species.

Domestication of NTFPs is a scientific conservation process which enhances the capacity of species to fulfil either or both services and production functions. Domestication of NTFPs is a human induced evolution bringing species into wider cultivation through farmer-driven or market-led approach (Simons, 1996). It includes a wide range of activities, viz., exploration and collection of natural populations, evaluation and selection of suitable species and provenance, breeding to develop suitable cultivars, development of propagation techniques, multiplication and dissemination of germplasm, development of management techniques, utilization and tree product marketing, development and dissemination of relevant technical information (Akinnifesi et al., 2006). Strategies for individual species vary according to their functional use, biology, target environments and the domesticators involved and researchers or farmers. Domestication is primarily a strategy and its techniques are selected on the basis of economic choices and initial preference for a given cropping system within a given farming system.

Domestication of NTFP provides sustainable income to farmers on one hand and on the other meets the commercial demand for the species, thereby reducing pressure on the wild resources. It is a way of responsible business wherein social equity and environmental concern are interwoven with economic prosperity. Domestication is therefore, a win-win scenario wherein both conservation and economic development go hand-in-hand. Three dimensions of domestication may be distinguished which are associated with specific socio-economic and ecological conditions (Wiersum, 1996):

(a) Stage 1: Uncontrolled exploitation to controlled collection from forest
(b) Stage 2: Purposeful cultivation of wild species outside forest
(c) Stage 3: Commercial cultivation of domesticated species in farmer's field

The recognition of these different dimensions is of particular importance with respect to the domestication of tree species, because commercial tree utilization usually starts with the exploitation of trees from the wild in natural forests.

Stage 1: The domestication of species in forest environment involves the change from uncontrolled collection of the wild tree products to their controlled collection. During this stage of domestication of species, the management practices are directed at defining user rights.

Stage 2: The second phase consists of purposeful cultivation of wild trees, either in a resource-enriched natural environment, or in an indigenous agroforestry system, such as forest gardens. In this stage, management of biological characteristics is undertaken. In the process of adoption, farmers grow species due to its commercial demand of particular product and due to that they try to improve the morphological and reproductive characteristics of the species.

Stage 3: The cultivation of domesticated trees in either agroforestry systems or commercial tree-crop plantations is the last stage of this domestication process. During this stage of plant domestication, the genetic characteristic of the tree population is modified under scientific projects.

Domestication process of NTFP tree species is linked with social and cultural characteristics as well as the economic needs of local inhabitants for the products of such plants. In places where the markets are easily available for certain NTFPs, the species are readily domesticated (Nair and Merry, 1995).

The study was carried out in two states, viz., Madhya Pradesh (MP) and Odisha (Od). Two districts were studied in each of the two states, viz., Hoshangabad and Sehore in Madhya Pradesh and Mayurbhanj and Koraput in Odisha. Ten villages were studied in each of the two states.The six species studied are *Annona squamosa, Buchanania lanzan, Chlorophytum borivilianum, Commiphora wightii, Emblica officinalis* and *Litsea glutinosa.*

1. *A. squamosa* The species is in Stage 2 of domestication in Sehore (MP), Rairangpur and Koraput (Odisha). The species is grown under traditional agroforestry system (home gardens) in these areas. In Hoshangabad, the fruits were unsustainably harvested from forest, and therefore, is in Stage 0 of domestication process. The stage of domestication of the species in different study sites is presented in Table 9.1.

The species, locally known as Sitafal, is a high value fruit crop and has low gestation period (4 years). Propagation of trees takes place through seeds and grows easily on wastelands. Easy availability of the market and maximum value capture at farm gate makes the species highly preferable for cultivation by the farmers. Still large-scale cultivation of species under commercial plantation has not been taken up in the study sites. However, in Sangareddy, Andhra Pradesh, a study on germplasm, collected from wild sources from all over India is carried out. Work is in progress to select seedlings having better yield, quality and size of fruit with lesser number of seeds. Attempts are also being made to induce seedless variety with hormonal treatment. A few promising cultivars have been isolated from the germplasm and their progenies have been propagated through budding. Amoo (2008) reported that the cultivar of Balanagar, Andhra Pradesh, produced maximum fruit weight (700 g). Large varietals were reported in the 'Red fruited' (*A. squamosa* var. *rubra*) and 'Green fruited' (Chadha, 2003).

Therefore, though in MP, the species was found to be in the second stage of domestication, the above discussion suggests that the actual domestication stage for the species is 3.

2. *B. lanzan* The species is in stage 1 of domestication in Baripada and Koraput (Odisha), whereas in MP the seeds of the tree were unsustainably harvested and is therefore, in Stage 0 (Table 9.1).

The species is also not preferred for commercial cultivation due to its long gestation period (10 years). Chironji is facing severe genetic erosion as a result of activities related to afforestation in tribal inhabited areas (Singh, 2007). Cultivation of Chironji trees is unorganized and fruits are directly collected from the natural wild trees occurring in the forest and marginal lands, therefore, the species is facing a great threat. Occasionally, the trees are cut partly or completely by the tribals, to collect the fruits. Many of the populations have been completely wiped off in the recent past due to developmental activities and want of agriculture land by farmers. There are no standard cultivars available in chironji since little work has been done to exploit genetic resources.

Genetic resources of Chironji have not been given due attention till date. Genetic diversity of Chironji has been collected from Gujarat, Uttar Pradesh, Maharashtra, Bihar, Chhattisgarh and Rajasthan by Central Horticulture Research Station (CHES), Central Institute of Arid Horticultural (CIAH), Godhra, Central Institute of Sub-tropical Horticulture (CISH), Lucknow, and National Bureau of Plant Genetic Resources (NBPGR), New Delhi. CHES (CIAH), Godhra collected 30 variable accessions from Gujarat (Singh, 2007). At CISH, Lucknow 8 accessions of Chironji have been identified and collected from Uttar Pradesh and Madhya Pradesh (Annon, 2008).

Recently NBPGR, New Delhi, has taken up specific exploration and collection missions in the diversity rich areas of Madhya Pradesh, Gujarat, Chhattisgarh and Rajasthan for the collection of Chironji and 74 accessions have been collected. Thirty accessions collected from Gujarat have been characterized for physico-chemical characters at CHES (CIAH), Godhra (Singh et al., 2006). Eight collections have been found promising for important horticultural traits and are being evaluated for field performance at CHES (CIAH), Godhra. Germplasm collected from the parts of Uttar Pradesh, Maharashtra, Bihar has been characterized for fruit weight, total soluble solids, acidity, protein content and earliness (Rai, 1982).

In the region of its natural occurrence it regenerates through seeds. Vegetative propagation through soft wood grafting and chip budding is successful but rarely tried as there is no demand of planting material for commercial cultivation. Presently, the trees are available only in the forest or marginal lands near the villages. It is a highly heterozygous, cross pollinated fruit crop and seedling selections are required to be identified with desirable characters.

Based on the observation from the study sites regarding the stage of domestication and the above discussion regarding the efforts undertaken for domestication from various studies, it is clear that the most prevalent status of *B. lanzan* is Stage 0, i.e., the species is harvested unsustainably in the wild and there have been no efforts regarding its domestication in situ or ex situ. Thus, Stage 0 is considered its overall stage of domestication.

3. C. borivilianum The species is one of the most important species required for Ayurvedic formulations. The tuberous roots of *C. borivilianum* are the economically important part which contains saponins and are used for the preparation of many Ayurvedic tonics. It fetches an attractive market price (Rs. 900–1200/kg of dry roots). It is in stage 2 of domestication in Hoshangabad and Sehore (MP) where the wild variety of the species was commercially cultivated. In Baripada and Rairangpur, the tuberous roots of Safed Musli were unsustainably harvested and thus, the species is in Stage 0 of domestication (Table 9.1).

The wild varieties are cultivated as no improved variety of the species has been developed till date. The increasing demand of the roots by the pharmaceutical companies has resulted in intense competition among collectors. This affected regeneration owing to collection of roots by the tribal/ local people before seed formation and dissemination. The species is a rare herb and hence its preservation and conservation carries a lot of importance. A serious effort is required for the collection, conservation and maintenance of its variability.

Based on the field observations and above discussion, it may be concluded that the species exhibits two types of domestication stages, viz., unsustainable harvesting in wild and ex situ cultivation of wild varieties in farmers' field. However, the cultivation of wild variety was found dominant and therefore, the overall stage of domestication for the species is considered as 2.

4. *C. wightii* The species is found in wild state in MP but is not found in Odisha. It was found to be domesticated in Hoshangabad and is in Stage 2 of domestication process (Table 9.1). It is not preferred for cultivation by the farmers, even though there is an ever increasing demand for guggal. The increase in commercial interest in plant and resultant excessive tapping has pushed the species into the list of Rare Endangered and Threatened plants in India. Central Institute of Medicinal and Aromatic Plants (CIMAP), Lucknow, has developed a good oleo-gum-resin-yielding cultivar called 'Marusudha'. Some studies by CIMAP also indicated the presence of high guggal sterones in the cuttings obtained from the areas of Ajmer in Rajasthan and Kot Lakhpat area of Bhuj in Gujarat. The germplasm from Nakoda in Rajasthan also gives a good yield (CIMAP, 1997). But all these are still at trial level and are yet to be transferred to farmers' field. Therefore, Stage 2 has been considered at its overall stage of domestication.

5. *E. officinalis* *E. officinalis* was found to be cultivated as regular cropping system in the study sites of MP. Therefore, it is in Stage 3 of domestication (Table 9.1). In Baripada and Koraput the fruits of the species were sustainably collected from the forest, whereas in Rairangpur it was traditionally grown in homegardens (Table 9.1). There are three main varieties of Aonla – Banarasi, Francis (Hathijhool) and Chakaiya, which are cultivated under regular cropping system.

National Research Centre for Agroforestry (NRCAF), Jhansi, India has advocated inter cultivation of *Vigna mungo* (black gram), *V. radiata* (Green gram/Mung) and *Lens culinaris* (Masur) in *E. officinalis* (Aonla) orchards. At present Aonla germplasm is being maintained at several places in the country in field gene banks. These germplasms were mainly collected from Narendra Dev University of Agriculture and Technology (NDUAT), Faizabad. Under the Indo-US project, National Bureau of Plant Genetic Resources (NBPGR) is maintaining National Active Germplasm Site (NAGS) of Aonla at Central Horticulture Experiment Station, Ranchi and Jharkhand in Bihar. Besides this, a total of 51 Aonla accessions are maintained at the NBPGR Regional Station, Trissur, Kerala (Malik et al., 2010). *E. officinalis,* being astringent in taste, is not consumed as a table fruit. The fruits are processed into several products, which are much in demand in Indian households. Demand for the species in the herbal medicine sector comes from Ayurveda, Siddha and Unani systems of medicine, which are the major wings of Indian System of Medicine and Homeopathy (ISM&H).

Based on the observations at the study sites (Table 9.1) and the above discussion it may be concluded that *E. officinalis* domestication stage is 3, i.e., it has achieved the highest order of domestication that is of commercial cultivation though improved varieties. Thus, the overall stage of domestication for *E. officinalis* is considered 3.

6. *L. glutinosa* In MP the species is in second stage of domestication only in Hoshangabad, whereas in Sehore the species was not found. MP Forest Department has established a nursery in Matkuli in Pipperiya range of Hoshangabad Forest Division. The species is propagated through root suckers in nursery. When the plant attains a sapling stage, it is transplanted in the forest area near the vicinity of natural *Litsea* plant. In Baripada, Rairangpur and Koraput Forest Divisions of Odisha, the bark of *Litsea* is unsustainably harvested by Khoria tribes, and is therefore, in stage 0 of domestication (Table 9.1). In Koraput the species is locally called 'Tangiaachina' (Franco and Narasimhan, 2009), whereas in Baripada and Rairangpur it is called 'Pojo'. In Hoshangabad it is called 'Maida chhal.'

The bark of *L. glutinosa* is used in making Jiggat (adhesive) for agarbatti (incense stick). The Agarbatti industry is a source of livelihood for thousands of women in India (Mahasuar, 2012). Illegal unsustainable harvesting of *Litsea* bark is widespread for preparation of *jiggat* used in Agarbatti industry. Very few initiatives have been taken for the propagation of the species. Thus, the overall stage of domestication for *L. glutinosa* is taken as 0.

Table 9.1 Stages of domestication of the selected species in the study sites

Study Sites	*Stages of domestication*					
	A. squamosa	*B. lanzan*	*C. borivilianum*	*C. wightii*	*E. officinalis*	*L. glutinosa*
Hoshangabad	0	0	2	2	3	2
Sehore	2	0	2	NA	3	NA
Baripada	NA	1	0	NA	0	0
Rairangpur	2	0	0	NA	3	0
Koraput	2	1	NA	NA	3	0
Mode	2	0	2	2	3	0

From the present study it can be concluded that among the six species studied, only *A. squamosa* and *E. officinalis* are in the third stage of domestication. Serious conservation measures are required for *C. borivilianum, C. wightii* and *L. glutinosa* to maintain their natural population in the wild. Efforts for domesticating the rest of the selected species are required. Ex situ conservation and sustainable harvesting protocols need to be developed and specific efforts are required for domestication of *C. wightii* and *L. glutinosa.*

REFERENCES

1. Akinnifesi, F.K., Kwesiga, F., Mhango, J., Chilanga, T., Mkonda, A., Kadu, C.A.C., Kadzere, I., Mithofer, D., Saka, J.D.K., Silashi, G., Ramdhani, T., and Dhliwayo, P. (2006). Towards the development of miombo fruit trees as commercial tree crops in Southern Africa. Forests, Trees and Livelihood. 16. pp. 103-121.

2. Amoo, I.A., Amenike, A.E., and Akpambang, V.O.E. (2008). Compositional evaluation of *Annona cherimoya* (Custard Apple Fruit). *Trends. Appl. Sci. Res.* 3(2): pp. 216- 220.
3. Annon (2008). Agrotechniques of selected medicinal plants. Published by National Medicinal Plant Board.Department of Ayush, New Delhi. Published by TERI, New Delhi. 1: p. 90.
4. Bhattacharya, P., and Hayat, S.F. (2004). Sustainable NTFP management for rural development: A case from Madhya Pradesh, *India. International Forestry Review.* 6(2): pp. 161-168.
5. Bhattarai, N.K., and Croucher, J. (1996). Viability of local commercialization of non-timberforest products as a strategy for promoting biodiversity conservation. In: Environment and Biodiversity in the context of South Asia. (Eds.) P.K. Jha, G.P.S. Ghimire; S.B. Karmacharya; S.R. Baral and P. Lacoul. Ecological Society, Kathmandu, Nepal. pp. 346-353.
6. Chadha, K.L. (Ed.) (2003). Handbook of Horticulture Published by Directorate of Information and Publications of Agriculture, Indian Council of Agricultural Research, Krishi Anusandhan Bhavan, Pusa, New Delhi.
7. Franco, M., and Narasimhan, D. (2009). Plant names and uses as indicators of knowledge patterns. *Indian J. Trad. Know.* Vol. 8(4): pp. 645- 648.
8. Lange, D. (2006). International trade in medicinal and aromatic plants actors, volumes and commodities. In: Medicinal and Aromatic Plants. (Eds.) R.J. Bogers; L.E. Craker and D. Lange. pp. 155-170.
9. Mahasuar, K. (2012). The journey of an incense stick: A rural success story. Indian Management.
10. Malik, S.K., Chaudhury, R., Dhariwal, O.P., and Bhandari, D.C. (2010). Genetic Resources of Tropical Underutilized Fruits in India. Published by the Director, NBPGR, New Delhi. p. 168.
11. Malhotra, K.C., Deb, D., Dutta, M., Vasula, T.S., Yadav, G., and Adhikari, M. (1991). Role of non-timber forest products in village economy: A household survey in Jamboni Range, Midnapore, West Bengal. Indian Institute of Bio-social Research and Development.
12. Mishra, M. (2011). Wild Harvesting and Management of some medicinal plants in natural forest of Central India. *Indian J. Fund. Appl. Lif. Sci.* 1(2): pp. 90- 97.
13. Nair, P.K.R., and Merry, F.D. (1995). Development of non-wood forest products through agroforestry: Issues and strategies. In: Non-Wood Forest Products. Report of the international expert consultation on non-wood forest products. Series 3. Rome: FAO.
14. Rai, Y.C. (1982) *Buchanania lanzan* Spreng. Studies on methods of propagation and estimation of fruit yield. Indian Forester. 108: pp. 501-511.

15. Rawat, R.B.S. (2002). Medicinal plants sector in India with reference to traditional knowledge and IPR issues. Paper presented at International Seminar for the Protection of Traditional Knowledge, New Delhi. http://r0.unctad.org/trade_env/test1/meetings/delhi/India/mik-094.doc. (accessed on December 11, 2011).
16. Simons, A.J. (1996). ICRAF's strategy for domestication of indigenous tree species. Pp. 8–22, in: R.R.B. Leakey, A.B. Temu, M. Melnyk and P. Vantomme (eds) Domestication and Commercialization of Non-Timber Forest Products in Agroforestry Systems, Non-Wood Forest Products 9. Food and Agriculture Organization of the United Nations, Rome, Italy.
17. Singh, S. (2007). Chironji. In: Underutilized and underexploited crops. (Ed.) K.V. Peter New India Publishing Agency, New Delhi. 2: pp. 295-309.
18. Singh, S., Singh, A.K., and Apparao, V.V. (2006). Genetic diversity in Chironji (*Buchanania lanzan*) under semi-arid ecosystem of Gujarat. *Indian J. Agr. Sci.* 76: pp. 695-698.
19. Ved, D.K., Goraya, G.S. (2007). Demand and supply of medicinal plants in India. NMPB, New Delhi; FRLHT, Bangalore.
20. World Health Organization (WHO)(2003). WHO guidelines for good agricultural and collection practices (GACP) for medicinal plants.(www. hqlibdoc.who.int/publications/2003/9241546271.pdf.)Accessed on 5/1/12.
21. Wiersum, K.F. (1996). Domestication of valuable tree species in agroforestry systems: evolutionary stages from gathering to breeding. In: Domestication and Commercialization of Non Timber Forest Products in Agroforestry Systems. (Eds.) R.R.B. Leakey, A.B. Temu, M. Melnyk and P. Vantomme. Non Wood Forest Products 9, FAO, Rome, Italy. p. 298.

10

Forest Products Utilization During Famine Conditions in Raigad District of Maharashtra: A Case of Thakur Tribes

Manish Mishra, PA Jadhav and Rajshree Sinha

Faculty area of Forestry,
Indian Institute of Forest, Bhopal (M.P.), India
E-mail: manishm@iifm.ac.in

ABSTRACT

Thakur tribes of Raigad district of Maharashtra, who form the majority in the area, inhabit the upland areas of the foothills. They have small landholdings and are skilled bamboo workers. They eat several NTFPs, medicinal plants, wild roots, rhizome, seeds, fruits and mushrooms at the time of famine and natural calamities like severe drought, flood, etc. They use stored wild tubers mostly Ratalu *(Dioscorea bulbifera)*, Akhur (*Dioscorea wallichii)*, Vazkand (*Dioscorea daemona*), Shivla (*Amorphophallus sylveticus*), etc., to fulfill their food, medicinal and nutritional requirements, particularly in harsh climatic conditions. They eat various dried and stored wild leafy vegetables like Kolu bhaji *(Chlorophytum tuberosum)*, Taklu bhaji (*Cassia tora*), Matth bhaji (*Amaranthus spinosus*), Kurudu bhaji (*Celosia argentea*) during famine conditions. Some of the fruit species like Tembru (*Diospyros melanoxylon*), Karanda (*Carrisa carandus*) etc., thrives during extreme droughts. They drink Kokum *(Garcinia indica)* sharbat (liquid), peg made of Mahua (*Madhuca indica*) flowers, etc., during scarcity of food. They also utilize NTFPs like amla (*Emblica officinalis*), mango (*Mangifera indica*) fruits, bamboo seeds and some wild mushrooms (*Agaricus* spp.) after drying and crushing during drought period. Indigenous practices of Thakur tribes to cope up with extreme environmental conditions and utilization of forest products during severe drought and famine etc., are discussed in the paper.

INTRODUCTION

Thakur or lord is the common Rajput title by which they are generally addressed. They have rich traditions of folk dance, songs and unique culture of their own. Thakurs live in small groups. Their villages are mostly situated in forest or on hilltop. Their houses are more like huts made up of mud and bricks. They are also called by other names like Thakur, Thakar, Ka Thakar, Ma Thakur. They come under scheduled tribe category. Thakurs are Indo-Aryans and speak a dialect of Marathi. They mostly use cereals of different types in their food. Raigad is located in the Konkan region in the state of Maharashtra. The district is bounded by Mumbai Harbour to the northwest, Thane District to the north, Pune district to the east, Ratnagiri district to the south, and the Arabian Sea to the west. The Thakurs live primarily in the hilly areas of Karjat, Khalapur, Panvel, Sudhagad and Pen talukas of Raigad district in the western Indian state of Maharashtra in India. They are also found in Dhule and Jalgaon. As per 2001 census, the population of the Thakurs in Maharashtra was 487,696 and the literacy rate was 50.9%. Thakur women's are fond of jewellery. Their main occupation is agriculture. They use traditional methods of cultivation. They use plants to cope up with the extreme environmental conditions.

Indigenous people in the world have used various strategies to respond and adapt to environmental change: diversified resource base chiefly to minimize the risk due to harvest failure, they grow many different crops and varieties, and they also hunt, fish, and gather wild food plants. Change in crop varieties and species and change in the timing of activities like crop harvests, wild plant gathering, fishing, changes in resources resorting to wild foods in the case of emergency situations such as droughts and floods are their coping strategies to combat adverse climatic condition (Salick and Byg, 2007). Plants are used as emergency food by the tribals of Rajasthan desert during periods of famine (Bhandari, 1974). The utilization of little known foods in times of acute crisis is recognized as a form of resilience. Several indigenous crop species are described which could be grown and utilized to prevent a great deal of suffering. The information could be an indicator of the extent to which a region may be suffering silent famine. These natural phenomena may be related to wind direction, cloud pattern, position of planets, behaviour of animals, birds and changes in plants, etc. (Pareek and Trivedi, 2011).

Local communities have always aimed to adopt variations in their climate. They possess enough knowledge about wild varieties of agricultural crops, edible food, tubers, rhizomes, vegetable crops, etc. Traditionally, they store and utilize wild tubers, yams, medicinal plants, etc., during the harsh climatic conditions (Mishra and Singh, 2009; Mishra, 2010 and Mishra et al., 2011). Later on Mishra et al. (2012a and 2012b) documented traditional knowledge of Baiga and Gond tribes of Central India and reported their coping strategies to fight adverse environmental conditions such as drought, famine and floods. They also reported a few wild tubers, NTFPs, agricultural crops utilized during drought and famine conditions. They concluded that the selected tribes have their own strategy to cope up with natural calamities by utilizing NTFPs, yams, roots, fruits, etc., after drying in particular drought situations.

METHODOLOGY

Thakur tribes of the District Raigad were surveyed thoroughly along with forest officials. Raigad district lies between latitude and longitude 18° 38′ 24″ North, 72° 52′ 48″ East. Six forest ranges of the study area were selected for the survey conducted in 2012 in all the three seasons. This period of study helped the team to record the indigenous knowledge of the Thakur tribe from a variety of contexts, such as community food gathering activities, community farming activities, etc. Each range was visited 2-3 times, and old aged persons were interviewed for collecting information. Four villages in each range were selected, i.e., in Uran range, villages Karanja, Dighode Nagopirwadi, Mora were selected. In range Neral matheran (Karjat east), villages Bhadwal-Takachiwadi,

Damat-Katkarwadi, Ambherwada were selected. In range Karjat west, Ashane-Thakurwadi, Akulwadi, Gorewadi, Palashdari wadi villages were selected. In Pen range Jamrajwadi, Katechiwadi, Katkarwadi and Fanaswadi villages were selected. In Khalapur range, Kandholi, Katkarwadi, Khapoli, Bihari-Thakurwadi villages were selected. In Raigad range, Varsoli, Navgaoun, Rewash and Chaul were selected.

The study documents local knowledge of the Thakur about the methods, management options. The data collection was mainly through transect walks and observation of the key informants. The key informants were selected after focus group discussions in the selected sites. The information about the utilization of wild products collected from the natural forests was obtained from the local inhabitants, local medicine men, *vaidyas* and forest officials. During field visits of the study area, efforts were made to collect information regarding agricultural vegetable and wild products which are used as food supplement in harsh climatic conditions and utilized for curing various human diseases.

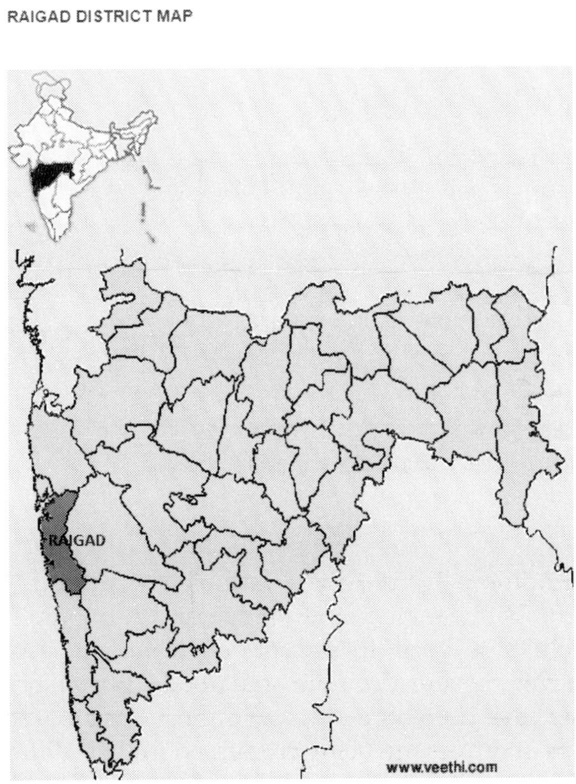

Fig. 10.1 Map of Maharashtra state showing study site, District Raigad

Random questionnaire-based survey of 20% households was done on selected forest ranges of Raigad. Information on wild plant parts (seeds, leaves, fruits, etc.) used to fight crop diseases and their curing methods was also collected. Focus group discussions along with forest department officials, villagers, *Vaidyas* were organized at field level to confirm various indigenous practices adopted by the tribes. Old aged persons (*Gunias*) of above fifty years of age in each village were interviewed separately to get idea about traditional coping strategies to fight various crop diseases using wild products and their mode of applications.

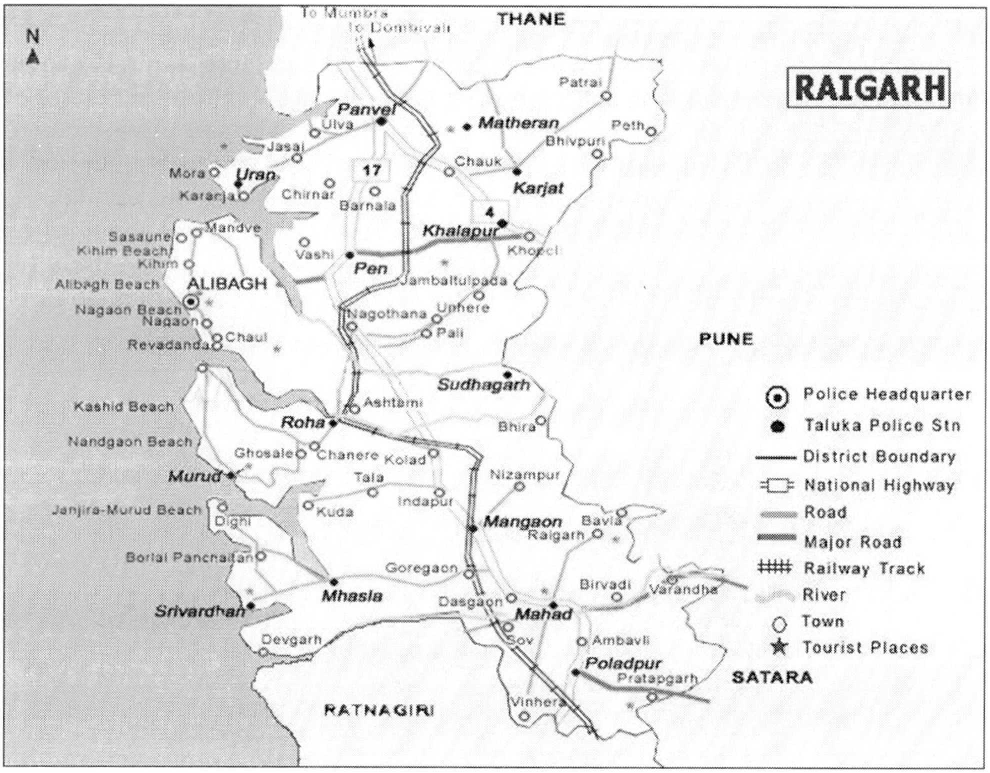

Fig. 10.2 Map of District Raigad

RESULTS

Thakur community traditionally conserves various agricultural crop seeds, vegetable, wild tubers, etc., to combat changing environmental conditions. Traditional crop seeds possess different qualities to grow in extreme environmental conditions and to fight climatic disturbances like severe drought and floods. Social practices of using local plants as famine foods have been observed as a regular practice (Singhi and Joshi, 2010). Different herbs, shrubs and trees are being consumed in their different preparations and incorporated in their diet during famine and drought conditions. Some of the traditional methods were described.

A list of different agricultural crops, cereals, etc., used by the Thakur tribes in severe drought and famine conditions is depicted in Table 10.1. In case of the continuous subsequent famine, crops like Nachani (*E. coracana*) are used whose seeds were used to make flour for chapati; Vari (*Panicum* sp.) chapatis are made from seeds; Makka (*Zea maize*) seeds after boiling, are eaten in form of aqueous paste; and Urad (*Vigna radiata*) seeds are boiled and utilized as cereals, etc. These crops were largely utilized in continuous famine conditions by the Thakur tribes after processing and can be stored for future use in harsh climatic conditions.

Table 10.1 Agricultural crops used in famine conditions by Thakur tribes of Raigad district

S. No.	*Agriculture crops in famine*	*Processing*
1.	Nachani (*Eleusine coracana*)	Seed is used for making chapatti
2.	Vari (*Panicum* sp.)	Chapattis are made from crushed seeds
3.	Red Makka (*Zea maize*)	After boiling eaten in form of aqueous paste
4.	Urad (*Vigna radiata*)	Boiled and utilized as cereals
5.	Bhadas (*Oryza sativa*)	Dried, boiled and aqueous paste
6.	Red jaw/lal milo (*Hordeum vulgare*)	Chapatti is made from its crushed seeds
7.	Uri (*Phaseolus* spp.)	Seeds were used to make chapatti
8.	Kulthi (black and red) (*Microtylom uniflorum*)	Grown as pulse for consumption

The information depicted in Table 10.2. shows various species of yams and tubers used as food supplement in famine and drought conditions by the Thakur tribes. Root, tubers and yams are important sources of food for survival in the region. Most of the tubers like Ratalu (*D. bulbifera*), Balkand (*D. pentaphylla*), Valkand (*A. companulatus*), and Wild alu (*Colocasia esculenta*), Geethkanda (*D. daemona*) were used as food supplements, medicinal and vegetable after thorough washing, boiling and skin removal. The tubers are eaten baked, boiled, or fried much like potatoes. Plants like Rankelikand (*M. superba*), Van karela /Kartoli (*M. dioca*), Sewal/Shivla (*A. sylveticus*), etc., were used as vegetables after thorough washing and boiling. Similarly, these yams were also used for curing various diseases such as for curing jaundice, stomach ache, malaria, fever, etc., after some modification and processing. Kokum (*G. indica*) fruit and its juice is used to drink in famine, and oil extracted is used for bone pain by the tribes. Most of the yams, tubers can be stored up to six months to one year without refrigeration, which makes them a valuable resource for the yearly period of food scarcity at the beginning of the rainy season.

Table 10.2 Wild yams, tubers used as food supplement in famine conditions by the tribes of Raigad district, Maharashtra.

S. No.	*Local name/ Botanical name of Yams/ tuber species*	*Processing*	*Medicinal uses by tribes*	*Utilization in drought/ famine conditions*
1.	Rankelikand (*M. superba*)	Raw tubers are eaten after thorough washing	Not used medicinally by tribes	Under stressed conditions, during vegetable shortage
2.	Balkand (*D. pentaphylla*)	Raw tubers are eaten after thorough washing and cleaning	Not used medicinally by tribes	During food shortage, lack of nutrition
3.	Ratalu (*D. bulbifera*)	Tubers are eaten baked, boiled, or fried like potatoes. After boiling used as vegetable	Tuber powder mix with butter is given to check diarrhoea.	Flowers used as vegetables. Tubers utilized during food shortage, crop failure and consecutive droughts
4.	Geethkanda (*D. daemona*)	After thorough washing making tuber chips, fried chips	Not used medicinally by tribes	Dried chips can be stored for long duration, famine situations
5.	Van karela/Kartoli (*M. dioca*)	Fruit used as vegetables.	Root is used with honey in curing gangrene.	At the time of severe drought, crop failure
6.	Wild alu (*C. esculenta*)	Tubers are boiled or roasted, fried much like potatoes.	Not used medicinally by tribes	During drought
7.	Karanda (*C. carandus*)	Eaten as vegetable and pickles making	Relieves burning sensation during urination, toothache	Used as pickle in drought
8.	Akhur (*D. wallichi*)	Eaten in raw form after thorough washing	Tuber juice used in jaundice	Tuber used in time of food shortage
9.	Valkand/wild suran (*A. companulatus*)	Edible, tubers are eaten, baked and boiled.	Not used medicinally by tribes	Tuber used in time of food shortage
10.	Dukkarkand (*D. alata*)	Tubers eaten as raw after cleaning, washing, etc.	Tuber powder is useful in piles, burning and eye diseases	Tubers and bulbils used as vegetables.
11.	Alambi/Chhattri (*Agaricus* spp.)	Edible /sundried and powdered	Mushroom powder is used to treat small pox and chickenpox	Edible during food shortage

12.	Sewal/Shivla (*A. sylveticus*)	Edible when cooked	Body pain	Rhizome is edible during famine conditions
13.	Kokam sharbat (*G. indica*)	Juice is made of its fruits, drinkable	Headache, oil is used to massage bones, body ache	Its juice is edible in harsh climatic conditions. It is highly nutritious.

Leaves are boiled, sun-dried and stored in bamboo made containers. These dried vegetables can be stored for six months to one year. For curing diseases, insects, etc., Kadu neem (*Azadirachta indica*) and Nirgundi (*Vitex nigundo*) leaves are used. Table 10.3. shows coping strategies of tribes using various wild edible vegetable species in harsh climatic conditions (Narayan and Kumar, 2007). After boiling, meshing and drying wild varieties of vegetables (bhaji) like Kolu (*Chlorophytum tuberosum*), leaves are cooked as vegetable, and roots are eaten raw; Taklu (*Cassia tora*) leaves are used as vegetables; Mokha (*Schrebera swietenoides*) young leaves are boiled and cooked as vegetable; Bhokar (*Cordia dichotoma*) fruits are used to make vegetable, and pickles/ripe fruits are edible; Karambel (*Dillenia pentagyna*) fruits are cooked as vegetable; Khurasani *(Hyoscyamus niger)* leaves are used as vegetable; Akhur (*D. wallichi*) tubers are edible etc. These are used in harsh climatic conditions as food material in absence of locally available vegetables. The plant leaves are largely sun-dried and sometimes boiled and stored in bamboo baskets for six months to one year. To remove insects, flies etc., dried neem, Nirgundi leaves are also spread in the locally made containers.

Table 10.3 Local coping strategies of Thakurs of Raigad district: Various wild edible vegetable species used in harsh climatic conditions

S. No.	*Name of Vegetable/ Bhaji*	*Part used*	*Habit*	*Processing*
1.	Kolubhaji (*Chlorophytum tuberosum*)	Leaves, root	Herb	Leaves are cooked as vegetable; also roots eaten raw.
2.	Taklu (*Cassia tora*)	Herb	Herb	Leaves are used as vegetables.
3.	Nachani (*Eleusine coracana*)	Seeds	Herb	Seeds flour is used to make chapattis, and stored in kanga (storage container) for years after drying in sunlight.
4.	Mokha (*Schrebera swietenoides*)	Leaves	Tree	Young leaves are boiled and cooked as vegetable.
5.	Bhokar (*Cordia dichotoma*)	Flower	Tree / large shrub	The fruits are used to make vegetable and also for pickles/ripe fruits are edible.

contd...

Contd...

6.	Khurasani (*Hyoscyamus niger*)	Leaf	Herb	Leaves are used as vegetable.
7.	Akhur (*D. wallichi*)	Tuber	Climber	Tubers are edible.
8.	Karambel (*Dillenia pentagyna*)	Leaf / fruits	Small Tree	Fruits are cooked as vegetable.
9.	Vankakdi (*Cucumis sativus*)	Fruit	Climber	Fruits are used as vegetable.
10.	Vankarela/kartoli (*Momordica dioca*)	Fruit	Climber	Fruit used as vegetables.
11.	Umbar (*Ficus racemosa*)	Flower and fruits	Tree	Inflorescence or young fruits were cut and prepared with little amount of lemon juice.
12.	Tembur (*Diospyros melanoxylon*)	Fruit	Tree	Ripe fruits are eaten raw.
13.	Kurudu bhaji (*Celosia argentea*)	Leaves	Herb	Leaves and twigs are cooked as vegetable.
14.	Kantoli (*Zizyphus oenoplea*)	Leaves	Herb	The berries are edible and the bark is used for tanning.
15.	Rankeli (*Musa superba*)	Fruit	Shrub	Whole inflorescence and rhizome is cooked as vegetable.
16.	Bharangi (*Clerodendrum serratum*)	Leaves/ root	Shrub	The leaves are used as a vegetable; the root is believed to have medicinal value.
17.	Jangli suran (*Amorphophallus companulatus*)	Tuber	Herb	The pungent corms after cutting them into small pieces put in boiling water with tamarind, and then fried.
18.	Pandhra kuda (*Holarrhena pubescence*)	Flower	Tree	Cooked as vegetable
19.	Matth bhaji (*Amaranthus spinosus*)	Leaves	Herb	Cooked as vegetable

Non-timber forest products (NTFPs) used by the tribes during famine conditions are shown in Table 10.4. Tribals utilize these NTFPs in different forms, directly or after processing, for their livelihood to cope up with the environmental disasters. Amla (*Emblica officinalis)* fruits are consumed as pickle and after sun-dried used in famine. *Harra (Terminalia chebula)* fruits are edible and Mahua (*Madhuca indica*) flowers are used to prepare a drink known as 'Peg' drink. Aam (*Mangifera indica*) seeds are dried and used as flour; Chinch (*Tamarindus indica*) seeds are used in extracting gum, and Bamboo (*Dendrocalamus strictus*) culms, seeds and flowers are used as food. Kosum *(Schleichera oleosa)* fruit pulp is used for drinking. These products are commonly used by the tribals, particularly in famine conditions.

Table 10.4 NTFPs used during famine conditions by Thakur tribes of Raigad District, Maharashtra

S.No.	*Name of species*	*Processing and utilization*
1.	Amla (*Emblica officinalis*)	Fruits are pickled or sun-dried and used in famines
2.	Harra (*Terminalia chebula*)	Fruits are edible and can be stored for long
3.	Bahera (*T. bellarica*)	Fruits are edible and after sun-drying can be stored for long
4.	Mahua (*Madhuca indica*)	Dried flowers are used for making chapattis and Peg drink in food shortage. Oil extracted from fruits used in absence of edible oil during famines
5.	Valkal (*Antiaris toxicaria*)	Fruits are edible
6.	Kokam (*Garcinia indica*)	Sharbat (liquid) is prepared from kokam
7.	Karanj (*Pongamia glabra*)	Oil extracted from seeds used in absence of edible oil during famines
8.	Bamboo (*Dendrocalamus strictus*)	Bamboo seeds and flowers are eaten in famine conditions
9.	Chinch (*Tamarindus indica*)	Seeds were used in extracting gum
10.	Aam (*Mangifera indica*)	Seeds were dried and flour is made from seeds
11.	Kosum (*Schleichera oleosa*)	Fruits pulp is edible for cooling effect
12.	Aalambi/ chhatri mushroom (*Agaricus* spp.)	After drying and powdering can be used for edible purpose

DISCUSSION

Wild agricultural crop seeds like Nachani (*E. coracana*), Red Makka (*Zea maize*), Uri (*P. mungo*), Vari (*Panicum* sp.) and Bhadas *(Oryza sativa)*, etc., possess qualities to grow in extreme weather situations and are prone to insect infestations. The crushed, powdered seeds of these crops were largely used for making chapatti by the tribes. Sometimes aqueous paste (Peg) was also made using Mahua flowers and Red maize seeds. Mishra et al., (2012a, 2012b) also reported utilization of wild crops relatives and NTFPs like Mahua and Bamboo seeds after processing flowers, particularly in harsh climatic conditions. They reported several wild relatives of crops like Kulthi, Baigani tuar, jhunjhura, adzuki bean, etc., used during extreme drought conditions by the Gond and Baiga tribes of Central India.

Some wild yams, tubers and rhizomes like Balkand (*D. pentaphylla*), Ratalu (*D. bulbifera*), Wild alu (*C. esculenta*), Rankelikand (*M. superba*) and Valkand/wild suran (*A. companulatus*), etc., are used as food material after processing and thorough washing and boiling. Most of the tubers are thoroughly washed in running water for a few hours and then baked, boiled to remove toxicity. Few tubers are cut and chips are made. After processing, these products made out of raw tubers are utilized during famine conditions. However, tubers have medicinal value and hence are used in curing various human ailments. Local tribes also used these wild yams, tubers during stressed, food shortage conditions whenever drought occurred. During floods and crop failure, most of the NTFPs available locally in the forest were used to cope with extreme environmental conditions.

Tribal people, through their traditional knowledge, infer what to eat and what not to eat. They are thoroughly acquainted with the methods of excluding the harmful substances from wild plants and preparing acceptable recipes for their meager meals. Corms and aerial bulbs of wild *Dioscorea*, eaten raw, cause a terrible itching sensation, so they are peeled, boiled in tamarind water and smeared with turmeric paste. The toxicity of *Cycas* seeds is removed by placing crushed and powdered seeds in water overnight. Next morning the water is run off and the seeds are again washed with pure water and decanted off (Yashodharan and Sujana, 2007).

Wild leaves are among the most widely consumed wild foods. Most of the leafy wild foods plants are locally referred to and classified as weeds, sprouting, and flourishing after rains. Tribes use them in soups, stews and relishes that add flavour to staples. Some wild leafy vegetables like Kolubhaji leaves are cooked as vegetable, and roots are eaten raw; Taklu leaves are used as vegetables; Tembur ripe fruits are eaten raw; Kurudu bhaji leaves and twigs are cooked as vegetable; and Kantoli berries are edible and the bark is used for tanning, etc. These vegetables are used as food in harsh climatic conditions like drought, famine, etc., after proper processing. Misra, et al. (2008) reported diversity of wild leafy vegetables of Nanda Devi Biosphere Reserves used by local inhabitants was around 21 species belonging to 14 genera and 11 families. They also described availability, use period and associated traditional knowledge. Similarly, Ratheesh and Kumar (2007) also reported changing trends in utilization of wild leafy greens in the Western Ghats. They reported that alien species invasion and modern agri-practices lead to local extinction of many greens and the erosion of traditional knowledge, especially among the youth, due to materialistic lifestyle affects sustainable use of many wild greens. Sasi, et al. (2011) studied the Nilgiri Biosphere Reserve and concluded that leafy vegetables are a good resource for the upliftment of the economy of rural people.

Thakur tribes of Raigad district also use some NTFPs as food material in drought or famine. They use either directly or after processing, like Harra, Bahera fruits are edible and after sun-drying can be stored for long duration. Karanj oil extracted from seeds is used in the absence of edible oil during famine. Bamboo seeds and flowers are eaten in famine conditions and Chinch seeds were used in extracting gum, etc. NTFPs play a vital role in the economy of a tribal, both in terms of sustenance and source of income.

Similar observations were also reported by Bhandari, 1974, while working in Rajasthan. He reported that many plants are used as famine foods in the Rajasthan desert. Some are of great importance as emergency foods, while others are of minor value, i.e., used only or rarely when better famine foods are not available. Kelbessa, 2001, reported storage of farmers of Sahel, South Africa while using emergency fodder in times of drought and calling of weak livestock for food during periods of drought. Similarly, Mishra et al., 2012a also reported traditional knowledge of stored yams, tubers and their utilization during harsh climatic conditions. They reported several NTFP species which were utilized after simple processing and stored for longer duration. These NTFPs were utilized during natural calamities like floods, frost, drought, etc. and fulfilling their nutritional requirements.

CONCLUSION

The study shows that wide ranges of uncultivated species are used by the majority of households as leafy vegetables. These herbs are harvested and used directly by the households and often without any form of trade especially because the majority of the households often engage in the harvesting within their local environments. Value addition through storage and commercialization could probably widen the livelihood base and thus draw attention of planners. Such strategies have been effectively used to combat vitamin and micronutrient deficiencies and thus should be replicated in all other regions.

In the above study it is seen that agricultural crops and wild plants are used by Thakur tribes in famine conditions. They use wild relatives of agricultural plants like Kurudu, Nachani , Vari, Uri etc. Wild tubers and yams are also used as food by the Thakur tribe, i.e., Rankeli, Balkand, Ratalu, Geethkanda, etc., in famine conditions in the study area. Wild vegetables like Kolubhaji, Taklu, Nachani, Mokha, Khadsing, etc., utilized after drying, processing, particularly in severe drought conditions. They are equally rich and nutritious like normal crops and help the tribals to cope up from scarcity of food.

These wild food plants are not only dietary staples but also provide nutritionally valuable supplements as they are rich in proteins, sugars, starch, vitamins, minerals, etc. Some edible products are highly priced for their delicacies, medicinal values and industrial uses. As such these are in great demand in the urban centers also.

This study, thus, suggests that academicians, planners and tribal communities have to work together to address the adverse impacts of climate change. Scientists and policy makers who want to address the impacts of global climate change can derive important lessons from tribal communities, locals and pastoralists' strategies for addressing weather-related risks, although indigenous strategies have been affected by various internal and external factors. It is important to incorporate indigenous knowledge and perceptions into the climate change forum in the country. It would always be useful to involve local communities in the design and implementation of climate coping strategies. Local communities should be encouraged to grow drought-resistant and early-maturing crops, protect forests, and improve empirically tested coping strategies. Documentation of traditional knowledge about plants used during famine, severe drought and flood situation will be helpful in coping with extreme environmental conditions by the locals. While indigenous communities will undoubtedly need much support to adapt to climate change, they also have expertise to offer on coping through traditional time tested mechanism.

Acknowledgements

Authors are thankful to Director, IIFM for his kind inspiration and Research Cell of IIFM, Bhopal for providing financial support. Authors are also thankful to Dr. Madhu Verma and Dr. Suprava Patnaik for their guidance and constant encouragement. We are extremely thankful to the Thakur tribal community and the State Forest Department of Raigad, Maharashtra, for providing valuable informations and field work.

REFERENCES

1. Bhandari, M.M. (1974). Famine foods in the Rajasthan desert. *Econ. Bot.,* 28(7). 73-81.
2. Kelbessa and Workineh. Traditional Oromo attitudes towards the environment: An argument for environmentally sound development. OSSREA Social Science Research Report Series, Addis Adaba: Commercial Printing Enterprise, 19, (2001). [http:/www.ossrea.net/ssrr/workneh/toc.htm.]
3. Mishra, M., and Singh, S.P. (2009). Indigenous community participation in India. *European Tropical Forest Research Network,* Special Issue on Climate Change, 50, 84-90.
4. Mishra, M. (2010). Climate change and its impact on forest and dependent local communities: A case of two tribal communities of Central India. Proceedings of the *National Seminar on Management of Natural Resources and Environment in India – MNRE,* (Guru Arjan Dev Development Institute, Amritsar, Punjab) 217-221.
5. Mishra, M., Gupta, H.S., and Bhangre, M. (2012). Climate change mitigation through participation of tribal communities: A case of central India, Proceed. *Natl. Conf. on Recent Advances in Plant Sciences.* NCRAPS, (Ed. Dr. P Shrivastava), 189-194.
6. Mishra, M., Gupta, H.S., and Shrivastava, M. (2012a). Documentation of Traditional Knowledge of Baigas of Dindori District, Madhya Pradesh with special reference to their strategies in coping climate change, *Life Science Leaflets,* 11, 61-72.
7. Mishra, M., Gupta, H.S., and Bhangre, M., (2012b). Climate variability and its impact on Non- Timber Forest Products (NTFPs): A Case of Baiga and Gond Tribes of Central India. A Report submitted to Regional Centre, National Afforestation & Eco-development Board, IIFM, Bhopal, Sponsored by RCNAEB, Bhopal, 108.
8. Misra, S., Maikhuri, R.K., Kala, C.P., Rao, K.S., and Saxena, K.G. (2008). Wild leafy vegetables: A study of their subsistence dietetic support to the inhabitants of Nanda Devi Biosphere Reserves, India, *J Ethnobiol Ethnomedicine,* 4, 15.
9. Narayan, RMK., and Kumar, A.N. (2007). Gendered knowledge and changing trends in utilization of wild edible greens in Western Ghats, India, *Indian J Tradit Knowl,* 6(1) 204-216.

10. Pareek, A., and Trivedi, P.C. (2011). Cultural values and indigenous knowledge of climate change and disaster prediction in Rajasthan, India, *Indian J Tradit Knowl,* 10(1): 183-189.
11. Salick, Jan., and Anja Byg., Indigenous Peoples and Climate Change, 2007 http: // tyndall webapp .uea.ac.uk/publications/Indigenouspeoples.pdf.] Website viewed on 12/4/12.
12. Sasi, R., Rajendran, A., and Sivalingam, R. (2011). Diversity of leafy vegetables in Nilgiris, Nilgiri Biosphere Reserve, Southern India, *Int. J. Biol. Tech.,* 2, 1-7.
13. Singhi, M., and Joshi, R. (2010). Famine food of arid Rajasthan: Utilization, perceptions and need to integrate social practices by bioresolution, *Ethnome.,* 4(2): 121-124.
14. Yashodharan, K., and Sujana, K.A. (2007). Wild edible plants traditionally used by the tribes in the Parambhikulum wildlife sanctuary, Kerala, India, *Natural Product Radiance,* 6(1): 74-80.

11

A Study on Sustainable Harvest of Wild Medicinal Plants of Uttarakhand with a Case Study on Lichens

Sudipto Chatterjee*, Prodyut Bhattacharya and Anand Kandya†**

**Conservation Sciences India, New Delhi*

*** University School of Environment Management*

GGS Indraprastha University, New Delhi-110078

†Department of Botany, Dr. H.S. Gour University, Sagar, Madhya Pradesh

E-mail: s.chatterjee@teriuniversity.ac.in & prodyutbhattacharya@yahoo.com

ABSTRACT

Medicinal plant species are obtained both from cultivation and from the wild. With increase in demand from the related industries, there has been an effort to bring such species under high demand into cultivation. Huge investments are being made on developing cultivation protocols. Government of India has set up a National Medicinal Plants Board to address the sector. There has been a directive by the government to integrate the medicinal and aromatic plants sector, with the Forest Working Plans. Such efforts in the country would require relevant datasets on ecological aspects of species, their existing and potential demand in trade, usage, present levels and methods of harvest, traditional ecological knowledge and awareness among different groups of stakeholders on the species.

Global conservation community has been struggling with the concept on sustainable use of biodiversity. There is a clarity in its meaning but it has not been possible to quantify sustainability. Hence, efforts have been made to develop indicators, good practices or standards. Newer approaches have been proposed, certifying forests for their management and the products extracted from them. Such initiatives however, require region-specific studies.

The present study addresses the key question: are the medicinal plant resources in Uttarakhand nested in the Indian biodiversity 'hotspots' of western Himalaya sustainably managed? What management regimes the wild medicinal plants of Uttarakhand are subjected to and what management interventions are presently needed? This frames the objective of the present study.

INTRODUCTION

Medicinal plants and their conservation have been increasingly receiving attention world over. They are important component of both modern and traditional systems of medicine. Globally traditional healers and scientists are on a look out for newer medicines for many of the ailments for which effective medications do not exist, and the diversity of medicinal plants holds the potential. The herbal industry, globally, has annual turnover of 60 billion US $ and has been consistently on the rise. Nations rich in medicinal plants find the same as a source of revenue, and India is no exception. In 2000, there were 7800 pharmaceutical units in India, and globally, the rate of growth in these units is 7% (Anon, 2000).

The demand for medicinal plants far exceeds what the forests which harbor these medicinal plants can provide and what could be procured from cultivation. Moreover, for many medicinal plants, the cultivation protocols and agro-technology is yet to be developed or under the development process. The pressure therefore is on the wild resources, and more than 95% of the medicinal plants are obtained from the wild. Of the 881 plant species reported to be used by the pharmaceutical industry in India, more than 82% are collected from the wild (Ved and Goraya, 2008). The scenario holds true for developed as well as developing countries. Of the 960 medicinal plants species that have been recorded under trade for raw drugs in the country, 176 are consumed in volumes in excess of 100 MT per year. Eighty per cent of the industrial demand is from domestic herbal industry. Thirty-four species (19%) are sourced from cultivation, 46 species (26%) are sourced from farm bunds and 5 species (3%) are imported. The remaining 91 species (52%) are sourced from tropical and temperate forests. Analysis of 91 species that are sourced primarily from forests shows that 21 are from Himalayan (temperate) forests and 70 from tropical forests. An estimated 4000 to 10,000 medicinal plants are facing local, national and global extinction. Of the 71 rare medicinal plants 92% are in active trade, 74% are traded nationally and 35% are traded internationally (Kala et al., 2006).

There has been great emphasis on conservation of medicinal plant resources world over including India (Shanley et al., 2005). These include their documentation of ethno-botanical uses, inventorization, assessment of conservation criteria, delineation of landscapes rich in medicinal plant resources and conservation areas, bio-prospecting and in situ and ex situ conservation. Non-timber forest products including medicinal plants are important sources of livelihood for local communities in India. A detailed account of the same has been provided by (Prasad and Bhattacharya, 2003 and Malhotra and Bhattacharya, 2010).

India is a mega-diversity country, and three of the global 'hot spots' viz., the 'Himalayas', 'The Western Ghats' and the 'Indo-Burma' are located in the Indian subcontinent as per Conservation International, a leading global conservation agency, USA.

The present study addresses the key question: are the medicinal plant resources in Uttarakhand nested in the Indian biodiversity 'hotspots' of western Himalaya sustainably managed? What management regimes the wild medicinal plants of Uttarakhand are subjected to and what management interventions are presently needed? This frames the objective of the present study.

Medicinal plant diversity of Uttarakhand is relatively well documented and inventorised. Based on the assessments made by institutions of repute in the state, plant species have been categorized as either rare, threatened or endangered. The IUCN (International Union for Conservation of Nature and Natural Resources) criteria, viz., Critically Endangered (CE), Endangered (E), Threatened (T), Vulnerable (Vu), Near Threatened (NT) and Low Risk (LR) have gained popularity through the Conservation Assessment Management Process (CAMP). However, in-depth studies particularly on the management practices are generally lacking which has been a severe constraint in formulating management plans for the plants. This is more so for the medicinal plants for which cultivation protocols do not exist or have not been developed and are continued to be extracted from the wild. Due to excessive extraction of medicinal plants species in high altitude areas, Uttarakhand has witnessed a serious depletion of its biological resources. The state government has introduced policies to promote the conservation of these species and encourage farmers to cultivate them and supplement their income. In spite of the efforts and the fact that certain species cannot be cultivated, a large proportion of the demand is met by collection from the wild. This is particularly true for plants found in high altitude areas, which are generally characterized by fragile ecosystems.

A recent effort by the Government of India has been to include the medicinal and aromatic plants sector in the forest working plans. Several efforts have been made in this regard by National Medicinal Plants Board (NMPB), Govt. of India. Forest Certification and developing standards for harvest of wild and aromatic medicinal plants are some such global initiatives (Chatterjee et al., 2002).

The study emanates from the authors' experience of working on relevance of forest certification in India, as certification of both forest management and its products is viewed as a tool to sustainable forest management (SFM). The origin of this thesis also draws from the authors' close working on the WWF supported 'Measures of SFM and its implementation, at International Center for Community Forestry (ICCF), Indian Institute of Forest Management (IIFM) Bhopal, working for a project commissioned by Uttarakhand Council of Science and Technology (UCOST), Govt. of Uttarakhand to develop standards for wild medicinal and aromatic plants from the state. The author has also implemented landscape level conservation initiatives in the biodiversity hot spots of Western Ghats and Indo Burma in India. During the course of the previous work and thesis it was realized that lower group of plant species, which serve as pioneers and colonizers in different ecosystems often remain unaddressed. For preserving and protecting habitats, we need to develop an understanding of the requirements of the species, their autecology and the threats these species are subjected to. Levels of harvest of the wild species, their impacts on the ecosystems and dependence of local communities on these resources need data collection, collation and analysis. Until then, developing conservation management plans and harvesting protocols for their sustainable use will remain a far cry.

Some of the global studies on sustainable use of wild medicinal plants include *Manilkarazapota* in Mexico (Shanely et al., 2005), *Bertholletia excelsa* and *Euterpeoleracea* in Brazil and Bolivia (www.bertholettia.org, Van Andel et al., 1988), *Harpagophytum procumbens* in Namibia (Schneider et al., 2000), *Euphorbia clivicola* and *Sclerocarya birrea* in South Africa (Pfab and Scholes, 2004 and Emanuel et al., 2005), *Iriartea deltoidae*, *Aquillaria* sps. and *Aloe pegerae* in Indonesia (Anderson, 1998, Soehartono and Newton, 2001 and Pfab and Scholes, 2004) and *Pterocarpus angolensis* in Tanzania (Schwartz et al., 2002). Indian studies are on *Emblica officinalis*, *Sterculia*sp and *Asparagus racemosus* (Bhattacharya, 2004 and Ganesan and Siddappa, 2004).

The state of Uttarakhand, which has declared itself as a herbal state, provides an opportunity to study the management practices, regulations, threats and pressures to medicinal plants. The present thesis titled 'Studies on the conservation and management practices of wild medicinal plant resources of Uttarakhand' is based on an interdisciplinary approach. It is based on traditional knowledge system and perceptions of local communities, scientific studies by other researchers, socio-economic studies and modern tools and techniques that aid biodiversity conservation. A baseline on wild medicinal plant resources in the state of Uttarakhand and the management practices has been created, on which further studies can be undertaken. This study would result in sustainable use of medicinal plants in the state.

OBJECTIVES OF THE STUDY

It was aimed to explore the relevance of forest certification in India through a study for the Uttarakhand Council of Science and Technology, Govt. of Uttarakhand, to develop standards for wild medicinal plants.

An alarming fact was revealed during the study that low-volume high-priced medicinal plants, like lichens and mosses which are considered indicators and pioneers to ecosystems are harvested and traded in huge quantity. Standards cannot be developed until and unless there is an understanding of ecological, economic and social aspects of species in concern. This led to genesis of this study on the conservation and management practices of wild medicinal plant resources of Uttarakhand. The duration of this research work also witnessed some of the relevant initiatives in India. Indian Institute of Forest Management, Bhopal has developed the principles, criteria and indicators for sustainable use of medicinal plants and had been developing ecological, economic and social standards for different regions in India. These standards were based on the International Standards for Wild Collection of Medicinal and Aromatic Plants (ISSC-MAP) which were separately promoted by WWF Germany and TRAFFIC India. These were being field tested in states of India like Karnataka and Uttarakhand which are extremely rich in medicinal plants.

The state provides a set of policy framework, a network of scientific institutions, landscapes rich in medicinal plant resources, local communities dependent on medicinal and aromatic plant resources. Uttarakhand, therefore, was found to be appropriate for this work. The author's familiarity with issues related to medicinal plants conservation in India, through Biodiversity Conservation Prioritization Process (BCPP), Forest Certification and experience on cultivation of medicinal plants in Uttarakhand, landscape level conservation in Southern Western Ghats and Eastern Himalaya was helpful in conducting the research.

APPROACH AND METHODOLOGY

At the outset a global search was undertaken to look at case studies both in India and overseas for sustainable use of plants. *Manilkara zapota* (Chicle), *Bertholletia excelsa* (Brazil nuts), *Harpagophytum procumbens* (Devil's claw), *Panax quinquefolium* (Amercan ginseng), *Sclerocarya birrea* (Marula fruits), *Euterpe oleracea* (Palm hears), *Iratea deltodae, Euphorbia clivicola, Aquilaria sps, Pterocarpus angolensis* and *Aloe pegerae* are some examples that provided global case studies. In India only *Sterculia urens* and *Asparagus racemosus* have been investigated.

Sustainable use of plants entails assessment of resources. Kumar (2010) provided a study on assessment of Lichen resources from the very same district where the present study was undertaken. The study provided valuable information on diversity of lichen species in Uttarakhand and amount of lichens that is available on the forest and can be marketed. A short term study from the Chopta forests of Uttarakhand suggests that around 6.4 kg of lichen/ha/yr is available to the collectors of this species.

The whole state of Uttarakhand contributes to medicinal plants. For identification of a landscape, and initiating studies related to conservation and management practices of wild medicinal plants, a step-wise approach was followed.

1. A series of consultations with official and non-official agencies.
2. Collection of data sets on wild medicinal and aromatic plants legally traded from three strategically located trade depots in Uttarakhand, viz., Bibiwala, Ramanagar and Tanakpur.

Using the criteria in all the three depots, traded quantity and rate of medicinal plant species like Lichens, *Pueraria tuberosa* (Bidarikand), *Ventilago denticulata* (Kalibel), *Asparagus racemosus* (Shatavar) *and Juglansregia* (Dandasha) were initially shortlisted.

The conservation status of the medicinal and aromatic plants of the state was ascertained through the CAMP (Conservation Assessment Management Plan) process undertaken in 2003, by ZOO Outreach Organization, Coimbatore, at Shimla. The CAMP process categorizes medicinal plant species into IUCN conservation categories, viz., Critically Endangered (CE), Endangered (E), Vulnerable (Vu), Threatened (T), Least Concern (LC) and Data Deficient (DD).

Sustainable use and conservation/management practices of medicinal plants are greatly influenced by the policies of the state. Uttarakhand, being a herbal state, has state level policies for conservation of medicinal plants. The state has set up institutions like, Forest Development Corporation, Herbal Research and Development Research Institute, Selaqui, Dehradun. Collection agencies like Bhesaj Samaj, Garhwal Mandal Vikas Nigam (GMVN) and Kumayun Mandal Vikas Nigam (KMVN) also exist in the state. The state has issued guidelines for collection of wild medicinal plants and regulations exist for Lichens and *Cordyceps sinensis*. Since 2005, the state introduced the rotational harvesting policy (*Chakravatpranali*) which alternatively opens and closes forest ranges, so that plant species get enough time to regenerate. Data was collected on the rotational harvesting policy for the years 2004-2007. Since it was difficult to procure range level maps for all the ranges in the state of Uttarakhand, administrative blocks where these ranges lie were mapped.

A suitable landscape had to be identified in the state of Uttarakhand for intensive study and data collection. Langha Forest range near Dehradun was explored for study of collection of *Pueraria tuberosa* and *Ventilago denticulata*. Kedarnath *chetra* was explored from where *Sassurea obvallata* is sourced for offerings at the Kedarnath shrine. A specific landscape could not be identified for *Juglansregia* as it was reported to be sourced from agricultural land in the state. *Asparagus racemosus* was not further studied as the species is now reported to be under cultivation.

Plate 11.1 Lichen—a consortium of different species harvested and traded in Western Himalaya

Chamoli district was explored and it was found that in addition to huge quantities of lichens, *Picrorhi zakurroa* and *Cordyceps sinensis* are collected from the wild. Of the three landscapes, therefore, District Chamoli was finally selected. A Lichen Conservation Area was delineated in the district as it contributed bulk of the Lichens traded in the state.

Based on the discussions with the harvesters at two villages, Waan and Kuling, community mapping and use of a freely available GIS, software DIVA GIS, a Landscape Conservation Area of 189 sq. km was delineated. Socio-economic status of the lichen harvesters was assessed through a sample size of 44 medicinal plants collectors in these two villages. Landscape boundaries were fixed up to the point the villagers reported to go for collection of lichens. The lichen conservation landscape spans through the administrative blocks of Naraynbagad, Dewal and Tharali.

Ten villages were identified for intensive study. Of these villages, Waan, Kuling, Mundoli, and Baank are present within, and villages Bursol, Bunga, Dungri and Ratagaon are outside the landscape. A total of 150 wild medicinal plant harversters were interviewed for assessment of collection of Lichens (120), *Picrorihi zakurroa* (20) and *Cordyceps sinensis* (10). Information on the forests from which the medicinal plants are sourced were collected from the medicinal plant collectors and harvesters (Fig. 11.1).

An assessment of marketable lichens available in the Lichen Conservation Area was made (Chatterjee et al., 2011). As per the traditional knowledge system that exists with the harvesters, lichens prefer niches and habitats that receive high insolation and rainfall. GIS techniques were used to identify lichen rich areas. Oak tree species ((*Quercus leucotrichophora, Q. floribunda* and *Q. semecaripfolia*), which are hosts to lichens were used as a surrogate because they form continuous patches of forests.

Maps of villages surveyed were superimposed over the favourable areas and assessment was made for these areas available to these villages for harvest of lichens. (Figs. 11.2 and 11.3).

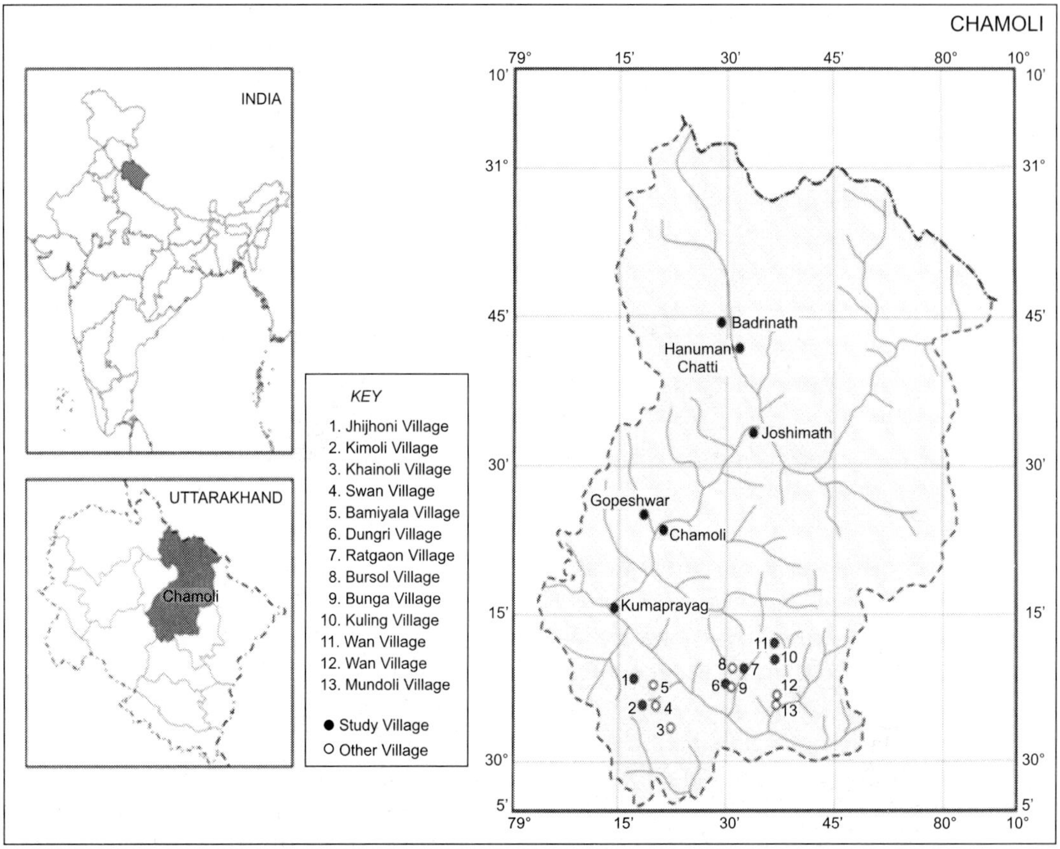

Fig. 11.1 The study area in village Chamoli with mapping locations of the villages surveyed

Based on the regulation by the State Forest Department that only lichens that have fallen on the ground and conclusions by Kumar (2010), that forest floors in Chamoli provide only 6.4 kg of lichens/ha/yr, calculations were made for lichens that should be collected from these villages and forests.

Bhattacharya et al. (2008) developed a 4-tier framework which includes principles, criteria and indicators for sustainable forest management. There has also been an effort by Govt. of India to integrate the medicinal and aromatic plants sector with the state working plans. A selected group of knowledgeable individuals, representing the state forest department, forest development corporation and Bhesaj Samaj were interviewed through a questionnaire for making an assessment of the readiness of the state vis-a-vis the aforementioned principles, criteria and indicators.

OBSERVATIONS

As per the CAMP workshop conducted in 2003, 7 sps, viz., *Aconitum heterophyllum*, *Arnebia benthemii*, *Dactylorhiza hattagirea*, *Gentiana kurroo*, *Lilium polyphyllum*, *Nardostachys-grandiflora* and *Picrorhiza kurrooa* were categorized as critically endangered, 18 species

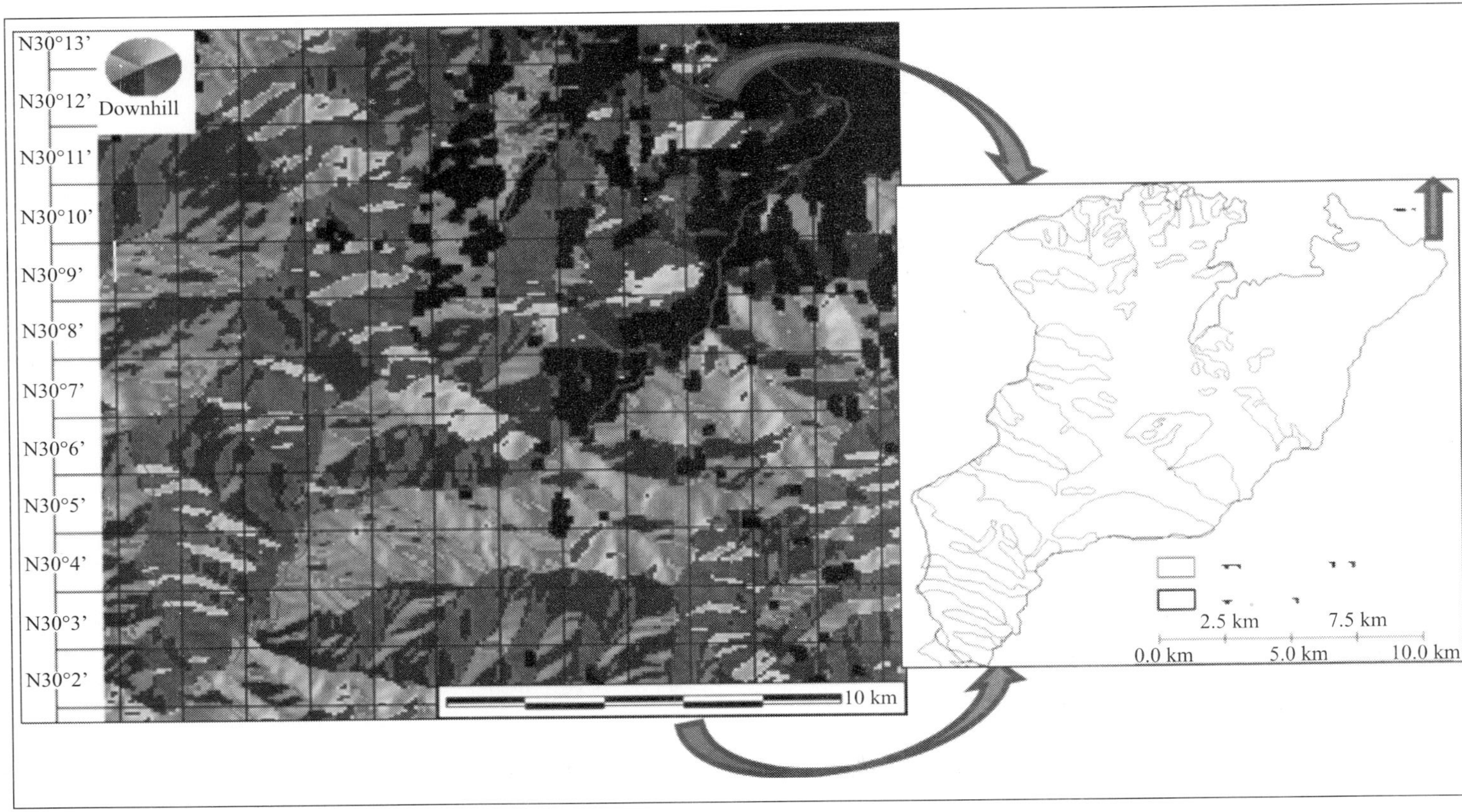

Fig. 11.2 Delineation of Lichen conservation areas

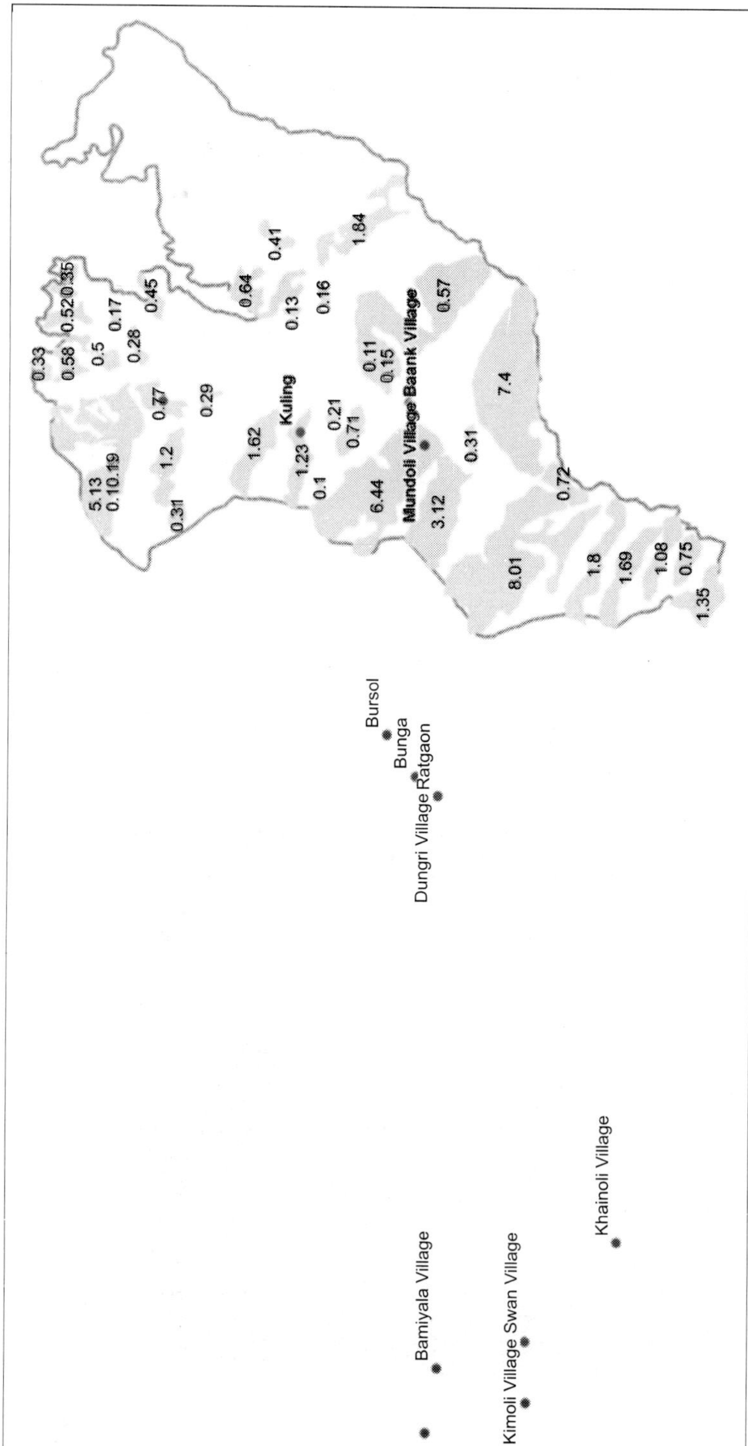

Fig. 11.3 Villages surveyed within and outside the landscape

were categorized as endangered and 20 as vulnerable. National Medicinal Plants Board, Govt. of India, assessed the annual demand and the rate of increase in the same for 32 medicinal plant species during the period 2001- 2005. According to Jain and Mishra (2007), unsustainable harvest practice of medicinal plants was realized during 1920s which led to creation of collection agencies like the Jila Aushadhiya Padap Sahakari Samiti (Bhesaj Samaj) and Kumayun Mandal Vikas Nigam. Since mid-2000s the Forest Development Corporation played an important role in collection of wild medicinal and aromatic plants from the state. Trade depots were set up to ensure that wild medicinal and aromatic plants are auctioned on fixed dates. Major trade depots are at Ramnagar, Bibiwala and Tanakpur. Data sets are available on medicinal plants traded at these depots since 2005. Tables 11.1, 11.2 and 11.3 provide the quantum of wild medicinal and aromatic plants traded through the depots in Uttarakhand and the trends in their prices respectively.

Table 11.1 Quantities of various medicinal plant species sold during 2005-2010 in Uttarakhand

MAP Species	*Quantities of MAPs sold (Quintals)*					
	2005-06	*2006-07*	*2007-08*	*2008-09*	*2009-10*	*2010-11*
Lichens	12785.6	14942.32	11433.5	14069.36	14221.14	12687.76
Julans regia	132.22	114.67	114.67	62.826	0	20
Asparagus racemosus	2.12	0.71	20.71	0	0	0
Pueraria tuberosa	0	0.88	0	12.508	0	0
Cordyceps sinensis	0	0	0	0.9624	0.075	0.27
Moss Grass	4179.171	2230.29	2242.46	3250.942	4359.956	2393.738

Table 11.2 Trends in quantities of MAPs traded in depots of Uttarakhand and revenue received in different years during 2005-2011

Year	*Quantity of MAPs collected (Quintals)*	*Revenue Earned (Rs)*
2005-06	22111.06	93415133
2006-07	18983.8	91629110
2007-08	16009.34	75876319
2008-09	22439.45	130514714
2009-10	19501.15	127657708
2010-11	15897.72	121108287

Table 11.3 Trends in selling rates of priority MAP species

Species	*Selling Rate of medicinal plants (Rs/Quintals)*					
	2005-06	*2006-07*	*2007-08*	*2008-09*	*2009-10*	*2010-11*
Lichens	6048.59	5607.7	5820.64	8140	8070	9002
Juglans regia	14990.73	13780.74	NA	15843	NA	10000

Contd...

Contd...

Asparagus racemosus	1264.62	600	1152.78	NA	NA	NA
Pueraria tuberosa	NA	1250	1152.78	700	NA	NA
Cordyceps sinensis	NA	NA	NA	18200087	1831075	17704000
Moss Grass	1803.89	1842.73	1995.83	1873	2230	2082

Community mapping revealed that villagers were able to name the forests from which they collect medicinal species. Records of these forests were made for the villages identified for the study for future reference. This also helped in delineating a Lichen Conservation landscape. Observations through questionnaire method were made on collection cycle, socio-economic status, methods and intensity of harvesting. Community perceptions on possible impact on forests and conservation measures that need to be taken were also recorded.

Observations on the policy environment included a listing of those species which are prohibited for harvesting and species for which no such restrictions exist. Detailed observations were made on the rotational harvesting policy and data was collected on forest ranges that were open and closed during the period 2004-2007. Some of the efforts of the state include establishment of institutions like Herbal Research and Development Institute (HRDI) at Selaqui, Dehradun, Medicinal Plants Conservation Areas (MPCAs) in the districts of Uttarkashi, Beageshwar, Chamoli, Champawat and Almora.

The policy environment of the state should synergize with some of the national efforts and facilitate implementation of the principles, criteria and indicators developed for sustainable use of medicinal plants. Observations were made on the readiness of the state on the four broad principles, viz., legal and policy framework, wild area conservation and management, responsible collection and use practices and benefit sharing and livelihood security.

Picrorhi zakurroa, a species assessed as critically endangered, was one of the plants found to be collected from the Lichen Conservation Landscape identified in this study. This species is presently now under cultivation in some villages in the study area. Several of the species which are high in demand and have registered more than 10% of growth like *Emblica officinalis, Aconitum heterophyllum, Berberis aristata, Tinospora cordifolia, Picrorhiza kurroa, Glycorrhiza glabra, Rauwolfia serpentina, Asparagus racemosus, Ocimum basilicum* and *Aconitum ferox* are sourced and traded in Uttarakhand, both legally and illegally.

The state has laid a lot of focus on cultivation of medicinal plants to reduce the pressure of collection from the wild. Herbal Research Development Institute at Selaqui, Dehradun, has registered farmers in the state and has provided facilities for laboratory testing.

The State Forest Department, Govt of Uttarakhand, has enlisted 100 medicinal plant species for conservation, of which 35 are of high priority. An analysis of this list against the list of medicinal plants traded through the depots reveals that many of the plants that are not traded are also prioritized. While species like lichens and *Tinospora malabarica* are included in the list of priority, species like *Bergenia ciliata, B. stracheyi* (Pashanbhed) which are consistently traded in the depots are low in priority. *Bergenia* sps therefore need conservation attention.

Analysis also revealed that of the 106 sps of lichens found, species like *Parmelia, Everniastrum, Usnea* and *Ramalina* are more in demand. Harvesters, while aware of species in demand, lop entire branches of oak trees and no segregation is done during the harvest.

Medicinal plant collectors are mostly engaged in agriculture and collect lichens for 3-6 months in a year. The state regulation, which says that only Lichens that fall on the ground should be collected is flouted. Lopped oak branches stand as a testimony to this. Illiteracy is high and the harvesters mostly belong to the age group of 35-45 years. Women are also engaged in medicinal plants collection. Both general category and scheduled castes are engaged in collection. Average land holding is low, to the tune of 0.40 acres per harvester. Most of the harvesters belong to the BPL (Below Poverty Line) category. In a small sample size of two villages, it was revealed that an average collector collects 0.5 – 1.4 quintals of lichens per year. However, a larger sample (120 individuals) revealed that the collection averaged to 11.3 quintals per year per harvester. The collection ranged from 1 to 20 quintals, therefore the amount varies for individual collectors. The difference is for two reasons. First a selected group of harvesters is engaged in collection of large amounts (15-20 quintals). It needs a mention that harvesters adulterate the collections with branches and twigs to add weight. An average collector collects 6 quintals of *Picrorhi zakurroa* and 630 gm of *Cordyceps sinensis* every year. Table 11.4 provides an account of village-wise collection of data on Lichens, *Picrorhi zakurroa* and *Cordyceps sinensis*.

Table 11.4 Assessment of Volume of extraction of Lichens

(Sample nos. 1-60, 71-90, 101-150 pertain to collection of Lichens, 61-70 and 91-100, collection of *Picrorhi zakurroa*, 101-110, collection of *Cordyceps sinensis*)

Sample No.	*Age (yrs)*	*Collecting since (yrs)*	*Harvest (Quintals) Annually/ kg for Cordyceps*	*Tools used*	*Impact on forest*	*Perception / Conservation Measure*
1.	35	35	10	Spade Axe	Declining	Forests be protected from fire
2.	32	7	5	Spade	Declining	Immature thalii should not be harvested
3.	38	6	10	Spade Axe	Declining	Protection and conservation
4.	45	4	9	Spade Wood	Declining	Immature thalii should not be harvested
5.	50	4	6	Spade	Declining	Only fallen thalii should be harvested
6.	46	8	12	Spade	Declining	Only fallen thalii should be harvested

Contd...

7.	48	8	13	Spade and wood	Declining	Only fallen thalii should be harvested
8.	25	5	5	Spade and wood	Declining	Only fallen thalii should be harvested
9.	36	6	15		Declining	Only fallen thalii should be harvested
10.	50	6	10	Spade and wood	Declining	Only fallen thalii should be harvested
11.	34	10	14	Spade	Declining	Protection to forests to be provided
12.	35	10	14	Spade	Declining	Training to harvesters
13.	39	10	14	Spade	Declining	Protection from fires
14.	26	10	12	Spade	Decline in quality	Protection to forests
15.	45	10	8	Spade	Decline in quality	Protection to forests
16.	55	20	20	Spade and wood	Decline in quality	Protection to forests
17.	28	10	10	Spade and wood	Decline in quality	Protection to forests
18.	28	10	15	Spade and wood	Decline in quality	Protection to forests
19.	30	10	13	Spade	Decline in quality	Protection to forests
20.	42	15	20	Spade	Decline in quality	Protection to forests
21.	45	2	4	By Hand	Production dependent on climatic conditions	Appropriate collection methods
22.	30	1	1	By Hand	Production dependent on climatic conditions	Thalli fallen by wind should be collected
23.	55	4	3	By Hand	Production dependent on climatic conditions	Thalli fallen by wind should be collected
24.	28	4	3	By Hand	Production dependent on climatic conditions	Thalli fallen by wind should be collected
25.	60	5	5	By Hand	Production dependent on climatic conditions	Thalli fallen by wind should be collected
26.	19	1	1	By Hand	Production dependent on climatic conditions	Thalii on forest floor should be collected

27.	18	2	2	By Hand	Production dependent on rainfall	Thalii on forest floor should be collected
28.	18	1	2	By Hand	Production dependent on rainfall	Thalii on forest floor should be collected
29.	18	1	1	By Hand	Production dependent on climatic condition	Thalii on forest floor should be collected
30.	27	3	3	By Hand	Production dependent on climatic condition	Thalii on forest floor should be collected
31.	43	2	4	By Hand	Production dependent on climatic condition	Thalli fallen due to wind are collected
32.	50	10	8	By Hand	Production dependent on climatic condition	Thalli fallen due to wind are collected
33.	52	2	3	By Hand	Production dependent on climatic condition	Thalli fallen due to wind are collected
34.	52	3	3	By Hand	Production dependent on climatic condition	Thalli fallen due to wind are collected
35.	50	4	4	By Hand	Production dependent on climatic condition	Thalli fallen due to wind are collected
36.	52	5	5	By Hand	Production dependent on climatic condition	Thalli fallen due to wind are collected
37.	55	4	4	By Hand	Production dependent on climatic condition	Thalli fallen due to wind are collected
38.	52	5	5	By Hand	Production dependent on climatic condition	Thalli fallen from the tree are collected
39.	45	5	10	By Hand	Production dependent on climatic condition	Thalli fallen from the tree are collected
40.	43	3	4	By Hand	Production dependent on climatic condition	Thalli fallen from the tree are collected
41.	45	2	2	By Hand	No impact	Thalli fallen from the tree are collected

Contd...

42.	55	3	2	By Hand	No impact	Thalli fallen from the tree are collected
43.	45	4	3	By Hand	Impacted by Rainfall	Thalli fallen from the tree are collected
44.	48	4	4	By Hand	Production dependent on Climate	Thalli fallen from the tree are collected
45.	32	3	3	By Hand	Production dependent on Climate	Thalli fallen from the tree are collected
46.	50	4	3	By Hand	Production dependent on Climate	Thalli fallen from the tree are collected
47.	52	2	2	By Hand	Production dependent on Climate	Thalli fallen from the tree are collected
48.	36	4	4	By Hand	Production dependent on Climate	Thalli fallen from the tree are collected
49.	34	5	5	By Hand	Production dependent on Climate	Thalli fallen from the tree are collected
50.	44	3	3	By Hand	Production dependent on Climate	Thalli fallen from the tree are collected
51.	28	15	15	By Hand	No impact. Entirely dependent on climate	Thalii fallen due to wind should be collected
52.	40	15	20	By Hand	No impact. Entirely dependent on climate	Some thalli should be left on the branches
53.	35	10	20	By Hand	No impact. Entirely dependent on climate	Thalli fallen due to wind should be collected
54.	33	6	12	By Hand	No impact. Entirely dependent on climate	Thalli fallen due to wind should be collected
55.	42	6	13	By Hand	No impact. Entirely dependent on climate	Some thalli should be left on the branches

56.	32	7	15	By Hand	No impact. Entirely dependent on climate	Some thalli should be left on the branches
57.	18	2	4	By Hand	No impact. Entirely dependent on climate	Some thalli should be left on the branches
58.	18	10	20	By Hand	No impact. Entirely dependent on climate	Some thalli should be left on the branches
59.	22	4	6	By Hand	No impact. Entirely dependent on climate	Some thalli should be left on the branches
60.	27	7	14	By Hand	No impact. Entirely dependent on climate	Some thalli should be left on the branches
61.	67	3	1	Digging by Kudal	No impact	A part of the root is left below ground
62.	60	2	2	Digging by Kudal	No impact	A part of the root is left below ground
63.	52	3	3	Digging by Kudal	No impact	A part of the root is left below ground
64.	35	3	2	Digging by Kudal	No impact	A part of the root is left below ground
65.	28	3	6	Digging by Kudal	No impact	A part of the root is left below ground
66.	40	4	5	Digging by Kudal	No impact	A part of the root is left below ground
67.	45	3	3	Digging by Kudal	No impact	A part of the root is left below ground
68.	45	4	4	Digging by Kudal	No impact	A part of the root is left below ground
69.	38	3	3	Digging by Kudal	No impact	A part of the root is left below ground
70.	42	4	4	Digging by Kudal	No impact	A part of the root is left below ground
71.	44	15	15	By Hand and Daranti	No impact	Some thalli are left for regeneration from location of harvest

Contd...

72.	24	7	7	By Hand and Daranti	No impact	Some thalli are left for regeneration from location of harvest
73.	48	15	15	By Hand and Daranti		Some thalli are left for regeneration from location of harvest
74.	42	15	15	By Hand and Daranti	No impact	Some thalli are left for regeneration from location of harvest
75.	42	15	25	By Hand and Daranti	No impact	Some thalli are left for regeneration from location of harvest
76.	55	20	25	By Hand and Daranti	No impact	Some thalli are left for regeneration from location of harvest
77.	42	10	15	By Hand and Daranti	No impact	Some thalli are left for regeneration from location of harvest
78.	45	10	12	By Hand and Daranti	No impact	Some thalli are left for regeneration from location of harvest
79.	38	10	20	By Hand and Daranti	No impact	Some thalli are left for regeneration from location of harvest
80.	36	15	25	By Hand and Daranti	No impact	Some thalli are left for regeneration from location of harvest
81.	25	15	20	By Hand	Lichens dependent on climate	Some thalli are left for regeneration from location of harvest
82.	40	10	10	By Hand	Lichens dependent on climate	Some thalli are left for regeneration from location of harvest

83.	40	12	12	By Hand	Lichens dependent on climate	Some thalli are left for regeneration from location of harvest
84.	32	12	15	By Hand	Lichens dependent on climate	Some thalli are left for regeneration from location of harvest
85.	40	10	10	By Hand	Lichens dependent on climate	Some thalli are left for regeneration from location of harvest
86.	40	10	10	By Hand	Lichens dependent on climate	Some thalli are left for regeneration from location of harvest
87.	30	10	10	By Hand	Lichens dependent on climate	Some thalli are left for regeneration from location of harvest
88.	25	5	5	By Hand	Lichens dependent on climate	Some thalli are left for regeneration from location of harvest
89.	35	15	15	By Hand	Lichens dependent on climate	Some thalli are left for regeneration from location of harvest
90.	35	12	15	By Hand	Lichens dependent on climate	Some thalli are left for regeneration from location of harvest
91.	36	7	7	Digging by Kudal	No impact	Root part left below ground
92.	40	8	5	Digging by Kudal	No impact	Root part left below ground
93.	55	10	10	Digging by Kudal	No impact	Root part left below ground
94.	70	15	15	Digging by Kudal	No impact	Root part left below ground
95.	36	6	7	Digging by Kudal	No impact	Root part left below ground
96.	40	12	12	Digging by Kudal	No impact	Root part left below ground

Contd...

97.	35	20	12	Digging by Kudal	No impact	Root part left below ground
98.	30	9	7	Digging by Kudal	No impact	Root part left below ground
99.	28	5	5	Digging by Kudal	No impact	Root part left below ground
100.	35	7	7	Digging by Kudal	No impact	Root part left below ground
101.	70	10	500gm	By hand	No damage but dependent on climate	NA
102.	25	5	400- 500 gm	By hand	No damage but dependent on climate	NA
103.	32	5	500-600 gm	By hand	No damage but dependent on climate	NA
104.	27	5	700-800 gm	By hand	No damage but dependent on climate	NA
105.	30	5	600-700gm	By hand	No damage but dependent on climate	NA
106.	25	5	500-600gm	By hand	No damage but dependent on climate	NA
107.	40	8	700-800gm	By hand	No damage but dependent on climate	NA
108.	35	5	300-500gm	By hand	No damage but dependent on climate	NA
109.	36	6	500-600gm	By hand	No damage but dependent on climate	NA
110.	55	7	600-700gm	By hand	No damage but dependent on climate	NA
111.	35	12	15	By hand	No damage but dependent on climate	Lichens fallen due to wind are collected
112.	40	10	13	By hand	No damage but dependent on climate	Lichens fallen due to wind are collected

113.	35	12	15	By hand	No damage but dependent on climate	Lichens fallen due to wind are collected
114.	28	5	8	By hand	No damage but dependent on climate	Lichens fallen due to wind are collected
115.	23	5	9	By hand	No damage but dependent on climate	Lichens fallen due to wind are collected
116.	25	10	15	By hand	No damage but dependent on climate	Lichens fallen due to wind are collected
117.	32	15	18	By hand	No damage but dependent on climate	Only 50% of the thalli should be removed
118.	22	5	10	By hand	No damage but dependent on climate	Only 50% of the thalli should be removed
119.	27	8	16	By hand	No damage but dependent on climate	All the thalii should not be removed
120.	30	10	18	By hand	No damage but dependent on climate	Thalii fallen due to wind are collected
121.	35	12	12	By hand	dependent on climate	Part of thalii left on the branches
122.	45	15	15	By hand	Dependent on rainfall	Part of thalii left on the branches
123.	50	20	20	By hand	dependent on climate	Part of thalii left on the branches
124.	36	16	16	By hand	Dependent on rainfall	Part of thalii left on the branches
125.	24	8	8	By hand	Dependent on rainfall	Part of thalii left on the branches
126.	23	6	7	By hand	Dependent on rainfall	Part of thalii left on the branches
127.	50	20	20	By hand	Dependent on rainfall	Thalii fallen by wind are collected
128.	29	9	9	By hand	Dependent on Climate	Thalli fallen due to wind are collected
129.	47	20	22	By hand	Dependent on rainfall	Thalli fallen due to wind are collected

Contd...

130.	53	22	22	By hand	Dependent on climate	Some thalli are left on the branches
131.	45	5	12	By hand	Dependent on climate	Some thalli are left on the branches
132.	45	4	8	By hand	Dependent on rainfall	Some thalli are left on the branches
133.	27	7	14	By hand	Dependent on climate	Some thalli are left in the branches
134.	27	7	14	By hand	Dependent on climate	Thalli fallen by wind are collected
135.	30	8	16	By hand	Dependent on climate	Thalli fallen by wind are collected
136.	55	12	20	By hand	Dependent on climate	Thalli fallen by wind are collected
137.	58	15	20	By hand	Dependent on climate	Thalli fallen by wind are collected
138.	42	10	20	By hand	Dependent on climate	Thalli fallen by wind are collected
139.	35	10	12	By hand	Dependent on climate	Some thalli are left on the branches
140.	40	8	15	By hand	Dependent on climate	Thalli fallen by wind are collected
141.	35	15	15	By hand	Dependent on climate	Some thalli are left on the branches
142.	50	15	20	By hand	Dependent on climate	Some thalli are left on the branches
143.	55	15	15	By hand	Dependent on climate	Some thalli are left on the branches
144.	45	12	15	By hand	Dependent on climate	Some thalli are left on the branches
145.	32	13	15	By hand	Dependent on climate	Some thalli are left on the branches
146.	18	2	4	By hand	Dependent on climate	Some thalli are left on the branches
147.	18	2	5	By hand	Dependent on climate	Thalli fallen by wind are collected
148.	40	15	15	By hand	Dependent on climate	Thalli fallen by wind are collected
149.	35	15	20	By hand	Dependent on climate	Some thalli are left on the branches
150.	35	12	15	By hand	Dependent on climate	Some thalli are left on the branches

Table 11.5(a) Average collection of lichens per year from the lichen conservation area in Uttarakhand

Sample No.	*Collection of Lichens/collector / year (Quintals)*
1	10
2	5
3	10
4	9
5	6
6	12
7	13
8	5
9	15
10	10
11	14
12	13
13	14
14	12
15	8
16	20
17	10
18	15
19	13
20	20
21	4
22	1
23	3
24	3
25	5
26	1
27	2
28	2
29	1
30	3
31	4
32	8
33	3
34	3
35	4
36	5
37	4
38	5
39	10
40	4
41	2
42	2
43	3
44	4
45	3
46	3
47	2
48	4
49	5
50	3
51	15
52	20
53	20
54	12
55	13
56	15
57	4
58	20
59	6
60	14
71	15
72	7
73	15
74	15
75	25
76	25
77	15
78	12
79	20
80	25
81	20
82	10
83	12
84	15
85	10
86	10
87	10
88	5

Contd...

89	15
90	15
111	15
112	13
113	15
114	8
115	9
116	15
117	18
118	10
119	16
120	18
121	12
122	15
123	20
124	16
125	8
126	7
127	20
128	9
129	22
130	22
131	12
132	8
133	14
134	14
135	16
136	20
137	20
138	20
139	12
140	15
141	15
142	20
143	15
144	15
145	15
146	4
147	5
148	15
149	20
150	15
N = 120	**1353**
AVG	**11.3** quintals

Table 11.5(b)

Sample No.	*Collection of Picrorhiza kurroa Quintals per collector per year*
61	1
62	2
63	3
64	2
65	6
66	5
67	3
68	4
69	3
70	4
91	7
92	5
93	10
94	15
95	7
96	12
97	12
98	7
99	5
100	7
N = 20	**120**
AVG	**6.0** quintals

CONCLUSION

Assessment of favourable areas for lichens revealed that Waan, Kuling, Bank and Mundoli villages have forest areas of 77, 123, 26.6 and 64.5 ha respectively. As per assessment made by Kumar (2010), these 4 villages should collect no more than 4.62, 7.38, 1.59 and 3.87 quintals per year.

OBSERVATIONS ON THE POLICY ENVIRONMENT

Uttarakhand, the herbal state in India, is rich in medicinal plant resources and has emphasized on their conservation. It has taken the right steps in right direction by promoting both in situ and ex situ conservation of the medicinal plants. *Picrorhi zakurroa* is now into cultivation and it is expected that cultivation will reduce the pressure of collection from the wild. The state has developed the necessary infrastructure and enunciated policies to ensure conservation of wild medicinal plants. The state has:

1. Issued guidelines for sustainable harvest of wild medicinal plants and has addressed priority species like *Cordyceps sinensis.*
2. Designed rotational harvesting policy that opens up and closes forest ranges to ensure regeneration of plants. A systematic record of the forest ranges opened and closed since 2004- 05 exists for analysis in the future.
3. Strategically located medicinal plant depots to ensure that wild medicinal plants are routed and, fair pricing is ensured through auctions. Trade data at major depots since 2005 will help in future studies related to sustainable use of wild medicinal plants in the future.
4. Making efforts towards integration of the medicinal plants sector with the State Forest Working Plans. Guidelines have been developed in this regard.
5. The state has been able to draw the attention of larger investments like the GEF (Global Environmental Facility) through UNDP (United Nations Development Programme) and State Medicinal Plants Board, Govt. of Uttarakhand on conservation and sustainable use of medicinal plants.

In spite of the regulations, harvest of wild medicinal plants is done unsustainably owing to the following reasons:

1. Present levels of regulations are highly generic; estimations for harvest made are based on ocular estimations. Species level standards are not available to the harvesters. Existing studies on levels of harvest of specific wild medicinal plant species are extremely limited in terms of duration and are site-specific. These would require further research, validation and consultation with local communities in particular. The present estimation of marketable lichens to the tune of 0.06 quintals of lichens per ha of forest floor needs further investigations. Local communities have to be communicated the quantum of harvests undertaken from the forests and these levels do not commensurate with the rate of regeneration of the species.

2. Enforcements are not adequate, moreover all the wild medicinal plants harvested are not necessarily traded through the trade depot. Illegal trade thrives, hence it is difficult to provide correct estimations. Harvesters confide the actual amounts harvested in the apprehension of the fact that collection of species might be banned by the government and their livelihoods will be impacted. At present, there is no assessment on the impacts of the rotational harvesting policy.
3. Due to the lack of awareness on sustainable harvesting techniques and the impact of non-availability of harvesting manuals and practically no investments on training the plant collectors and harvesters on scientific methods of collection, grading and value addition of the harvested material, collectors of lichens are unaware of the end product, grading and packaging requirements to minimize waste in collection.

RECOMMENDATIONS

Need for Resource Assessment and Impacts of Harvest

As mentioned in the preceding paragraph it is essential that species-specific guidelines be developed for all medicinal plant species traded for commercial use. This has been done for species like Lichens, and state guidelines suggest collection of species only from the forest ground, however sustainable harvesting modeling is the need of the day. This would require information on the autecology of the species, climatological and socio-economic parameters. Special emphasis has to be provided on regeneration of species. Use of newer tools and techniques like DIVA GIS and sustainable harvesting modeling studies can aid in undertaking resource assessment. A thorough data-set could meet such modeling requirements for sustainable harvest of lichens as a case study.

Ecological Studies on Impact of Harvests as well as Regulations

It is important to make an assessment of the impacts of harvest of wild medicinal plants. Wild plants play an important role in nutrient cycling and are also food sources for many species of invertebrates. Lichens, for example, are important food sources for *Pica* sps, the tailless mice. They are also used by birds in their nests for unknown ecological reasons. It has been observed that oak species are host for lichen colonization. It is essential to quantify the after effects of harvest of lichens on oak regeneration patterns. Information of this nature will be of much help in developing species level management plan.

Emphasis on Harvesting Techniques, Prepare Guidelines on the same and Raise the Level of Awareness Amongst the Harvesters

An integral component of the management programme for sustainable harvest of the medicinal plants is the continuing research on the harvesting techniques and tools. For this purpose, it is proposed here that efficiency of the current methods and equipment being used for harvesting the prioritized species should be tested. It is very likely that the equipment may lie more on the inefficient side. As far as the situation on the field is concerned, most of the techniques are primitive and not specialized for the harvesting purpose. What is more disturbing is the fact that most of the lichens harvest is being taken from the standing crop and less from the forest floor because of the desiccation at the floor. The destructive method of harvest leads to a total disrupt in minimal amount of lichen regeneration that is going on.

Apart from the equipment, harvesting techniques and schemes also need to be defined in order to achieve the sustainable harvest goals till the time modeling becomes an alternative. There have been numerous examples of sustainable harvesting techniques by mapping the resource and designating areas for regeneration and harvesting separately. This can be a feasible option of a rotational policy as in case of Devil's Claw.

This presents an urgent need to conduct research on the non-destructive methods of harvest as well as the training modules to be followed while training the harvesting professionals. As far as the applicability of this concept is concerned, some regions of the Lichen Conservation Areas can be devoted for testing the efficiency of the proposed equipment and techniques of harvest. A comparison can be drawn between the conventional and the proposed techniques. Apart from the efficiency of the equipment, the feasibility and cost-effectiveness of the alternatives also needs to be assessed.

Findings of academic research of this nature have to be communicated to the local communities. With lichens, as a case in point, it needs to be communicated to both the scientific community and users that requirements of trade in case of Lichen species like *Everniastrum, Parmelia and Parmotrema* are justified. But, harvest of all species of Lichens that have little value in trade needs to be spared.

Value Addition Techniques

A common observation during the study was that most of the harvesters as well as the traders did not know about the end product which is being made from lichens (and other medicinal plants as well). This indicates the void in the supply chain and the benefit sharing mechanism. As the supply chain becomes clearer with the help of questionnaires, an assessment of the value addition techniques can be done by surveying the markets where medicinal plants are sold in packaged form. For example, lichens are an important ingredient in a popular sherbet called Rooh-afza™ manufactured by Hamdard Laboratories, and *Pueraria tuberosa* pulp is used in Dabur Chyawanprash™. A survey can be conducted to assess the quality and form of lichens which the industry requires. Keeping this in mind, transparency in the supply chain can be enhanced by training the harvesters or the post-harvesting personnel for selective harvesting.

The harvesters can themselves be taught the value addition techniques and proper storage techniques in order to reduce the hierarchical dominance in the supply chain and making the product more acceptable to the end user.

Payment for Environmental Services

Interactions between biotic and abiotic components of the ecosystem result in ecosystem services, which in turn sustain human life and without them economics of the world would come to a halt (Singh, 2007). It is thus ideal to amalgamate economics with biodiversity conservation. The monetary value of global ecosystem services has been estimated to be at least US $ 33 trillion (Singh, 2007). The total value of forest ecosystem services flowing from Uttarakhand has been estimated to be about 2.4 billion US $/year. Because of the river connections, the ecosystem services flowing from Uttarakhand have played a principal role in shaping the rise of culture in the Gangetic plains inhabited by nearly 500 million Indians. It is thus important to conserve these forests and provide the people of the Himalayan regions with appropriate economic incentives.

Community forestry has been strongly institutionalized in the state of Uttarakhand, and issues related to ecosystem services can directly affect their economic condition and processes leading to the empowerment of local people. It is proposed to incorporate the concept of environmental incentives in the Lichen Conservation Areas. The concept has been well received and appreciated during the initial talks with the *Van Panchayats* of Waan and Kuling villages. A mechanism to provide economic incentives to the collectors for sustainably harvesting medicinal and aromatic plants needs to be worked out. The income of the collectors will also be boosted with the implementation of other value added practices. Provision of economic incentives will ensure the maintenance of the Lichen Conservation Areas. Upreti (2008) has already mooted the idea of lichen gardens and sanctuaries. Lichen Conservation Landscape identified in Chamoli District can be a good location for a network of such lichen gardens. These gardens could also serve as stations for scientific research.

This mechanism of financial support, once extended to all the Himalayan states, will eventually help in (i) maintaining the flow of ecosystem services to the rest of the country on a sustainable basis, and (ii) improve the quality of life without jeopardizing the national ecological security. Sustainable harvesting of wild medicinal plants has huge prospects in India. We have to build upon the success stories available in the country and abroad.

Ministry of Environment and Forests, Government of India, has launched the 'The Economics of Ecosystem Services and Biodiversity (TEEB)' process. The ecosystem function and importance of wild medicinal plants should be highlighted in such initiatives.

Assessment of External Drivers to Demands on Wild Medicinal and Aromatic Plant Species

Lichens from the state of the Uttarakhand find their way to neighbouring state of Uttar Pradesh, for further exports. Perfumes are extracted from lichens for export to Arabian countries, where the product is in high demand as these are sprayed before prayers in the mosques. It is important to understand the present level of demand outside the country. Lichens were also collected by the Sholiga tribals in Karnataka where the consumption is more as spices and therefore they are sold in the local markets. It will be interesting to know why lichens are not extracted heavily from the Eastern Himalayan states like Arunachal Pradesh.

Understanding Habitat Degradation and Land Use Land Cover Change

It must be mentioned that conservation of medicinal plant species will not happen by regulating trade. There are other major drivers which influence wild population of medicinal and aromatic plants, that need data collection, collation and analysis. Impact of land degradation and change in land use land cover, which are major drivers of deforestation and degradation, is not only on flagship species that receive attention, but also on lower group of species like lichens and mosses, which play an important role in ecosystem functioning. These species are dependent on their hosts which provide them their ecological niche. Dangwal (2009) has provided an account of degradation of ecosystem in Uttarakhand.

Scaling Up of Study to Other Wild Medicinal Plant Species in India

The study can be scaled-up to many other wild medicinal tree species of India. These include *Garcinia* which is an endemic genus to India, and extracted not only locally but also for commercial use of hydrocyanic acid in Western Ghats in India; *Rhododendrons arboretum,* flowers of which are extracted heavily from the Western Himalaya for preparation of squashes; *Swertia chirata,* extracted from Eastern Himalaya, particularly the states of Arunachal and Sikkim; *Madhuca latifolia* from the central India's states of Chattisgarh and Madhya Pradesh. These studies will help our policy makers and the State Forest Departments. In addition, the studies will also support the present effort of Govt. of India in integrating the medicinal and aromatic plants in State Forest Working Plans and to the proposed Indian Council for Forest Certification which will require standards and good management practices for sustainable harvesting.

REFERENCES

1. Anderson, P.J. (1998). Using ecological and economic information to determine sustainable harvest levels of a plant population. In incomes from the Forest - Methods for development and conservation of forest products for local communities. (Eds.) E. Wollenberg and A. Ingles. Occasional paper No. 33, Center for International Forestry Research, Bogor, Indonesia.
2. Anonymous (2000). Report of the task force on conservation and sustainable use of medicinal plants, Planning Commission of India, Govt. of India, March p. 194.
3. Bhattacharya, P. (2004). Sustainable harvesting and value addition of Satavar (*Asparagus racemosus*): A case of participatory livelihood generation in Chambal ravine of Madhya Pradesh, India. Paper presented at the International conference on 'Medicinal herbs and herbal product: Livelihood and trade options'. New Delhi, 26-27 March.
4. Bhattacharya, P. Prasad, R. Bhattacharya, R. and Asokan, A. (2008). Study on certification of wild medicinal and aromatic plants in four Indian states. Unasylva, 59: 35-44.
5. Chatterjee, S.J. Singh, A. Acharya, A. Jha, R. and Prasad, R. (2011). Setting standards for sustainable harvest of wild medicinal plants in Uttarakhand: A case study of Lichens. 2011. In: Community based biodiversity conservation in the Himalayas. Eds: Yogesh Gokhale and Ajeet K Negi. The Energy Research Institute. New Delhi. pp. 102-123.
6. Chatterjee, S. Pande, M. Semwal, R.L. Kumar, K.S. Dutta, P. Sudha, S. and Pangging, G. (2002). Relevance of Certification to the Wood Carving Industry in India. WWF-India, New Delhi, p. 56.

7. Dutt, D.D. (2009). Himalayan degradation. Colonial Forestry and Environmental Change in India. Foundation Asia. New Delhi, pp. 19-50.
8. Emanuel, P.L. Shackleton, C.M. and Baxter, J.S. (2005). Modeling the sustainable harvest of *Sclerocaryabirrea* subsp. caffra fruits in the South African lowveld. *Forest Ecology and Management*, 214: 91-103.
9. Ganesan, R. and Siddappa Setty, R. (2004). Regeneration of Amla, an important non timber forest product from southern India. Conservation and Society, 2(2): 365-375.
10. Jain, P. and Mishra, M. (2007). Policy and laws for medicinal and aromatic plant sector – A guide for Uttaranchal Villagers. National Medicinal Plants Board, New Delhi.
11. Kala, C.P. Dhyani, P.P. and Sajwan, B.S. (2006). Developing the medicinal plants sector in Indian states. challenges and opportunities. *J. Ethnobiol. Ethnomed.* 2(32): 1-15.
12. Balwant, K. (2010). Ecological, social and commercial role of lichens in India with special reference to Garhwal Himalayas, Acdemia arena, Supplement 0201, pp. 1-117.
13. Malhotra, K.C. and Bhattacharya, P. (2010). Forest and Livelihood. Status paper (2010). Research Unit for Livelihoods and Natural Resources, Center for Economic and Social Studies. Hyderabad, p. 246.
14. Pfab, M.F. and Scholes, M.A. (2004). Is collection of Aloe peglerae from the wild sustainable? An evaluation using stochastic population modeling. *Biol. Conserv.* 118. 695-701.
15. Prasad, R. and Bhattacharya, P. (2003). Sustainable harvesting of medicinal plant resources. In: Contemporary studies in Natural resource management in India. Ed. S.B. Roy. Inter India Publication. New Delhi, pp. 168-198.
16. Schneider, E. (2000). Handbook for Sustainable Harvest in Semi-Wild Populations of Harpagophytum procumbens (Devil's Claw, Grapple plant, Sengaparile) and Preparation of Good Quality Devil's Claw Root Tubers, Kalahari Harpagophytum Project, GTZ.
17. Schwartz, M.W., Caro, T.M. and Banda Sakala, T. (2002). Assessing the sustainability of harvest of Pterocarpusangolensis in Rukwa region, Tanzania, *Forest Ecology and Management*, 170: 259-269.
18. Shanley, P. Gullien, A. Laird Sarah, A. and Pierce Alan, R. (2002). Tapping the Green Market, Certification and Management of Non-Timber Forest Products, People and Plants Conservation Series, Earthscan Publications Ltd.
19. Shanley, P. Pierce, A. Laird, S. (2005) Beyond timber: Certification of non-timber forest products. Forest Trends/CIFOR.
20. Singh, S.P. (2007). Himalayan Forest Ecosystem Services-Incorporating in national accounting. Central Himalayan Environment Association (CHEA), Uttarakhand.
21. Soehartono, T. and Newton, A.C. (2001). Conservation and sustainable use of tropical trees in the genus Aquillaria II. The impact of Gaharu harvesting in Indonesia. *Biol. Conserv.*, 97: 29-41.

22. Upreti, D.K. (2008). Need for creation of lichen gardens and sanctuaries in India. *Curr. Sci.*, 94(8): 976-978.
23. Van Andel, T.R. Huyskens, P.E. and Broker, K.C.A. (1998). Palm heart harvesting in Guyana's North-West District: Exploitation and regeneration of Euterpeoleraceae Swamp Forests. Tropenbos–Guyana Interim Report; Utrecht University, p. 98.
24. Ved, D.K. and Goraya, G.S. (2008). Demand and supply of medicinal plants in India. National Medicinal Plants Board. Govt. of India, New Delhi and Foundation of Rehabilitation of Local Health Traditions, Bangalore, pp. 1-18.

12

Richness and Variability of Tree Species in Achanakmar and Amarkantak Forests of Central India

R.V. Shukla*, Kamalesh Shukla and R. Sukumar†**

**Department of Botany, C.M.D. College Bilaspur, Chhattisgarh*

***School of Studies in Biotechnology, Pt. R.S.S. University Raipur, Chhattisgarh*

†Center for Ecological Science, Indian Institute of Science, Bengaluru

ABSTRACT

The Achanakmar and Amarkantak forests in Central India include a large variety of tree species of tropical semi-ever green type belonging mostly to the families Combretaceae, Dipterocarpaceae, Papilionaceae. Physiographically, this is a sub-arid forest zone divided into a lowland of Achanakmar and upland of Amarkantak, which is characterised by the uniform density of *Shorea robusta.* The forest area is clothed with dominating indigenous Sal tree found in different percentages of distribution in different altitudinal gradients. The mixed Sal forest is situated at an elevation of 450 metres in Achanakmar. It has a long history of its biome types. The area is diverse and seems to be a cradle for the generation of new taxa or even a museum for the preservation of existing tree diversity. This is an area of higher speciation and greater variance in species ages. The present study, a preliminary investigation, sensitizes us about the endemicity of tree species against the land-use changes.

INTRODUCTION

In Central India, the Satpura Maikal landscape spreads across Madhya Pradesh, Chhattisgarh, Maharashtra, Jharkhand and Andhra Pradesh, of which the major part is covered by the tropical forests. This is the richest tract in wildlife as well as a myriad of indigenous people are living in the forests since time immemorial. The Maikal range in Amarkantak (AMK) and Achanakmar (ACH) is an extension of mountain Satpuda and it is gifted with a large variety of tree species. This sub-arid area of Sal and associated species (members of Combretaceae, Dipterocarpaceae, etc.) includes Achanakmar Amarkantak Biosphere Reserve (ABR), the 14th biosphere reserve in the country that focuses on the uniform dominance and intact tree cover over the forest areas.

ABR is one of the most crucial hotspots that shows endemicity and unusual longevity of tree species in medium height mountain (400-800 m) ranges over the Gondwana land. The biosphere consists of environmental heterogeneity that differs with the neighbouring northern state of the country. This reserve denotes a typical forest ecosystem where we find the floristic density, species evenness and their concentration of dominance at different altitudinal ranges. The soil composition and its quality in metalliferous rocks are quite different in continuum to understand and to explain numbers of analytical and synthetic parameters against the variability and richness of tree species.

ABR, includes an important eco-region, where bulk of tree species persist to keep the forest floor moist and cool (Tiwari et al., 2009). Within the uninterrupted local conditions, a tree plant has a longer period of survival and chances for the species interactions leading to niche specialization. Thus, this biosphere with higher speciation rates as well as a longer history has a rich banquet of tree species that are confined to moist lowland stratum in ACH. As per the age and hypothesis the occurrence of tree species in ACH represents a potential source of relevant evidence as they are older than an average relative to species of upland in AMK area.

In ABR, the geographical differences are more pronounced for natural selection of the tree species as they have shown close affinities with those of moist lowland in ACH and upland in AMK. The ACH forest area in ABR has a higher speciation as well as a longer history of its biome showing greater variance in species' ages than on higher latitudes in AMK. As per the age and hypothesis the occurrence of tree species in ACH represents a potential source of relevant evidence as they are older than on an average relative to species of upland in AMK area. This contradicts dry deciduous forests in the AMK and other neighbouring forests in Central India (Sukumar et al., 1992; Umashankar et al., 1998), where lengthy leaf-drop period renders the canopy almost naked and forest floor gets dry and hot during summer months.

Schemske (2002) conducted a comparative study on the temperate and tropical population to explain adaptive capability against the abiotic and biotic factors. He found out that abiotic interactions have impact on the temperate population whereas biotic interaction such as rapid evolution, coevolution , adaptation and speciations are prominent in tropics. Farrell et al. (1992); Latham and Ricklefs (1993) explained the physiological adaptation of plant species against the broad range of temperature and their invasion in temperate zone. Ravindranath and Sukumar (1996) explained the impact of climate change and stated that a shift in vegetation type is expected from lower elevation to higher altitude.

Umashankar (2001), Singh and Kushwaha (2005), studied floristic composition of Sal dominated lowland forest of the Eastern Himalayas. Neelakandan et al. (2006) emphasized on the importance of biogeographical information system for monitoring biodiversity and its sustainable use in site-specific and region-specific areas. In ABR, Shukla (2003), Tiwari et al. (2009), Shukla et al. (2010), Tiwari et al. (2010) conducted an extensive work on mushroom diversity. Mishra (1956), Sen Gupta (1977), Singh et al. (1986), studied floristic composition and Hewetson (1956), Gadgil & Meher-Homji (1982) had contributed summaries on physiography of vegetation of Madhya Pradesh and Chhattisgarh. But no one has explained the tree composition of forests that vary even at short distances in ACH and AMT.

Umashankar (2001) made a comparative analysis of the richness of Sal tree, its dispersal and regeneration, with that of deciduous forests and reported that the tree richness is greater in lowland forests of the Eastern Himalayas than that of Sal forests in the Central Himalayas. Jha and Singh (1990), Prasad and Pandey (1990) were of the opinion that the richness of large tree species is greater in Central India than that in the Himalayas. Singh and Kushwaha (2005) suggested that wide leaf phenological response in Sal tree is more prominent due to its semi-evergreen nature that helps extensive distribution in the tropics, than the moist regions (e.g., Central Himalayas) to the dry regions (e.g., Vindhyan). Therefore, latitudinal gradients and moist conditions are enough to explore the question why ACH mixed Sal forest represents far greater number of tree species than that of true Sal forest in Amarkantak.

The richness of tree diversity in ACH range is due to the local environmental conditions that have a direct physiological impact on the growth and development of tree species. Many tree species of the area are highly specific likely to be an obligate group that exist in a narrower range of temperatures and high level of precipitation. Stevens (1989) stated the 'Rapoport's rule' to explain the narrow range of climatic conditions which facilitate variety of species to coexist and the extent of latitudinal ranges, that declines species diversity in the tropics. Shafer et al. (2001) suggested counting mean temperature of the winter, annual precipitation, evapo-transpiration and the moisture index.

Regional variation in species richness is the subject of long standing debates in ecology and biogeography (Piank (1966), Huston (1994), Lomolino (2001), Whittaker et al., (2001)). The variation of species richness along elevation gradients has been documented for a variety of taxa and geographical areas (Stevens (1992), Rahbek (1995, 1997), Brown (2001), Heaney (2001), Bhattarai & Vetaas (2003, 2005, 2006), Carpenter (2005)). Bhattarai and Vetaas (2003) found maximum tree species richness around 800 m a.s.l. Suderpandian and Swamy (2000) studied altitudinal range of Dipterocarpaceous members and reported their absence at higher altitudes (above 1150 m).

Climate is an obvious explanation for the richness of tree species. Odland and Birks, (1999), Bhattarai and Vetaas (2003) reported that at lower elevational ranges tree species richness can be predicted as per the linear function of rainfall – a parabolic function of energy. Although rainfall is not governed by the mountain elevations, as there are variations in rainfall over the same elevation ranges in Satpuda mountain of Central India. The vegetational composition of ABR is more complex which can be apparently seen in respect to elevational gradient. This is an excellent site to explain the richness and variability of tree species all along the elevational gradient from lowland to upland.

Despite several explanations, how ABR differs in tree diversity than other forests, our aim is to explain the geographic variation in light of the documentation of tree species. We speculate the most important factors, i.e., the climate and the soil type that showed positive relationship mostly with the old tree types. Since the tree species are found more accurate in observation and are not influenced by the climate change as much the herbaceous species are influenced, therefore we opted a preliminary study on documentation of the tree species in both the mixed Sal forest and true Sal forest areas.

METHODOLOGY

Achanakmar-Amarkantak Biosphere Reserve (ABR)

The Achanakmar-Amarkantak Biosphere Reserve (ABR) is named after the Achanakmar (ACH) forest village in Bilaspur district in Chhattisgarh and Amarkantak (AMK) an edge of Anuppur district in Madhya Pradesh. Both the areas of ACH and AMK collectively were declared as the 14th biosphere reserve in 2005 by the Government of India. The topography of the biosphere falls under the biogeographic zone of Deccan Peninsular a central highland (6A) of the country. Physiographically, the area is divided in the lowland and upland representing various ecological niches for natural availability of a large spectrum of biological species.

ABR can be divided into the core and the buffer zones. The Achanakmar tiger reserve a core zone is located in Bilaspur district and rest of the area categorized as buffer zone and transitional zone of which major part falls in Chhattisgarh state.

Fig. 12.1 Map showing study area Achanakmar-Amarkantak Biosphere Reserve

STUDY SITE

Achanakmar

Achanakmar sanctuary, a core zone located in Bilaspur district, has a total geographical area of 551.1/5 km^2, within the area of the biosphere. Apart from the sanctuary there are Lamni and Chhaparwa forest ranges which are 62-70 km away at the north-west of Bilaspur forest division.

In the present investigation ACH wild sanctuary, a core zone of ABR, was selected to record the tree diversity. Apart from the sanctuary 56 km in north-west of Bilaspur, there are Chhaparwa (62 km), Lamni (70 km) and AMK 90 km where the higher temperature in summer (April-June) at altitudinal ranges (400 = 1100 m) is not similar. An average rainfall of 146 cm was recorded in the period of 2000-2008. The average relative humidity was 60% and 52% in the morning and evening respectively. The soil characteristic is acidic reddish brown and dark yellowish brown. The area is endowed with natural primary and manmade secondary forests.

Amarkantak

The agro topography of the biosphere varies from the rice fields below in Bilaspur and Anuppur areas and from the wheat fields in Dindori to the hills of Maikal range of Satpudas. The mineral topography of the soil in the Amarkantak plateau is of bauxite rocks. The area is one of the major watersheds of peninsular India. Several streams and *nallas* are flowing through the reserve and many of them are perennial. It separates the rivers that drain into the Arabian Sea by the Bay of Bengal. The reserve is also unique for being the source of three major river systems like Narmada, Johilla and Sone of the Ganga basin.

The above three study sites were selected to keep track of changes in biodiversity. An area of 500 × 5 metres was selected in each site (Site I, II and III; Tables 12.1, 12.2 and 12.3) to record the occurrence and distribution of tree species (Shukla 2003). The tree species were sampled at three levels, namely, adult, sapling and seedling.

Result and discussion

Study site I

The study site in ACH situated at an elevation of 450 metres is characterised by the presence and the prominence of Sal, a key-stone tree species of the biosphere. Despite, dominance of Sal tree the percentage of other tree species, i.e., *Anogeissus latifolia* (Dhaora), *Buchanania lanzan* (Char), *Butea monosperma* (Palas) *Diospyrous melanoxylon* (Tendu) *Eugenia jambolana* (Jamun) *Terminalia tomentosa* (Saja), etc. (Table 12.1) determine the combined effect on the humus, soil pH, degree of waterlogging, light, and the degree of human disturbances in the biosphere (Shukla 2003; Tiwari et al., 2009).

The thickness of tree species is almost uniform; climatic conditions with least deviation throughout the year promoting varieties of microorganisms, mushrooms, lower plants and insects. The warm, wet, seasonal climate within the site is apparently favourable for coexisting varieties of tree species that exhibit a great diversity in structure and in composition of tree species. The pattern and magnitude of the tree species richness shows an interrelationship between a large spectrum of biological species (ground flora, insects, birds, reptiles, and wildlife), of national importance.

Plate 12.1 An open necked wasteland exhibits soil texture

Table 12.1 List of tree species recorded in Sal dominated undisturbed habitat Site - I

Name of Tree species	*Adult*	*Sapling*	*Seedling*	*% Total*
Aegle marmelos	1	1	–	0.2
Anogeissus latifolia	5	–	–	0.5
Adina cordifolia	1	–	–	0.1
Acasia catechu	1	–	–	0.1
Annona squamosa	1	–	–	0.1
Bridelia retusa	2	–	–	0.2
Butea monosperma	2	–	–	0.2
Boswellia serrata	1	–	–	0.1
Bombax malabarica	1	–	–	0.1
Buchanania lanzan	17	–	–	1.8
Bauhinia malabarica	2	–	–	0.2
Cassia fistula	5	4	–	1.0
Dendrocalamus strictus	1	–	–	0.1
Diospyros melanoxylon	56	16	–	7.7
Dillenia pentagyna	–	1	–	0.1
Eugenia jambolana	23	10	–	3.5
Emblica officinalis	1	–	–	0.1
Feronia elephantopus	1	–	–	0.1
Ficus glomerat	6	–	–	0.6

Contd..

Contd..

Lannea coromandelica	3	–	–	0.3
Lagerstroemia parviflora	4	5	–	1.0
Madhuca longifolia	5	–	–	0.5
Pterocarpus marsupium	4	2	–	0.6
Phoenix acaulis	2	–	–	0.2
Saccopetalum tomentosum	6	–	–	0.6
Schleichera oleosa	2	–	–	0.2
Semecarpus anacardium	7	–	–	0.7
Shorea robusta	174	291	75	57.8
Terminalia belerica				
Terminalia tomentosa	30	2	–	3.4
T. chebula	2	–	–	0.2
Tamarindus indica Zizyphus jujuba unidentified 1	1	159	–	17.1
unidentified 2	1	–	–	0.1
unidentified 3	2	–	–	0.2
unidentified 4	–	–	–	-
unidentified 5	2	–	–	0.2

This site includes thick cover of tall Sal and Sal type semi-evergreen tree species; *Adina cordifolia* (Haldu), *Albizzia lebbek* (Siras), *Anogeissus* latifolia (Dhaora), *Bridelia retusa (kasai), Boswellia serrata* (Salai), *Buchanania lanzan* (Char), *Dillenia pentagyna* (Kalla), *Eugenia jambolana* (Jamun), *Ficus glomerata* (Gular), *Lagerstroemia parviflora* (Sej), *Mitragyna parvifolia, Pterocarpus marsupium* (Bija), *Saccopetalum tomentosum* (Kari), *Schleichera oleosa* (Kusum), *Tamarindus indica* (Imali), and many *Terminalia* species of the family Combretaceae, which are the promising source of minor forest produce: honey, gum, resin, oil seed, mushrooms, silkworm, and herbaceous ground flora, of medicinal importance. There are woody vines (*Butea superba, Tinospora cardifolia*), orchids and many others; lianas, herbaceous epiphyte, existing all along the moist lowland forest site.

The large-tree *Bombax* (Semal), *Elephantopus scaber*, (Kaitha), *Madhucac longifolia* (Mahuwa), *Mangifera indica* (Aam), and the medium small tree species; *Acacia arabica* (Babool), *A. catechu* (Khair), *Eagle marmelos* (Bel), *Zijuphus xylophora* (Ber), and Palas, Tendu, showed a distinct group which are found almost entirely leafless in summer months. Some species such as Babool, Palas, Tendu could succeed with a relatively small stock of juveniles, while others are found dependent on a large stocking of juveniles to compensate the mortality.

The temperature ranges in between January and June month impacting regeneration of many tree species in lowland areas. There are various understory tree plants but growth of shrubby group is found almost missing. However, this is an identified site for greater appearance of herbaceous flora due to moist soil condition. Shukla and Diwaker (1973), and Shukla and Jain (1984) found the abundance of weed flora of forest origin in moist and dry open land. Suderpandian and Swamy (2000) also identified and made their opinion about the higher percentage of monocotyledons.

In response to soil nutrients and high energy exchange the occurrence of a variety of tree species (Table 12.2) is quite positive as the species composition (large and medium size) maintains a very high humidity, moisture, wet and the warm climate. The tree plants return substantial amount of leaf litter to protect thin top soil for their seed germination. However, Sal, Jamun, Tendu, showed good-to-fair regeneration in upland plateau. The Sal tree with its viviparous seed begins to germinate on the tree and shows its sign of vegetational recovery mostly in plane surface area, but its seedlings and saplings fail to attain the adult state.

Table 12.2 List of tree species recorded in Sal dominated disturbed habitat site-II

Species	*Adult*	*Sapling*	*Seedling*	*% Total*
Aegle marmelos	1	1	–	0.3
Anogeissus latifolia	2	2	–	0.5
Adina cordifolia	–	–	–	–
Bridelia retusa	–	–	–	–
Butea monosperma	–	–	–	–
Boswellia serrata	–	–	–	–
Bombax malabarica	–	–	–	–
Buchanania lanzan	4	1		0.6
Bauhinia malabarica	–	–	–	–
Cassia fistula	1	4		0.6
Dendrocalamus strictus	–	–	–	–
Diospyros melanoxylon	23	64		11.0
Dillenia pentagyna	–	–	–	–
Eugenia jambolana	2	38		5.0
Emblica officinalis	–	–	–	–
Lannea coromandelica	–	–	–	–
Lagerstroemia parviflora	8	9		2.1
Madhuca indica	–	–	–	–
Pterocarpus marsupium	2	1		0.4
Phoenix acaulis	9	2		1.4
Saccopetalum tomentosum	–	–	–	–
Schleichera oleosa		2		0.3
Semecarpus anacardium	–	–	–	–
Shorea robusta	57	300	79	55.0

Contd..

Contd..

Tectona grandis	–	–	–	–
Terminalia tomentosa	16	2		2.3
T. chebula	–	–	–	–
Unidentified 1	15	143		19.9
Unidentified 2	–	–	–	–
Unidentified 3	–	–	–	–
Unidentified 4	–	–	–	–
Unidentified 5	5			0.6

Site II

This is a true Sal forest site in most of plane surface area of the ABR, where Sal tree dominates over other tree species such as Dhawara, Tendu, Jamun and Saja, trees (Table 12.2). There are different variables in the density of tree species in the same region and at the same altitude. Sites I and II are initially identical in everything, but density of Sal tree is significantly different in the same contiguous site. Although, the Sal tree is capable of maintaining its uniformity in a part of the area where it grows but, at some places it showed uniformity due to suitable habitat and in other places showed the non-uniformity exhibiting inadequate regeneration.

Plate 12.2 Sal trees in ABR

The occurrence of Sal tree is determined by the hygroscopicity of rocks and the soil. The overall population of tree species and their canopy layers decreasing population density of seedling to adult stage. No matter if density of saplings of Sal tree is good as in site-I (Table 12.1). The saplings and shrubby vegetation of the site indicate that tree plants are logged, but not in uniform pattern. The site showed uniform dominance of Sal tree which showed strong relationship for the recurrence of some of the mushrooms, i.e., *Geasrtum fimbriatum, Russula delica, Scleroderma texense, Termitomyces* spp., etc. (Shukla 2003; Tiwari et al., 2009, 2010; Shukla et al., 2010).

Site III

The site of Teak dominated secondary forest in ABR including Shivatarai showed increased level of local disturbance (fire, cattle grazing, etc.) attributing variation in microclimatic conditions such as temperature, humidity, and light intensity. The site reflects negative response to light soil texture which is prone to leach nutrients in excess. The leaf litter of Teak without microbial degradation is an overburden as it is not consumed by wildlife and can be lost by fire. The soil without water-holding capacity hardenes with unbound amount of iron, aluminum, manganese metals, causes toxic effects to the soil and waterbodies. The surface flora and the decomposer group are almost missing, however there is regeneration of seedlings, saplings and copping in stumps of Tenu, Sal, and *Cassia fistula,* etc. (Table 12.3) but their growth is limited due to mineral deficient soil (Shukla 2003).

Plate 12.3 Secondary forest: Teak plants of more than 60 years of age in ABR

Table 12.3 List of tree species recorded in Teak dominated habitat nearby Achanakmar Wildlife Sanctuary – site-III

Name of the Tree species	*Adults*	*Saplings*	*Seedlings*	*%Total*
Aegle marmelos	–	–	–	–
Anogeissus latifolia	–	1	–	0.1
Adina cordifolia	–	–	–	–
Bridelia retusa	–	–	–	–
Butea monosperma	–	2	–	0.2
Boswellia serrata	–	–	–	–
Bombax malabarica	–	–	–	–
Buchanania lanzan	–	3	–	0.3
Bauhinia malabarica	–	–	–	–
Cassia fistula	3	60	–	6.5
Dendrocalamus strictus	–	54	91	15.0
Diospyros melanoxylon	–	170	23	20.0
Dillenia pentagyna	–	–	–	–
Eugenia jambolana	–	–	–	–
Emblica officinalis	–	–	–	–
Lannea coromandelica	–	–	–	–
Lagerstroemia parviflora	–	1	–	0.1
Madhuca longifolia	–	–	–	–
Pterocarpus marsupium	–	–	–	–
Phoenix acaulis	–	–	–	–
Saccopetalum tomentosum	–	–	–	–
Schleichera oleosa	–	2	–	0.2
Semecarpus anacardium	–	–	–	0.0
Shorea robusta	–	100	16	12.0
Tectona grandis	255	166	–	43.6
Terminalia tomentosa	–	1	1	0.2
T. chebula	–	–	–	0.0
Unidentified 1	–	10	–	1.0
Unidentified 1	1	1	–	0.2
Unidentified 1	–	–	–	0.0
Unidentified 1	–	4	–	0.4
Unidentified 1	–	–	–	0.0

The Teak plantation site doesn't represent a net positive effect on the soil condition rather it shows lack of scientific understanding of the soil type in order to promote tree growth. The shifting of a calcicolous Teak tree to the habitat of calcifuge Sal tree is totally undesirable as the soil type has already determined the distribution of these two magnificent tree species (Teak and Sal). The rocks and the soil texture with high calcium content and high soil pH carried Teak tree and the acidic soil with low calcium carried Sal tree (Buch 1991).

Site IV

AMK of Madhya Pradesh is located in 22°40′′N, 81°45′′E / 22.67°N 81.75°E , at an average elevation of 1048 metres (3438 ft). The plateau of this area is made up of bauxite rocks that were mined out for commercial use. This is one of the pristine habitats in the country, which separates the rivers Narmada, Johilla and Sone, to drain into the Arabian Sea and to flow in the Bay of Bengal.

Plate 12.4 Sal tree cover aside origin stream line of Narmada river in Amarkantak

The vast upland area in AMK includes the uniformity of true Sal forest. In the past decades the area was also identified for planting other types of tree species such as *Pinus, Eucalyptus,* etc. (Plates 12.5 and 12.6), that requires more water to absorb all the nutrients from the soil without returning back to the soil. The large canopy gaps in plantation area allow penetration of too much light accelerating vertical movement of water and energy (nutrients). The area is comparatively important to ascertain the impact of vigorous growth of exotic weeds, viz., *Lantana* and *Parthenium.*

Plate 12.5 *Eucalyptus* tree cover aside stream line of Narmada river in Amarkantak

Plate 12.6 *Pinus* tree cover on way to Kapil-dhara in Amarkantak

Discussion

In the moist lowland forest of Achanakmar the indigenous semi-evergreen Sal and Sal-type tree species, seem to be older than dry upland forest in Amarkantak area. The structural and functional behaviour of large biomass high-grade timber species maintains ecological equilibrium to regulate energy cycle so that their own larger seed can also germinate. Still many tree species are found with their considerable numbers of seedlings and saplings population, however many of them do not mature in enough numbers to reach the juvenile and young stages (Table 12.1).

The *Anthocephalus cadamba, Bombax ceiba, Ficus bengalensis, F. religosa, Mangifera indica, Pithecolobium dulce, S. oleosa and Tamarindus indica,* are present in both the lowand and upland forests, but the core areas show less species as their numbers are consistent across the tree diversity. This attributes to either poorly regenerating capacity or failing to gain adult stage. Nevertheless many species such as *M. indica* and *Ficus* spp. rely mostly on biological factors and *A. pendula, Dalbergia sissoo, Tectona grandis* are almost missing in primary mixed forest as they are found to rely on edephic factor. *A. cadamba, Sterculia urence, Pterocarpous marsupium might* reach the verge of extinction.

The area with smaller elevation gradient in Achankmar, Chhaparwa and Lamni showcases mixed and monotonous species types, in which mixed forest area has plenty of species that have never been documented in respect to species area relationship. The tree species are more specific to narrower range of temperature and the high rainfall indicated how tree trunk density responded to the ground flora, epiphytes, climbers and tuberous forms. Virtually, the intact tree canopy cover maintains warm temperatures, moisture in soil to promote regeneration process of plant species. And a tree plant has to return back substantial amount of leaf litter to the nutrient deficient soil. The leaf litter enhances population of decomposer group that activate rapid decomposition and mineralization process. This is a kind of short circuited energy that runs consistently in between the soil and tree types.

Austin et al. (1996) explained richness of tree species positive to soil nutrients as the litter falling on ground surface holds water for its rapid decomposition. Leite & Furley, (1985) narrated that tree cover maintains higher level of organic matter to protect the soil from intense radiation and water run-off. In Central India, both the high temperature due to intense radiation, the water run-off due to lesser organic content, exist in a great way to realize that pioneer species are more suitable to non-pioneer ones. Still a very large number of native tree species (Table 12.1) are altogether better to survive against extremes of high temperature and to sustain the ecological equilibrium and agricultural production. Many pioneer tree species, i.e., Babul, Ber, Bel, Bamboo, Neem, Arjun, Tendu, Palas, etc., are capable of colonization rapidly even in degraded forest land and protect the soil surface from the radiation.

The area in between ACH and AMK includes a specific habitat of specific semi-evergreen tree species, Dhawra, Haldu, Babool, Pals and Saja tree as they are competent to grow in association with a tall, straight Sal tree which is competitively more superior. The Sal tree has extra dispersal adaptability to colonize and monopolizes monotonically with increasing elevation up to 1150 m (Swamy (2000). The nature of Sal tree in terms of leaf phenology is questionable due to evergreen v/s deciduous paradox (Troup 1921), while some workers have described Sal as a deciduous tree (Kirtikar and Basu (1975); Tiwari (1995), or semi-deciduous (Bor 1953), others have viewed it as an evergreen tree (Borchert 2000; Singh and Singh 1992). Joshi (198h0) has variously described Sal a deciduous tree or as border line between evergreen and deciduous plants. It is interesting to note that in annual cycles the entire population of the plant never became leafless, only about one-fifth individuals (variant), situated in relatively dry and shallow soil, became leafless for a short period in March.

Sundarapandian and Swamy, (1997); Ghate et al., (1998) stated that Sal, a semi-evergreen species, is hardly ever without green leaves, and if so only for a relatively short period. Singh et al. (2005), Agrawal et al. (2010) have described the forest of ACH as a deciduous forest type, under which the tree species hve to change their entire foliage every year. Probably, they have ignored uniformity of semi-ever greenness in Sal and Sal types trees and their extensive distribution from moist to the dry forests areas. Obviously, they were more concerned with numerical value of single Teak plant, which is of other type representing the secondary forest in some patches of ABR (Tiwari et al., 2009). Because, the small trees Palas and Tendu, the large deciduous trees Semal, Mahuwa, are not in uniform dominance and in intact canopy as Sal and Sal tree types representing the forest areas.

Although Teak is found in a variety of soils (alluvial, riverine , sandstone-derived) of natural forest and observed for variability in wood quality even within the Madhya Pradesh. As per the growth rate and quality of wood the Teak tree develops better in the deep, well-drained and fertile soils, especially on volcanic substrata such as igneous and metamorphic soils in certain parts of Central India. Seth and Yadava (1959) claimed that plantations of Teak tree completely failed when the tree was planted on low-lying, poorly drained land with clay soils. Bryce, (1966) alleged that planted Teak tree gives a light, weak and spongy wood of inferior quality than that of wood obtained from the natural primary forest.

Jose and Koshy (1972) pointed out that conversion of natural forest to Teak alters the physical soil conditions. They have pointed out that large canopy gaps allow penetration of too much light accelerating vertical movement of water and energy. In fact, land-use changes have harmed the forest areas more than the climate change. In this reference the Teak plantation site seems a new line thought of short-term gain like agricultural crops rather than a long-term thought of ecological sustenance. The fragmented site seems to be an ephemeral component which is not only against the utility of wildlife but also is more 'negative' than the protective values of native tree species of ABR. The variation in microclimatic conditions, i.e., soil moisture, humidity, light intensity and temperature described as to which type of tree species are being described in response to biodiversity and land-use changes.

The topographic heterogeneity also influences humidity, soil moisture and pattern of tree cover because, both the lowland and upland areas suffer with an excessive climatic demand of water (evapo-transpiration). Janzen (1967) narrated that species in lowland environments experience narrower range of temperature may reduce opportunity for dispersal across the unfavourable climatic conditions in upland regions. Odum (1971) pointed out the organic accretion influencing the tree species, but is less significant in reference to variation and the species richness.

In ABR, the local environmental conditions that have a direct physiological impact on the growth and development of plant species are assumed for the richness of tree species. Pitman et al. (2002) were of the opinion that the local species richness of an individual species is controlled by abiotic environmental conditions. O'Brien (1993, 1998) demonstrated a linear relationship of tree species richness with annual rainfall and with potential evapo-transpiration. The biosphere experiences much more severe drying in summer months. This may drive some species locally extinct in lowland, but capable of surviving there now have not yet returned due to slow migration rates. We see no reason why tree species in the site with their much less dramatic climatic history, should be any less at equilibrium with climate than tree species elsewhere.

Virtually, the biotic interactions are unexplored which facilitates the co-evolutionary diversification of populations and community in ABR. Despite the higher levels of defence mechanism in tree plants, there are higher rates of herbivory in the tropics (Coley & Barone 1996). Jeanne (1979) measured ant predation of wasp larvae and found a greater risk of predation at lower latitudes. Cornell and Lawton, (1992); Richardson et al., (2000b) were of the opinion that biotic interactions remain stronger in tropics than in the temperate regions. They have suggested that local biotic interactions such as competition, predation, and regional processes such as dispersal and speciation determine the species richness. Shukla (1997) evoked to determine the magnitude of biotic interactions against the mass tree felling in the name of insect infestation *Hoplocrambxy spinicornis* (Sal borer).

Takhtajan (1969), Wu (1980), Lidgard and Crane (1990) and many others expressed that angiosperm plants originated and diversified in tropics. Latham and Ricklefs (1993) demonstrated that many plant species of temperate regions are closely related with the tropical species. The tree species originated in tropical lowland area tend to subsequently expand their geographical ranges to the high latitudes without losing their tropical presence. Goldberg et al. (2005); Jablonski et al. (2006) were of the opinion that range expansions in many cases support determination of the latitudinal diversity gradient and its nature and strength.

Since the tree species have greater chances for evolution and physiological specialization in lower latitudes, therefore, these are older on an average relative to their species of temperate and polar areas. Wiens et al. (2006) made a detailed analysis of hylid frogs and narrated their origin and richness in South America. The abundance of an amphibian frog *Hyla* and an older tree species of Sal named Ram and Luxman for its giant height of 47 m and girth 4.85 m at Lamni range are tightly related with the time of their species colonization and explain the transitional linkage in between tropical and temperate latitudes.

CONCLUSION

In Maikal range, the topography and the local environmental conditions such as soil type, annual rainfall and potential evapo-transpiration showed a linear relationship against the richness of tree species. The tree species found in primary forest of ACH area are more specific to the temperature and the rainfall conditions as they stand at a narrow temperature range (10-35°C). The tree species in moist lowland forest in ACH leading more tree diversity than AMK is primarily due to accumulation of tree species. They are potential source relevant to the hypothesis of ancient origin. The Sal and its associated types of semi-evergreen tree species of primordial in origin have their own older variants on an average in ACH than the average relative to species in higher elevation of AMK. The ACH area has more than hundreds of endemic tree species of medicinal importance and for the richest collection of biological species particularly angiosperm plants.

We found that changes in land-use have worsened more rather than climate changes. Moreover, the changes over the total area of primary forests more or less are stable then the changes correspond to secondary vegetation. In reference to the biosphere biodiversity in general, the separation of tree species from its original environment (mixed and true Sal forest) area is not found significant. The removal of Sal and its associated types of tree plants and planting other types of tree types not only altered the species composition but, has also led to impoverishment of the area, where still energy is locked up in giant size tree plants and larger wild animals. Lastly, unless augmentation for regeneration of the constituent species is not maintained most of the species may not be available for extraction in future with sustainable forest management principle. It may be too late to turn back the clock for the trees, but doing nothing might cost us the eco-economics.

REFERENCES

1. Austin, M.P., Pausas, J.G. and Nicholls, A.O. (1996). Patterns of tree species richness in relation to environment in south-eastern New South Wales. *Aust. J. Ecol.* 21: 154-164.
2. Bal, S.S. (2002). Some Productivity Aspects of Ten Keystone Species of Similipal Biosphere Reserve. Ph.D. Thesis, Utkal University, Bhubaneswar.
3. Bor, N.L. (1953). *Manual of Indian Forest Botany*, Oxford University Press, Bombay, p. 441.
4. Borchert, R. (2000). Organismic and environmental controls of bud growth in tropical trees. In *Dormancy in Plants: From Whole Plant Behavior to Cellular Control* (eds Viemont, J.D. and Crabbe, J.), CAB International, Wallingford, pp. 87-107.
5. Bruijnzeel, L.A. (2004). Hydrological functions of tropical forests: not seeing the soil for the trees? Agriculture, Ecosystems and Environment 104: 185-228.
6. Bryce, JM. (1966). Mechanical properties of Tanzania-grown teak. Tech. Note Util. Sch. For. Div. Moshi No. 34.
7. Buch, M.N. (1991). The Forests of Madhya Pradesh, M.P. Madhyam, Bhopal.
8. Cornell, H.V. and Lawton, J.H. (1992). Species interactions, local and regional processes, and limits to the richness of ecological communities: A theoretical perspective. *J. Anim. Ecol.* 61, 1-12.
9. Champion, H.G. and Seth, S.K. (1968). A Revised Survey of the Forest Types of India. - Manager of Publication, Govt. of India Press, New Delhi.

10. Chakraborty Luna, Pawar, S.K. and Shukla, R.V. (1991). Effect of closure on soil properties and its fungus population in Sal forest. *Jour. Tropi. Forestry*, 7: 51-61.
11. Coley, P.D. and Barone, J.A. (1996). Herbivory and plant defenses in tropical forests. *Annu. Rev. Ecol. Syst.*, 27: 305-335.
12. De Bie S., Ketner, P. Paase, M. and Geerling, C. (1998). Woody plant phenology in the West African Savanna. *J. Biogeogr*, 25: 883-900.
13. Farrell, B.D., Mitter, C. and Futuyma, D.J. (1992). Diversification at the insect-plant interface: insights from phylogenetics. *Bioscience* 42: 34-42.
14. Gerhardt, K. and Hytteborn, H. (1992). Natural dynamics and regeneration methods in tropical dry forests - An introduction. *J. Vegetation Sci.*, 3: 361-364.
15. Gadgil, M. and Meher Homji, V.M. (1982). Conserving India's Biological Diversity: Indo US Binational workshop on Conservation of Biological Diversity, Department of Environment, Government of India, New Delhi.
16. Ghorpadé Kumar (2012). Butterflies (Lepidoptera—Rhopalocera) of the Achanakmar-Amarkantak Biosphere Reserve, in Chhattisgarh and Madhya Pradesh, with a synopsis of the recorded butterfly fauna of the eastern Central Highlands in India. Colemania, Number 26, pp. 1-38
17. Goldberg, E.E., Roy, K., Lande, R., and Jablonski, D. (2005). Diversity, endemism, and age distributions, in macro evolutionary sources and sinks. *Am. Nat.*, 165: 623-633.
18. Hewetson, C. (1956). Observations on the bird life of Madhya Pradesh. *J. Bombay Nat. Hist. Soc.* 53: 595-645.
19. Janzen, D.H. (1967). Why mountain passes are higher in the tropics. *Am. Nat.* 101, 233-249. doi:10.1086/282487. Cross Ref Web of Science
20. Jablonski, D. Roy, K. and Valentine, J.W. (2006). Out of the tropics: Evolutionary dynamics of the latitudinal diversity gradient. *Science*, 314: 102-106.
21. Joshi, H.B. (1980). *The Silviculture of Indian Trees*, Controller of Publications, New Delhi, p. 471.
22. Jose, A.I. and Koshy, M.M. (1972). A Study of the morphological, physical and chemical Characteristics of soils as influenced by teak vegetation: 1 *Indian For.* 6: 338
23. Jeanne, R.L. (1979). A latitudinal gradient in rates of ant predation. *Ecology*, 60: 1211-1224.
24. Jha, C.S. and Singh, J.S. (1990). Composition and dynamics of dry tropical forest in relation to soil texture. *Journal of Vegetation Science* 1: 609-614.
25. Kirtikar, K.R. and Basu, B.D. (1975). *Indian Medicinal Plants, Jayyed Press, New Delhi*, Vol. I, p. 838.
26. Krishnaswamy, V.S. and Mathauda, G.S. (1954). Phenological behaviour of a few forest spp. at New Forest, Dehra Dun. *Indian For.*, 80: 3-4.
27. Latham, R.E. and Ricklefs, R.E. (1993). Continental comparisons of temperate-zone tree species diversity. In: Species Diversity in Ecological Communities: Historial and Geographical Perspectives (Eds. Ricklefs R.E. and Schluter D.). University of Chicago Press, Chicago, IL, pp. 294-314.

28. Laurance, W.F. et al. (2002). Ecosystem decay of Amazonian forest fragments: a 22-year investigation. *Conserv. Biol.* 16: 605-618
29. Leite Laercio, L. and Peter A. Furley (1985). Land development in the Brazilian Amazon with particular reference to Rondônia and the Ouro Preto colonization project. In Hemming, John (Ed.) Change in the Amazon Basin: the frontier after a decade of colonization (pp. 119-139). Vol. II. Manchester University Press.
30. Lidgard, S. and Crane, P.R. (1990). Angiosperm diversification and Cretaceous floristic trends: a comparison of palynofloras and leaf macrofloras. *Paleobiology* 16: 77-93.
31. Mandal, S.C. (1984). Soil research in retrospect and prospect- acid soils of India: In soil Science in India Bull. No. 14 Indian Society of soil science, New Delhi, pp. 158-68.
32. Mishra, R. (1956). The vegetation of Amarkantak. Bulletin Botanical Society, University of Saugar, 8: 1-2.
33. Mishra, R. (1968). Ecology Work Book, Oxford & IBH Publishing Co., New Delhi.
34. Mishra, R.K., Upadhyay, V.P. and Mohanty, R.C. (2008). Vegetation ecology of the Similipal Bioshpere Reserve, *Applied Ecology and Environmental Research* 6(2): 89-99.
35. Mishra, R.K., Upadhyay, V.P. and Mohanty, R.C. (2003). Vegetation diversity of Similipal Biosphere Reserve, *E-Planet* 1(1): 4-9.
36. Uma Shankar, K.S. Murali, R. Uma Shaanker, K.N. Ganeshaiah, and K.S. Bawa, (1996). Extraction of non-timber forest products in the forests of Biligiri Rangan Hills, India. Economic Botany 50: 270–279.
37. Myers, N. (1990). The biodiversity challenge: expanded hotspots analysis. *The Environmentalist*, Vol. 10, No. 4, pp. 243-256.
38. Neelakandan, V.N., Mohan, C.N. and Sukumar, B. (2006). Development of bio-geographical information system: for conservation, monitoring, of biodiversity. *Curr. Sci.* Vol. 90, No. 3, pp. 444-450.
39. O'Brien, E.M. (1993). Climatic gradients in woody plant species richness towards an explanation based on an analysis of southern Africa's woody flora. *J. Biogeogr.* 20: 181-198.
40. O'Brien, E.M. (1998). Water-energy dynamics, climate, and prediction of woody plant species richness, an interim general model. *J. Biogeogr.* 25: 379-398.
41. O' Brien, E.M., Field, R. and Whittaker, R.J. (2000). Climatic gradients in woody plant (tree and shrub) diversity: waterenergy dynamics, residual variation, and topography, *Oikos* 89: 588-600.
42. Odum, E.P. (1971). Fundamentals of Ecology, W.B. Saunders Co., Philadelphia.
43. Overpeck, J.T., Bartlein, P.J. and Webb, T. (1991). Potential magnitude of future vegetation change in eastern North America: comparisons with the past. *Science* 254: 692-695
44. Pimm, S.L. et al. (1995). The future of biodiversity. *Science* 269: 347-350
45. Pimm, S.L. et al. (2006). Human impacts on the rates of recent, present, and future bird extinctions. *Proc. Natl. Acad. Sci. U.S.A.* 103: 10941-10946

46. Pitman, N.C.A., John W. Terborgh, Miles R. Silman, Percynu Nezv., David A. Neill, Carlos E. Cero´ N, Walter A. Palacios, Milton Aulestia (2002). A comparison of tree diversity tree species diversity in two upper Amazonian forests. *Ecology*, 83(11), pp. 3210-3224, Ecological Society of America
47. Prakash, R. (1992). Wildlife Working Plan. IInd Preliminary Proposal, Forest Department, Govt. of M.P., Bhopal.
48. Prasad, M.N.V. and Freitas, H. (2003). Metal hyperaccumulation in plants - *Biodiversity Prospecting for Phytoremediation Technology* Vol. 6, No. 3.
49. Ricklefs, R.E. and Miller, G.L. (2000). Ecology, W.H. Freeman & Company, New York.
50. Qian, H. and Ricklefs, R.E. (1999). A comparison of vascular plant taxonomic richness in China and the United States. *Am. Nat.* 154: 160-181.
51. Qian, H. and Ricklefs, R.E. (2000). Large-scale processes and the Asian bias in temperate plant species diversity. *Nature* 407: 180-182.
52. Rao, R. Shyama Prasad (2004). Water storage in Terminalia tomentosa, *Curr. Sci.* (87) 4, p. 416.
53. Ravindranath, N.H. and Sukumar, R. (1996). Impact of climate change on forest cover in India. *Commonwealth Forest. Review* 75 (1), pp. 76-79.
54. Ravindranath, N.H., Joshi, N.V. Sukumar, R. Saxena, A. (2006). Impact of climate Change on forests in India. *Curr. Sci.* Vol. 90, No. 3, pp. 354-361.
55. Richardson, D.M. Allsopp, N. D'Antonio, C.M. Milton, S.J. & Rejmanék, M. (2000). Plant invasions - *The Role of Mutualisms. Biol. Rev.*, 75: 63-93.
56. Saxena, H.O. and Brahmam, M. (1989). The Flora of Similipahar (Similipal), Orissa. Regional Research Laboratory, Bhubaneswar.
57. Schemske, D. (2002). Tropical diversity: patterns and processes. In: Ecological and Evolutionary Perspectives on the Origins of Tropical Diversity: Key Papers and Commentaries (Eds. Chazdon R. and Whitmore T.). University of Chicago Press, Chicago, IL, pp. 163-173.
58. Sengupta, G. (1977). A resume of botanical exploration and floristic studies in the Central Indian State of Madhya Pradesh. *Bull. Bot. Surv. India.* 19: 71-88.
59. Shafer, S.L, Bartlein, P.J. Thompson, R.S. (2001). Potential Changes in the distributions of western North America tree and shrub taxa under future climate scenarios. *Ecosystems* 4: 200-215.
60. Shukla, R.V. (1997). Why previously innocuous Sal borer insect now called Pest? M. P Chronicle - Sunday special Dec. 21, 1997.
61. Shukla, R.V. (2003). Biodiversity of Sal forest of Chhattisgarh with special reference to mushrooms. Project Report, Centre for Ecological Science, Indian Institute of Science Bangalore.
62. Shukla, R.V., Tripathi, A.M., Shukla, K., Tiwari, M. (2010) Diversity of Mushroom Research, Vol. 19 (1).

63. Agrawal, R., Dixit, B., Singh, L., and Ojha, B.M. (2010). Comoposition, Structure and floral diversity of forest communities of Achanakmar – amarkantak biosphere reserve: A comparison and conservation implication, International forestry and environment symposium University of Sri Jayewardenepura Sri Lanka
64. Singh, L. and Singh, J.S. (1991). Species structure, dry matter dynamics and carbon flux of a dry tropical forest in India. *Annals of Botany* 68: 263-273.
65. Singh, L. et.al., (2005) Diversity and Dominance of a Tropical Moist Deciduous Forest in Achanakmar Wildlife Sanctuary, Bulletin of the National Institute of Ecology 15: 1-9, Gupta et al. (Editors): Ecology and Environmental Management: Issues and Research Needs.
66. Singh, K.P. and Kushwaha, C.P. (2005). Paradox of leaf phenology: Shorea robusta is a semi-evergreen species in tropical dry deciduous forests in India, *Curr. Sci.*, 88: 1820-1824.
67. Singh, S.P. Tewari, J.C. Yadav, S. and Ralhan, P.K. (1986). Population structure of tree species in forests as an indicator of regeneration & future stability. *Proceeding Indian Academy of Science* 96(6): 443-455.
68. Singh, J.S. and Singh, S.P. *Forests of Himalaya*, Gyanodaya Prakashan, Naini Tal, 1992, p. 257.
69. Singh Lalji, Bhawana Sharma, Rashmi Agrawal, and Sunil Puri (2005). Diversity and Dominance of a Tropical Moist Deciduous Forest in Achanakmar Wildlife Sanctuary Gupta et al. (Ed.): Ecology and Environmental Management: Issues and Research Needs Bulletin of the National Institute of Ecology 15: 1-9, 2005 © National Institute of Ecology, New Delhi & Jaipur.
70. Specht, R.I. and Specht (1989a). Canopy structure in Eucalyptus dominated communities in Australia along climatic gradients *Acta Oecologia, Oecologia Plantrarum*, 191-213.
71. Specht, A. and Specht, R.I. (1993). Species richness and canopy productivity of Australion plant community. *Biodiversity and Conservation* 2: 152-167.
72. Stevens, G.C. (1989). The latitudinal gradient in geographical range: how so many species coexist in the tropics. *The American Naturalist*, 133, 240–256.
73. Sukumar, R. Dattaraja, H.S. Suresh, H.S. Radhakrishnan, J. Vasudeva, R. Nirmala, S. and Joshi, N.V. (1992). *Curr. Sci.*, 62, 608-616.
74. Sukumar, R. Suresh, H.S. Dattaraja, H.S. and Joshi, N.V. (1997). In *Forest Biodiversity Research, Monitoring and Modeling: Conceptual Background and Old World Case Studies*, (Eds. Dallmeier F. and Comiskey J.A.), Parthenon Publishing, Vol. I, pp. 529-540.
75. Sundarapandian, S.M. and Swamy, P.S. (1997). *Int. J. Ecol. Environ. Sci.*, 23, 363-379.
76. Sunderpandian, S.M. and Swamy, P.S. (2000). Forest ecosystem structure and composition along an altitudinal gradient in the Western Ghats, South India. *Journal of Tropical Forest Science* 12(1): 104-123.
77. Takhtajan, A. (1969). Flowering plants: origin and dispersal Smithsonian Inst. Press (translated by C. Jaffrey). Edinburgh: Oliver and Boyd, 310 pp.

78. Tilman, D. (1982). Resource competition and community structure. Princeton University Press, Princeton, NJ.
79. Tiwari, M., Shukla, K., Shukla, R.V., Tripathi, A.M. (2009). Prevalance of tribal mushroom in Sal forests of Chhattisgarh. Phytopathological Research Jour. of Mycopathological Research, 47 (2): 111-117.
80. Tiwari M., Shukla, R.V., Shukla, K.K. (2010). Taxonomic studies on agarics-the genus Amanita in Chhattisgarh Mushroom Research Accepted Mushroom Research Vol. 20(2) pp. 27-32
81. Tiwari, D.N. (1995). *A Monograph on Sal* (*Shorea robusta* Geartn. f.), Vedams eBooks (P) Ltd, New Delhi, p. 277.
82. Troup, R.S. (1921). *The Silviculture of Indian Trees*, Clarendon Press, Oxford, Vol. I, p. 336.
83. Uma Shaanker, R., Ganeshaiah, K.N., Murali, K.S., Uma Shankar and Bawa, K.S., 1998, Extraction of Non-Timber Forest Products in the forest of Biligiri Rangan Hill, India. 4. Impact on floristic diversity and population structure in a thorn scrub forest. Economic Botany., 52 (3): 302-315.
84. Uma Shankar (2001). A case of high tree diversity in a sal (Shorea robusta)-dominated lowland forest of Eastern Himalaya: Floristic composition, regeneration and conservation *Curr. Sci.* 81: 776-786.
85. Whitmore, T.C. (1984). Tropical Rain Forest of Far East. (2nd edition). Claredon Press, London.
86. Wiens, J.J., Graham, C.H. Moen, D.S. Smith, S.A. and Reederm, T.W. (2006). Evolutionary and ecological causes of the latitudinal diversity gradient in hylid frogs: tree frogs unearth the roots of high tropical diversity. *Am. Nat.*, 168: 579-596.
87. Wright, S.J. and Muller Landau, H.C. (2006). The future of tropical forest species *Biotropica* 38: 287-301.
88. Wu, Z.Y. (1980). The vegetation of China. Science Press, Beijing China.

13

Assessing Carbon Sequestration Potential of Bamboos from Manipur

Asha Gupta

Centre of Advanced Study, Department of Life Sciences, Manipur University, Manipur, India

E-mail: ashaguptamu@gmail.com

ABSTRACT

The world's forest ecosystems are estimated to store some 638 billion tonnes of carbon (C); more than the amount of C in the entire atmosphere. Deforestation is considered a major contributor to global warming and is often cited as one of the major factors responsible for the enhanced greenhouse effect. Tropical forests have the greatest potential to sequester carbon, primarily through reforestation and conservation of existing forests. Green India Mission puts the "greening" in the context of climate change adaptation and mitigation meant to enhance ecosystem services like carbon sequestration and storage in forests and other ecosystems, along with provisioning services like fuel, fodder, small timber and NTFPs. Bamboo is an environment-friendly plant and net carbon sink.

Bamboo plants grow fast, propagate rapidly and can be easily regenerated, hence bamboo can help in an increase in C stock in forests if plantations are spread over a large deforested area. Plantation and natural expansion of bamboo forests can contribute significantly to C storage. Bamboo can play a significant role in linking climate change mitigation to sustainable economic development in the developing world. Steps should be taken to protect the bamboo forest from being a carbon source. Application of modern engineering techniques can make bamboo products more durable with better storage of carbon for longer periods. Our study from Manipur state provides evidence that the bamboo plantation helps improve carbon absorption in vegetation and soil, reduces the emission of carbon and extends the storage time of carbon in soil.

INTRODUCTION

The world's forest ecosystems are estimated to store some 638 billion tonnes of carbon (C); more than the amount of C in the entire atmosphere. The overall C stored in living biomass decreased from 1990 to 2005, mainly due to deforestation in South and Southeast Asia, Western and Central Africa, and South America, although it remained relatively constant in Oceania, and even increased in East Asia, Europe and North and Central America. Deforestation is considered a major contributor to global warming and is often cited as one of the major factors responsible for the enhanced greenhouse effect.

Forests cover about one-fifth of India's land mass and comprise single largest land-based resource that has the potential of reducing poverty of the people due to the high dependence of people on forest resources, ensuring environmental sustainability. Tropical forests have the greatest potential to sequester carbon primarily through reforestation and conservation of existing forests.

Green India Mission puts the "greening" in the context of climate change adaptation and mitigation meant to enhance ecosystem services like carbon sequestration and storage in forests and other ecosystems, along with provisioning services like fuel, fodder, small timber and NTFPs. The mission aims at addressing climate change by enhancing carbon sinks in sustainably managed forests and other ecosystems; enhancing the resilience and ability of vulnerable species/ecosystems to adapt to the changing climate; and enabling adaptation of forest dependent local communities in the face of climatic variability.

NAPCC addresses the urgent and critical concerns of sustainable development and identifies the close linkage of the economy with its natural resource base, and cautions that the climate-sensitive sectors such as forestry may face a major threat and will have repercussions for livelihoods of people, in general, and forest dependent communities, in particular.

Climate impact assessments using BIOME-3 model and climate projections for the year 2085 show that 77% and 68% of the forested grids in India are likely to experience shift in forest types under A2 and B2 scenario. Indications show a shift towards wetter forest types in the NE region and drier forest types in the NW region in the absence of human influence. Increasing atmospheric CO_2 concentration and climate warming could also result in doubling and nearly 70% increase in net primary productivity under A2 and B2 scenario respectively.

Bamboos are some of the most productive and fastest growing species on the planet. Taxonomically, bamboo is a grass, with characteristics like fast growth, adaptability, resilience and substantial biomass production. Bamboo is an environment-friendly plant and net carbon sink. Bamboo plantation helps to reestablish forestry, reduces the emission of carbon and enhances the SOC and extends the storage time of carbon in soil. Many species of bamboo put on biomass at rates, which are far higher than for most tree species including important timber species like teak, have significantly higher carbon sequestration potential when measured over a rotation period.

BAMBOO DISTRIBUTION

Bamboo with 60 to 90 genera and over 1200 species is mostly distributed in tropical and subtropical zones over 37 million ha. between 46 North and 47 South latitude in Asia, Africa and America. Over 80% of the total area covered by bamboo with about 192 species of bamboos is located in Asia, and over 63% of bamboo resources are privately owned (FAO/INBAR). 15 genera and 115 species of bamboos from India have been reported. According to Tripathi (2008), the North-Eastern region holds more than two-thirds of the total growing stock of the country.

India is rich in genetic resource of bamboo present over 12.5% of forest area. National Mission on Bamboo Application (NMBA), National Bamboo Mission (NBM) and Department of Biotechnology (DBT) Govt. of India, are promoting cultivation and improvement of 15 industrially important bamboo species in the country (Swarup and Gambhir, 2008). India is second after China in bamboo resource and has 125 species in 23 genera spread over 8.96 million hectare area in 21 states and union territories, which is 12.8% of total forest area (Anonymous, 2003). Demand of bamboo is estimated to be 26.6 million tonnes and supply is only 13.47 million tonnes per year (Anonymous, 2003).

BAMBOO PRODUCTS

Bamboo culm with high strength, flexibility, and hardness compared with timber, is an ideal raw material for construction, handicraft, agriculture and fish industry. Longevity of bamboo products may determine the carbon sink function. It is important to produce durable and medium-longevity products during bamboo processing. Culm modification and bamboo preservation may improve the longevity of bamboo products. New uses of bamboo, particularly as a substitute of wood, housing sector and value added product will increase further demand of bamboo (Pandey, 2008). Bamboo can be used as a substitute for commercial timber, and by this India can save approximately 10,000 crore rupees annually that is currently used for importing timber.

BAMBOOS TO BE CONSIDERED AS FOREST TREES

Though bamboo is a grass, the following are the reasons to consider bamboo as forest tree:

- Bamboos are recognized as integral parts of forests in most countries including India and China. In India this recognition has also been accorded national legal validity under the Indian Forest Act.
- Bamboos sequester significantly higher amount of carbon than the most productive of tree species and are often used in long life products that can keep the carbon away from atmosphere for a long time.
- Bamboos have significant food value and thus have the potential of making CDM projects enhance food security in contrast to other forestry crops that tend to reduce food security.

To increase the possibility of carbon re-absorbtion, many people are directing their attention to bamboo as an effective alternative to reduce carbon dioxide.

BAMBOO'S ECONOMIC AND ENVIRONMENTAL POTENTIAL

Widely known as a "poor man's timber", about 1.5 billion people depend on bamboo for their daily lives. Over 20 million tonnes of bamboo are collected and utilized annually. Global bamboo trade is estimated to be between 1.5 and 2.5 billion USD. Bamboo contributes between 4-7% of the total tropical and subtropical timber trade. Known more for their uses in the industry, bamboos have an additional usage in utilization of their young shoots as food. Besides, several industrial uses the young shoots of bamboos are used as food. Juvenile shoots have high nutritive value, besides they are very delicious. Most bamboo species produce edible shoots, fibre and timber, but less than 100 species are commonly grown or utilized for their edible shoots (Midmore 1998; Collins and Keilar 2005). In India, the consumption of tender shoots is confined mainly to the North-Eastern states of the country where it is a part of the traditional cuisine. Bamboo shoots are nutritionally rich in proteins, carbohydrates, dietary fibre, and amino acids and low in fats and cholesterol; the juvenile younger shoots being much richer in nutrients than the fermented and canned form of shoots (Nirmala et al., 2008). The freshly emerging juvenile shoots are more nutritionally superior than the older emerged shoots (Nirmala et al., 2007). There are a large number of species whose shoots possess an acrid taste due to the presence of cyanogenic glucosides and have to be boiled in water for varying periods of time to make them fit for consumption. The species employed for commercial production of edible bamboo shoots belong to a few genera like *Bambusa, Dendrocalamus* and *Phyllostachys* (Kleinhenz et al., 2000), which are consumed in different forms as fresh, fermented, pickled, dried or canned in East and South East Asia,including the major countries of Japan, China, Korea and Indonesia.

More people are getting interested in planting bamboo for various reasons (e.g., bambusetum development and tourism culm production, landscape planning, control of soil erosion, carbon sequestration, food production, etc.). Consequently, the demand for planting materials has increased tremendously. There is a great need to establish more bamboo nurseries to produce propagules in large quantities for many ecological services through plantation establishment, watershed protection, carbon sequestration, land and water resource management, riverbank stabilization and the reduction of the incidence of landslides. Bamboo plants grow fast, propagate rapidly and can be easily regenerated, and hence they can help in increasing C stock in forests if plantations are spread over a large deforested area. Plantation and natural expansion of bamboo forests can contribute significantly to C storage. Most of the industrially important bamboo species have long flowering cycle (30 to 120 year) which is a limiting factor for planting programme. In addition, sporadic flowering is uncertain, and short viability of seeds further restrict the availability of seeds as and when required for planting programmes. Vegetative propagation through offset cutting, rhizome splitting, culm and branch cuttings is successful in bamboo species (Banik, 1994; Seethalakshmi et al., 2008).

BAMBOO CARBON FOR DEVELOPMENT

Bamboo is very efficient in the conversion of solar energy and can also be easily managed. Bamboo forests contribute greatly to C stock as an important C sink on earth. In total, there are 22 million ha of bamboo forests in the world (making up about 1% of the tropical and subtropical woody forests in the world). It is apparent that bamboo is an excellent wood substitute, with increasing production, while the world forest is shrinking.

More than 1500 applications of bamboo have been documented worldwide of which major ones include use as building material, agriculture implements, furniture, musical instruments, food items, handicraft, raw materials for large bamboo based industries like paper and pulp packing (Kishwan and Goyal, 2006). It has potential to provide benefits over time and space for market products such as medicine, food and charcoal; there should be understanding of constraints and trade-offs to planting and maintaining the plantations and look for possibilities and challenges to, engaging with carbon markets. In addition, the bamboo forest is included in the list of eligible afforestation and reforestation projects under the Clean Development Mechanism (CDM) as per Lobovikov (2010).

Because of bamboo's faster growth and shorter rotation compared with most tree species, conversion of bamboo into biochar can increase the residence time of carbon in the soil and other products (to more than 1000 years) and can reduce emissions of other greenhouse gases such as methane and nitrous oxide from the soil (Lehmann and Joseph, 2009; Yanai et al., 2007). This means that bamboo biochar is a potential carbon sink and that creating bamboo biochar can effectively store large quantities of carbon.

Bamboo in Manipur

In Manipur (latitude 23° 50′ N to 25° 42′ N latitudes and 92° 58′ E to 94° 45′ E) bamboo culture remains an essential part of daily life and civilization. Bamboo is an unique and important component of homestead gardens and landscapes. With better physical (mechanical) properties like high tensile strength, high flexibility, and light weight than many timbers, it is widely used as a building material in rural Manipur. Bamboo buildings are usually cheaper, lighter, and more earthquake-resistant. Bamboo-based cottage industry provides jobs and supplements income for the poor people of the state (Devi and Singh, 2007). Bamboo shoots have always been a common ingredient in Manipuri delicacies.

According to the State of Forest Report, 2009 by Forest Survey of India, Dehra Dun, the forest cover of Manipur is 17,280 km^2. which is 77.40% of the total geographical area of the State. At present about one lakh people depend on the forests for their survival. From the point of view of composition, forests can be classified into three types: coniferous (16.11 per cent of total forested area), non-coniferous (62.32 per cent of forest area) and bamboo brakes (21.57 per cent of total forest area). While the bamboo brakes and half of the coniferous forests (covering 2442 km^2) are accessible to human interventions and pressures. Bamboo brakes and bamboo mixed with scrubs of large extent occur in the western, south-western and north-western parts of the state. Smaller areas of bamboo brakes occur almost all over except at altitudes 1700 m above MSL. Most of the Cachar tropical semi-evergreen type of forests in the western part of the state have been hacked down for jhum and the trees have given way to a dense growth of Melocana bamboo interspersed with same clump forming bamboos and rare trees here and there. Bamboos also occur as an understory in tree forests. The percentage occurrence of bamboo in different districts of Manipur varies as: 4%(Senapati), 2.8% (Ukhrul), 29.8%(Chura chandpur), 28.4% (Tamenglong), 80.4% (Jiribam), 13.4% (Chandel), 1.9% (Thoubal), 1.6% (Bisnupur) and 0.7% (Imphal East and West) of their respective geographical area.

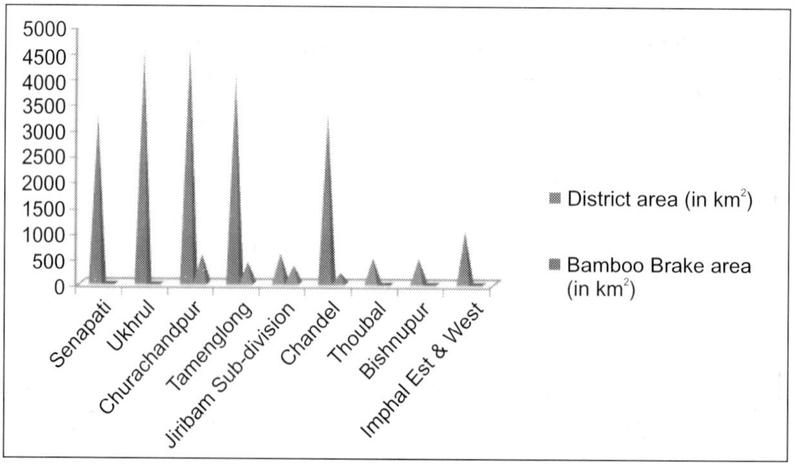

Fig. 13.1 Distribution of bamboo in Manipur

Bamboo Diversity of Manipur

A number of reports (Bhaumik et al., 2006; Deb, 1961; Devi and Sharma, 1993; Devi and Singh, 2007; Karthikeyan, 1983; Naithani et al., 2010; Sharma, 1982 and Shukla, 1996) are available on bamboo resources of Manipur. As per Devi and Sharma (1993), there are 53 species of bamboo under 10 genera which are distributed in various climatic zones of Manipur. Our recent study points out the occurrence of 45 species of bamboo growing in hills and plains of Manipur. These include 13 species of *Bambusa* along with one variety, 4 of *Cephalostachyum,* 1 of *Chimonobambusa,* 1 of *Chimonocalamus,* 9 of *Dendrocalamus,* 2 of *Drepanostachyum,* 2 of *Gigantocloa,* 1 of *Himalayacalamus,* 2 of *Melocalamus,* 1 of *Melocanna,* 1 of *Neomicrocalamus,* 1 of *Pseudostachyum,* 3 of *Schizostachyum,* 1 of *Thyrsostachys* and 2 of *Yushania.*

Geographical location, physiographical features like altitudinal variation, varied climatic zones and edaphic factor lead to bamboo diversity. During 2012-2013, NBM (National Bamboo Mission) plantation sites were visited and species composition and their performance in three different hill districts of Manipur recorded. Analysis of growth characteristics of selected species by established vegetational analysis norms was done. Recording of species specific variability in growth variables like: (i) no. of culms per clump, (ii) height of culms, (iii) nature of nodes, (iv) length of internode, (v) branching pattern and (vi) sheath characteristics were tabulated. Harvest method for biomass study in the age series was used.

Dendrocalamus hamiltonii

Bambusa tulda

Bambusa balcooa

Bambusa nutans

Bambusa bamook

Melocanna baccifera

Fig. 13.2 Bamboo species found in Manipur

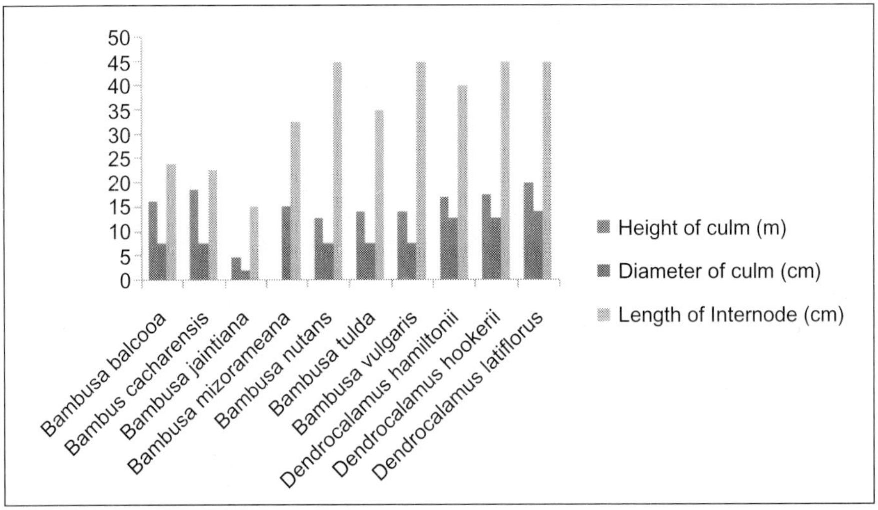

Fig. 13.3 Bamboo species found in Manipur Common bamboo species showing average growth characteristics in NBM plantations of Manipur

Bamboo offers a good material for forestry options in Manipur, specially its role in carbon sequestration and mitigation of climate change is being worked out. At present, the information on carbon capture and storage in the forest ecosystems of North-East India, in general and Manipur, in particular, is limited.

Carbon stock under seven species from Manipur:

1. *Bambusa tulda*
2. *B. cacharensis*
3. *B. mizorameana*
4. *Dendrocalamus hamiltonii*
5. *Bambusa balcooa*
6. *Melocanna baccifera*
7. *D. strictus*

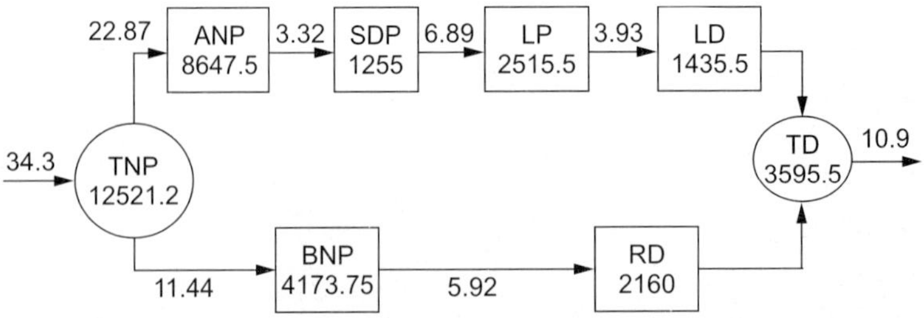

Fig. 13.4 Annual Carbon dynamics of bamboo plantation at Scenapati, Manipur. Compartments – $kgha^{-1}$, Accumulation and disappearance rate – $kgha^{-1}$ day^{-1}.

The carbon input from the total net primary production contributed 66 and 34% from the above and below ground parts respectively in the bamboo plantation ecosystem. The root respiration contributed 24% to the soil CO_2 flux and 76% by microbial respiration. The soil respiration was positively correlated with biomass, rainfall, air temperature, soil temperature and soil moisture. Annually 71.3% carbon still stored in the soil is not only utilised by soil microorganisms but it also maintains the suitability of the soil for sustaining the plant growth in the next generation of bamboo plantation ecosystem. AGC and SOC exhibited different patterns under different species management options.

Microbial biomass estimated under three management regimes, viz., bamboo forest, bamboo plantation and traditional bamboo grove from Bethel (at longitude 93° 49′ 21.2″ and latitude 24° 53′ 22.1″ at 764 m elevation) located in Senapati District of Manipur, NE India revealed that microbial biomass represents 83.33 to 85.15% of total soil carbon, 9.69 to 10.19% of total nitrogen and 4.16 to 5.13% of available phosphors up to 30 cm depth. The trend in microbial ratio showed considerable oscillations and followed significant seasonal variations ($p<0.05$) with increase in concentrations in the surface layer of soils. However, on an average, with remarkable little variation across three management regimes, the result of this preliminary investigation becomes important in that it demonstrates that the plantation of bamboo in denuded hill slopes of Manipur presents its ameliorative role by providing assimilable input of organic matter and moistening of dry soil, triggering a flush of microbial activity.

Our study provides evidence that the bamboo plantation helps improve carbon absorption in vegetation and soil, reduces the emission of carbon, and extends the storage time of carbon in soil.

Bamboo furniture *Bamboo basket*

Bamboo handicraft *Bamboo fishing trap*

Fig. 13.5 Bamboo products of Manipur

CARBON TRADING AND CREDITS

Bamboo can play a significant role in linking climate change mitigation to sustainable economic development in the developing world. As bamboo is a part of the world's forests and forest industries, it does qualify for carbon credits that may trigger creation of otherwise marginal bamboo plantations for processing, jobs and wealth generation. However, bamboo forests constitute an important livelihood source for rural people; the current extent of bamboo forests and area of potential distribution justifies amending the IPCC guidelines and additional methodological tools to allow for the inclusion of bamboo in carbon schemes (Lobovikov et al., 2009). To make this happen, more insights are needed in the potential contribution of bamboo to mitigating climate change. Due to its fast growth, bamboo is widely regarded as an ideal plant to sequester carbon, and is expected to play a bigger role in mitigating the impact of future climate change.

CONCLUSION

Bamboo, an indigenous grass, has proved to be a successful agroforestry plant for Manipur. The nutritive leaves can be used as fodder or as mulch. It tends to increase carbon stocks both above-ground and in the soil and improves soil water retention and nutrient status. Its short growth cycle makes it a great replacement for our slow growing forest that is being exploited for various reasons, and thus bamboo reduces our carbon footprint and helps fight global warming. Bamboo is an excellent species for agroforestry with high food and carbon values, thus it has significant potential of making CDM projects and enhance food security. The proliferating bamboo roots enhance below-ground biomass and soil porosity and, consequently, aeration, percolation, humidity and soil biodiversity leading to faster accretion to the below-ground biomass and soil organic carbon, and resulting climate change mitigation advantage.

It would open up opportunities for CDM projects for small-scale farmers, and for the first time the CDM projects can be "rural lifestyle projects" in which the plantations provide many necessities of rural life of the plantation owners besides providing carbon services to the globe. In Manipur, one of the key drivers of deforestation and landscape degradation is the demand for cheap energy. Firewood and charcoal account for between 60% and 85% of primary energy demand according to a survey. A farmer may require only 0.25 ha of land to sustainably maintain his household's primary energy demand, hence incentives to reforest or maintaining bamboo plantations may be widely deployed as an efficient and cost effective way of mitigating climate change. In addition, converting bamboo into charcoal as a substitute for fossil fuels could also provide additional opportunities to mitigate climate change. The carbon sequestration potential of the bamboo forest also depends on the life span of bamboo products. The carbon stored in these products will be released into the atmosphere when these products are either biologically degraded or burnt. Thus, producing more durable bamboo products will be a good way to prolong carbon storage. In addition, converting bamboo into charcoal as a substitute for fossil fuels could also provide additional opportunities to mitigate climate change. Bamboo harvesting and processing is mostly carried out by hand in cottage or village industries with technologies requiring little fossil fuel input, thus increasing the net carbon sink.

Steps should be taken to protect the bamboo forest from being a carbon source. Application of modern engineering techniques can make bamboo products more durable with better storage of carbon for longer periods. Thus, bamboo can contribute greatly to a future low-carbon society and also help in conserving and restoring the forest resources.

Manipur's long history of bamboo cultivation and use and its vast bamboo forest area give the country a unique opportunity to develop bamboo plantations and bamboo industry. With state government and the public paying more attention to bamboo plantations, improving bamboo's already high productivity and developing more sustainable management practices and products will allow bamboo to play an increasingly important role in mitigating climate change.

REFERENCES

1. Anonymous (2003). National Mission on Bamboo Technology and trade development. Published by Planning Commission, Govt. of India. p. 176.
2. Banik, R.L. (1994). Review of conventional propagation research in bamboo and future strategy. INBAR technical report No. 5-constraints to production of bamboo and rattan, INBAR, New Delhi, pp. 115-142.
3. Bhaumik, M., Mao, A. and Naithani, H.B. (2006). A note on the sporadic flowering of *Sinarundinaria Rolloana* (Gamble) Chao and Renv. (Poaceae-Bambusoideae) from Dzuko valley, Nagaland. *Indian Forester*. (13297): 906-907.
4. Collins, R.J. and Keilar, S.M. (2005). The Australian Bamboo Shoot Industry: A supply chain approach. Kinston, ACT: Rural Industries Research and Development Corporation.
5. Deb, D.B. (1961). Monocotyledonous plants of Manipur Territory. *Bull Bot Survey of India* 3: 115-138.
6. Devi Dhanapati L. and Devi Th. Sobita (1995). Ethnobotanical study of bamboos in Manipur. Keishamthong, Irom Pukhri Mapal, Imphal, Manipur. p. 69.
7. Devi, S.T and Sharma, G.S. (1993). Chromosome numbers in some bamboo species of Manipur. *BIC India Bulletin* 3(1): 16-21.
8. Devi, S.T. and Singh, T.B. (2007). Bamboo Resources of Manipur in biodiversity conservation and legal aspects (Ed.) A.K. Kandya and Asha Gupta. Aavishkar Pub: 90-100.
9. Karthikeyan, S. (1983). Endemic grasses of India with an emphasis on rare, indeterminate and extinct species. S.K. Jain and R.R. Rao (Eds.) Threatened Plants of India. 240-250, BSI, Howrah.
10. Kishwan, J. and Goyal, A.K. (2006). Gregarious flowering of bamboos in North-east India, need for Intensive Management. *Indian Forester*. 132(5): 525-533
11. Kleinhenz, V., Gosbee, M., Elsmore, S., Lyall, T.W., Blackburn, K. Harrower, K. Midmore, D.J. (2000). Storage methods for extending the shelf-life of fresh bamboo shoots [*Bambusa oldhamii* (Munro)]. *Postharvest Biology and Technology*, 19, 253-264.

12. Lehmann, J. and Joseph, S. (2009). Biochar for environmentalmanagement: Science and Technology. London: Earthscan.
13. Lobovikov, M. (2010). Bamboo: its potential role in climate change. *Non-wood News*, 20: 12-14.
14. Lobovikov, M., Paudel, S. Piazza, M. Ren, H. and Wu, J.Q. (2007). World bamboo Resources: A thematic study prepared in the framework of the Global Forest Resources Assessment 2005. FAO, Rome.
15. Midmore, D. (1998). Culinary bamboo shoots. In Hyde, K.W. (Ed.) The New Rural Industries. Canberra, Rural Industries Research and Development Corporation. pp. 188-196.
16. Naithani, H.B. Bisht, N.S. and Singsit, S. (2010). Distribution of bamboo species of Manipur. Forest Department of Manipur.
17. Nirmala, C. David, E. and Sharma, M.L. (2007). Changes in nutrient components during ageing of emerging juvenile bamboo shoots. *Int. J. Food Sci. Nutr.*, 58, 612-618.
18. Nirmala C., M.L. Sharma and E. David (2008). A comparative study of nutrient components of freshly harvested, fermented and canned bamboo shoots of *Dendrocalamus giganteus* Munro. Bamboo Science and Culture: *J. Amer. Bamboo Soci.*, 21(1), 33-39.
19. Pandey, S. (2008). New generation value added products of bamboo. In: proceeding of the international conference on improvement of bamboo productivity and marketing for sustainable livelihood, April 15 to 17th, 2008, New Delhi, pp. 76-91, Published by the Cane and Bamboo Technology Centre, Guwahati.
20. Seethalakshmi, K.K. Jijeesh, C.M. and Unni, K.K. (2008). Traditional methods for bamboo propagation in nursery. In: proceeding of the international conference on improvement of bamboo productivity and marketing for sustainable livelihood, April 15 to 17th, (2008), New Delhi, pp. 28-36, Published by the Cane and Bamboo Technology Centre, Guwahati.
21. Sharma, Y.M.L. (1982). Some aspects of bamboos in Asia and Pacific. FAO Bangkok, Thailand.
22. Shukla, U. (1986). Grasses of North-Eastern India, pp. 1-404. Scientific Publ, Jodhpur.
23. Swarup, R. and Gambhir, A. (2008). Mass production, certification and field evaluation of bamboo planting stock produced by tissue culture. In: proceeding of the international conference on improvement of bamboo productivity and marketing for sustainable livelihood, April 15 to 17th, (2008), New Delhi, pp. 22-27, Published by the Cane and Bamboo Technology Centre, Guwahati.
24. Tripathi, Y. (2008). Bamboo entrepreneurship – Opportunities for rural employment. *Indian Forester*. 134(9): 1199-1210.
25. Yanai, Y. Toyota, K. and Okazaki, M. (2007). Effects of charcoal addition on N_2O emissions from soil resulting from rewetting air-dried soil in short-term laboratory experiments. *Soil Sci. Plant Nutr.* 53(2): 181-188.

14

Vegetation Analysis of Teak-Dipterocarpus Forests from Indo-Burma Hotspot with a Note on Distribution and Diversity

N. Ranita[*], A.K. Kandya[**] and Asha Gupta[†]
[*]Department of Life Sciences, Manipur University, Canchipur-795003, Imphal, Manipur, India
[**]Department of Botany, Dr. H.S. Gour University, Sagar, Madhya Pradesh, India
[†]Centre of Advanced Study, Department of Life Sciences, Manipur University, Manipur, India
E-mail: ashaguptamu@gmail.com

ABSTRACT

Forest degradation is usually accompanied by species extinction, reduction in biodiversity and decrease in primary productivity. Consequently, there is a growing interest in quantifying habitat characteristics like forest structure, floristic composition and species richness in Indian forests. The analysis of tree community structure and diversity is still challenging for researchers in tropical ecology.

The present study focuses on variation in vegetative parameters in forest sites and diversity measures in them from Indo-Burma border forests, at present under much anthropogenic pressure. Species diversity is an important attribute of a natural community that influences functioning of an ecosystem. The number of tree species was found to be lower than the number of species reported by several workers in other tropical forests. Importance Value Index (IVI) of plant species indicates the richness of species in a community. Higher IVI indicates high ecological significance and it is a reflection of better natural regeneration status of plant species. Information on distribution and abundance of tree species and biodiversity is highly needed for the conservation and management of tropical forest.

The study is of much significance as Indo-Burma hotspot has a major contribution to the mega-biodiversity of North-East India and people here rely on jhumming or shifting cultivation depending upon forest productivity to maintain livestock and soil fertility. Sustainable management of valuable forest resources can only be ensured by the safeguard and maintenance of the valuable biodiversity present in this biodiversity hotspot.

INTRODUCTION

Tropical forests are often referred to as one of the most species diverse terrestrial ecosystems, and generate a variety of natural resources to help sustain the livelihood of local communities (Kumar et al., 2006). These forests are disappearing at alarming rates owing to deforestation for extraction of timber and other forest produce (Murphy & Lugo, 1986; Raghubanshi & Tripathi, 2009). Trees form the major structural and functional basis of tropical forest ecosystems and can serve as robust indicators of changes and stresses at the landscape scale. Tree species diversity in the tropics vary dramatically from place to place (Pitman et al., 2002)

Tropical forests cover only 7% of the earth's land surface but harbour more than half of the world's species. However, these species are disappearing at an estimated annual rate of 0.8–2.0% (May and Stumpf, 2002).

The exploitation of natural resources by the local populations has resulted in depletion of the biodiversity of forest communities (Ramakrishnan, 2003). Forest degradation is usually accompanied by species extinction, reduction in biodiversity and decrease in primary productivity. Consequently, there is a growing interest in quantifying habitat characteristics like forest structure, floristic composition and species richness in Indian forests (Nirmal Kumar et al., 1999, 2000, 2001, 2002; Yadav & Yadav, 2005).

The analysis of tree community structure and diversity is still challenging for researchers in tropical ecology (Bawa et al., 2004).

As Indo-Burma hotspot has a major contribution to the mega-biodiversity of North-East India, including Manipur, the present study focuses on variation in vegetative parameters in forest sites and diversity measures in them from Indo-Burma border forests, at present under much anthropogenic pressure.

STUDY AREA

Teak-Dipterocarpus forests, distributed on Indo-Burma border, about 110 km from the Imphal city, are located between 94°15′ and 94°20′E longitudes and 24°15′ and 24°25′N latitudes. They have an altitudinal range between 300 – 1200 m (average 350 m MSL) and experience a tropical climate.

The maximum temperature ranged from 23.5°C to 35.6°C and minimum varied from 8.6°C to 23.5°C in the study area. The maximum rainfall recorded was 195.20 mm (June) in the hydrological year 2007–2008. The relative humidity during the study period ranged between 55% and 84%. Figure 14.1 shows the location of the study area, and ombrothermic diagram is presented in Fig. 14.2.

MATERIALS AND METHODS

Intensive survey was done during July 2007 to June 2009 for plant collection and observations.

The phytosociological investigation was done by laying quadrats of size 10 × 10 m^2 for trees, and 5 × 5 m^2 for shrubs. Ten quadrats were placed randomly in both the sites for each category. Girth at breast height (1.37 m from ground level) of all the trees in each quadrat was measured and recorded. The vegetation data were quantitatively analysed for abundance, density, frequency and basal cover (Curtis and McIntosh, 1950, and Misra, 1968). Importance Value Index (IVI) for all species was worked out by summing their relative values. Dominance-Diversity Curves for trees and shrubs at two forest sites were drawn after Whittaker (1969). The ratio of the abundance to frequency was calculated to interpret the distribution pattern of the species (Curtis and Cottam, 1956).

Diversity Index H was also calculated from density data after Shannon and Wiener, (1963) whereas, Index of co-dominance (λ) was given after Simpson's Index (1949). Hill Diversity Numbers and Evenness indices were calculated after Pielou (1966).

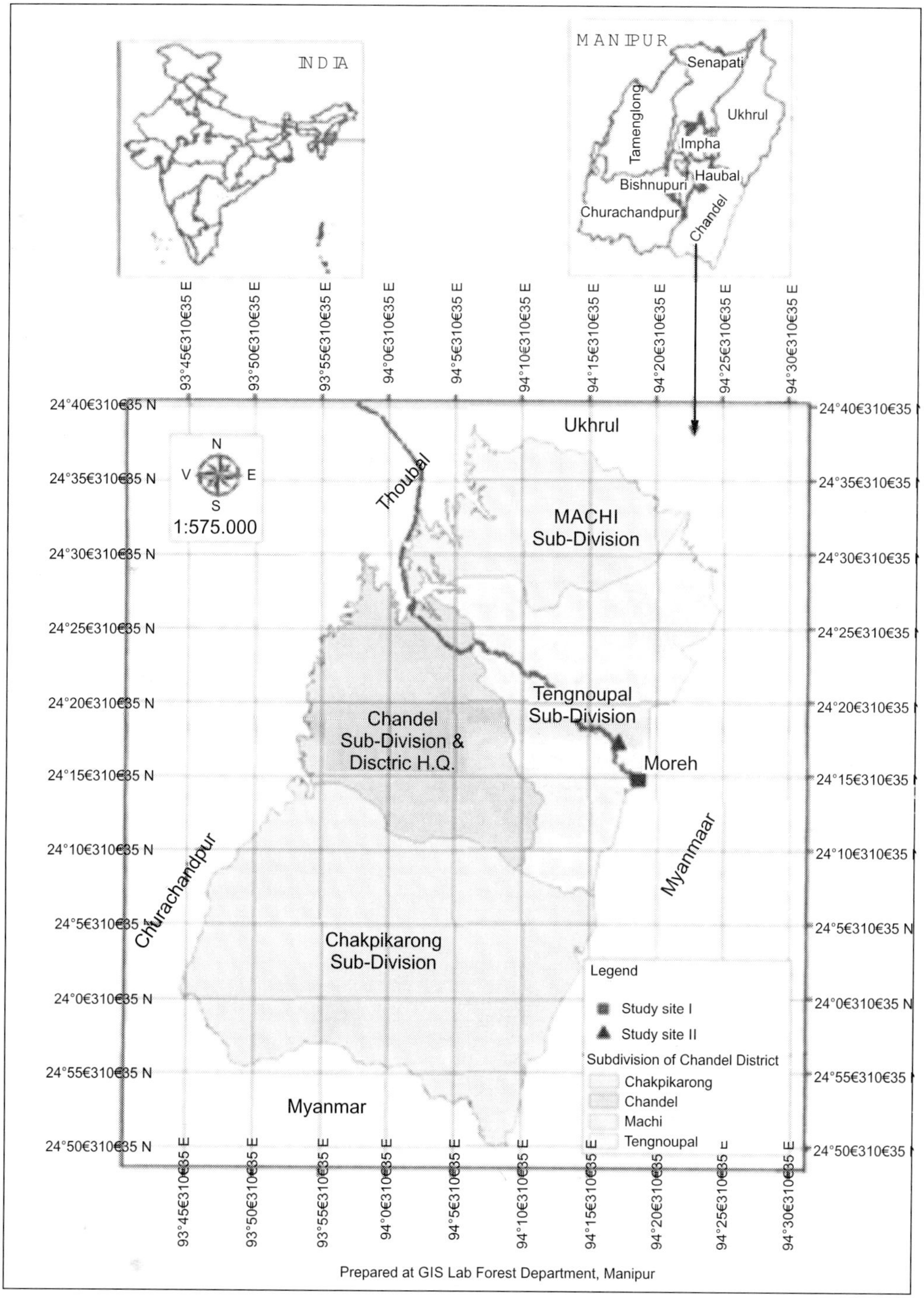

Fig. 14.1 Location of the study sites in Tengnoupal sub-division of Chandel district of Manipur.

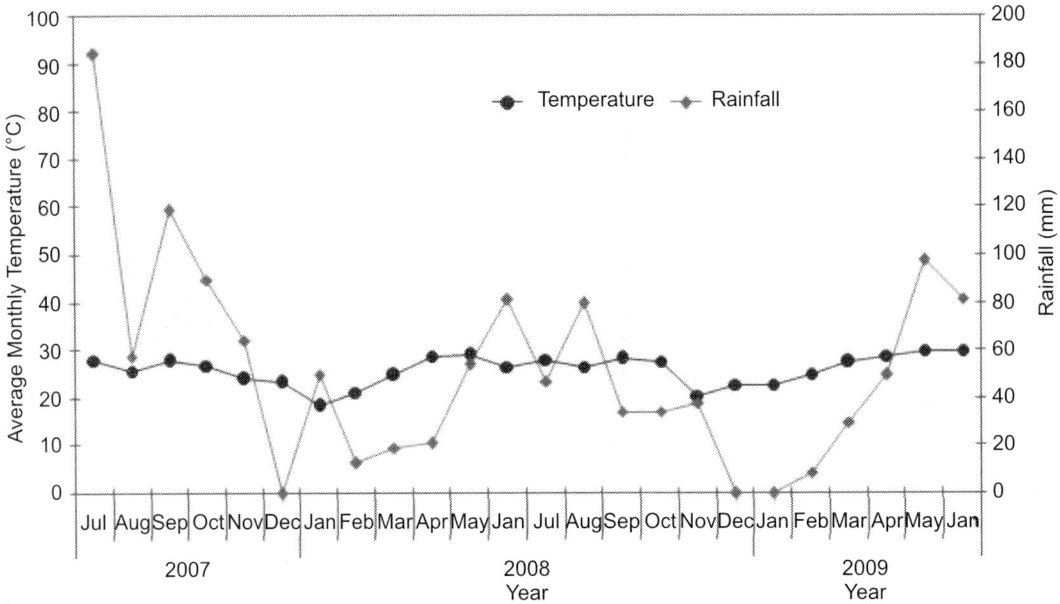

Fig. 14.2 Ombrothermic diagram for Thiubal district (neighbouring Chandel district) of Manipur for the study period (July 2007 to June 2009)

Table 14.1 Abundance/frequency ratios, basal area (m^2/ha) and distributional pattern of different tree species in two study sites of Tengnoupal subdivision of Manipur

Sl. No.	*Name of species*	*Abundance frequency ratio*		*Basal area m^2 ha^{-1}*		*Distributional pattern*	
		Site I	*Site II*	*Site I*	*Site II*	*Site I*	*Site II*
1.	*Aquilaria agallocha*	0.028	0.1	3.78	0.52	Random	Clumped
2.	*Artocarpus integrifolia*	0.05	0.05	0.34	0.37	Random	Random
3.	*Bauhinia variegate*	0.05	0.05	0.45	0.26	Random	Random
4.	*Cassia fistula*	0.05	–	0.67	–	Random	–
5.	*Castanopsis tribuloides*	–	0.05	–	0.28	–	Random
6.	*Celtis australis*	–	0.05	–	0.30	–	Random
7.	*Dipterocarpus tuberculatus*	0.018	0.021	13.86	3.01	Regular	Regular
8.	*Dipterocarpus turbinatus*	0.05	0.025	2.66	0.67	Random	Random
9.	*Erythrina indica*	–	0.05	–	0.30	–	Random
10.	*Ficus elastica*	0.1	–	0.42	–	Clumped	–
11.	*Ficus hispida*	–	0.05	–	0.19	–	Random
12.	*Ficus cunia*	–	0.05	–	0.45	–	Random
13.	*Gmelina arborea*	–	0.021	–	2.39	–	Regular

14.	Lagerstroemia speciosa	0.025	–	2.30	–	Random	–
15.	Mangifera indica	0.025	0.025	6.74	1.37	Random	Random
16.	Parkia roxburghii	0.025	0.05	1.15	0.19	Random	Random
17.	Quercus spp	0.037	0.025	2.37	3.18	Random	Random
18.	Salix tetrasperma	0.1	–	1.67	–	Clumped	–
19.	Tectona grandis	0.018	0.018	12.25	5.51	Regular	Regular
20.	Terminalia citrina	0.05	–	1.36	–	Random	–
21.	Xylosma longifolium	0.05	–	0.86	–	Random	–
22.	Terminalia myriocarpa	–	0.05	–	0.45	–	Random
23.	Toona ciliate	–	0.05	–	0.25	–	Random

RESULTS AND DISCUSSION

A study of floristic composition of the two study sites indicated that in all 22 plant families were reported. However in study site I, 16 plant families and in study site II, 15 families were recorded. Out of these, total 14 families were possessing trees and 10 families were possessing shrubs. Total number of species in both the sites is 36, but at study site I, 25 species, and at study site II, 25 species were observed.

When separated according to their habits, study site I showed 15 tree species and 10 shrub species, whereas study site II exhibited 17 tree species and 8 shrub species. Out of the 24 families recorded from the study sites, 2 were repeated, therefore the actual number of plant families was 22.

The highest basal area among the tree species in site I was observed in *Dipterocarpus tuberculatus* (13.86 m^2 ha^{-1}) and the lowest basal area was shown by *Artocarpus integrifolia* (0.34 m^2 ha^{-1}) Table 14.1. Among the shrub species in site I, maximum basal area was observed in *Cestrum nocturnum* (0.37 m^2 ha^{-1}) whereas the minimum basal area was recorded in *Adhatoda vasica* (0.02 m^2 ha^{-1}) as shown in Table 14.2.

Table 14.2 Abundance/frequency ratios, basal area (m^2/ha) and distributional pattern of different shrub species in two study sites of Tengnoupal subdivision of Manipur

Sl. No.	Name of species	Abundance frequency ratio		Basal area m^2 ha^{-1}		Distributional pattern	
		Site I	Site II	Site I	Site II	Site I	Site II
1.	Adhatoda vasica	0.05	0.05	0.02	0.024	Random	Random
2.	Bongainvillea spectabilis	0.05	–	0.09	–	Random	–
3.	Calotropis gigantea	–	0.033	–	0.47	–	Random
4.	Cassia occidentalis	0.05	0.05	0.22	0.09	Random	Random
5.	Cestrum nocturnum	0.03	–	0.37	–	Random	–

Contd...

Contd...

6.	*Duranta repens*	0.025	0.037	0.19	0.27	Random	Random
7.	*Hibiscus cannabinus*	–	0.05	–	0.07	–	Random
8.	*Jasminum amplexicaule*	–	0.05	–	0.08	–	Random
9.	*Mussaenda frondosa*	0.016	0.021	0.33	0.29	Regular	Regular
10.	*Mussaenda roxburghii*	0.1	–	0.06	–	Clumped	–
11.	*Nyctanthes arbortristis*	0.05	–	0.05	–	Random	–
12.	*Pogostemon benghalensis*	0.05	–	0.05	–	Random	–
13.	*Vitex negundo*	0.025	0.025	0.15	0.21	Random	Random

Of the total 15 tree species recorded at site I, the Importance Value Index (IVI) ranged from 6.36 to 54.79. Maximum IVI was exhibited by *Dipterocarpus tuberculatus* (54.79) followed by *Tectona grandis* (51.63), *Aquilaria agallocha* (42.19), *Mangifera indica* (24.62) and *Dipterocarpus turbinatus* (21.31). Minimum IVI was recorded by *Artocarpus integrifolia* (6.36) as shown in Table 14.3. In site I, a total of 10 shrub species were recorded. IVI for shrubs ranged from 13.7 to 59.16. Maximum IVI was exhibited by *Mussaenda frondosa* (59.16) followed by *Cestrum nocturnum* (55.08), *Duranta repens* (37.41), *Vitex negundo* (34.8) and *Cassia occidentalis* (26.77). Minimum IVI was recorded by *Adhatoda vasica* (13.7) as indicated in Table 14.4.

Table 14.3 Relative values of frequency, density, dominance, IVI and diversity indices of different tree species in two study sites of Tengnoupal subdivision of Manipur

Sl. No.	*Name of species*	*Relative frequency*	*Relative density*	*Relative dominance*	*Importance Value Indices*	*Diversity Index*
1.	*Aquilaria agallocha*	13.7 (3.5)	20.9 (5.2)	7.59 (2.64)	42.19 (11.34)	0.32 (0.15)
2.	*Artocarpus integrifolia*	3.4 (3.5)	2.3 (2.6)	0.66 (1.87)	6.36 (7.97)	0.08 (0.09)
3.	*Bauhinia variegate*	3.4 (3.5)	2.3 (2.6)	0.88 (1.32)	6.58 (7.42)	0.08 (0.09)
4.	*Cassia fistula*	3.4 (–)	2.3 (–)	1.31 (–)	7.01 (–)	0.08 (–)
5.	*Castanopsis tribuloides*	– (3.5)	– (2.6)	– (1.42)	– (7.52)	– (0.09)
6.	*Celtis australis*	– (3.5)	– (2.6)	– (1.52)	– (7.62)	– (0.09)
7.	*Dipterocarpus tuberculatus*	13.7 (14.2)	13.9 (18.4)	27.19 (15.28)	54.79 (47.88)	0.27 (0.31)
8.	*Dipterocarpus turbinatus*	6.8 (7.1)	9.3 (5.2)	5.21 (3.40)	21.31 (15.70)	0.22 (0.15)

9.	*Erythrina indica*	– (3.5)	– (2.6)	– (1.52)	– (7.62)	– (0.09)
10.	*Ficus elastica*	3.7 (–)	4.6 (–)	0.82 (–)	8.82 (–)	0.14 (–)
11.	*Ficus hispida*	– (3.5)	– (2.6)	– (0.96)	– (7.06)	– (0.09)
12.	*Ficus cunia*	– (3.5)	– (2.6)	– (2.28)	– (8.38)	– (0.09)
13.	*Gmelina arborea*	– (10.7)	– (10.5)	– (12.13)	– (33.33)	– (0.23)
14.	*Lagerstroemia speciosa*	6.8 (–)	4.6 (–)	4.51 (–)	15.91 (–)	0.14 (–)
15.	*Mangifera indica*	6.8 (7.1)	4.6 (5.2)	13.22 (6.95)	24.62 (19.25)	0.14 (0.15)
16.	*Parkia roxburghii*	6.8 (3.5)	4.6 (2.6)	2.25 (0.96)	13.65 (7.06)	0.14 (0.09)
17.	*Quercus* spp.	6.8 (7.1)	6.9 (5.2)	4.64 (16.15)	18.34 (28.45)	0.18 (0.15)
18.	*Salix tetrasperma*	3.4 (–)	4.6 (–)	3.27 (–)	11.27 (–)	0.14 (–)
19.	*Tectona grandis*	13.71 (14.2)	13.9 (23.6)	24.03 (27.98)	51.63 (65.78)	0.27 (0.33)
20.	*Terminalia citrina*	3.4 (–)	2.3 (–)	2.66 (–)	8.36 (–)	0.08 (–)
21.	*Terminalia myriocarpa*	– (3.5)	– (2.6)	– (2.28)	– (8.38)	– (0.09)
22.	*Toona ciliate*	– (3.5)	– (2.6)	– (1.26)	– (7.36)	– (0.09)
23.	*Xylosma longifolium*	3.4 (–)	2.3 (–)	1.68 (–)	7.38 (–)	0.08 (–)

Values in parentheses are for study site II

Of the total 17 tree species recorded at site II, the IVI ranged from 7.06 to 65.78. Maximum IVI value was exhibited by *Tectona grandis* (65.78) followed by *Dipterocarpus tuberculatus* (47.88) and *Gmelina arborea* (33.33), whereas the minimum IVI value was recorded by 2 species, viz., *Parkia roxburghii* and *Ficus hispida* (7.06) as shown in Table 14.3. For the 8 shrub species recorded in site II, the IVI ranged from 13.6 to 84.2. Maximum IVI was exhibited by *Calotropis gigantea* (84.2) followed by *Mussaenda frondosa* (61.75) and *Duranta repens* (47.9). Minimum IVI was recorded by *Adhatoda vasica* (13.6) as shown in Table 14.4.

Table 14.4 Relative values of frequency, density, dominance, IVI and diversity indices of different shrub species in two study sites of Tengnoupal subdivision of Manipur

Sl. No.	Name of species	Relative frequency	Relative density	Relative dominance	Importance Value Indices	Diversity Index
1.	*Adhatoda vasica*	6.6 (7.1)	5.8 (5.2)	1.30 (1.3)	13.7 (13.6)	0.16 (0.15)
2.	*Bougainvillea spectabilis*	6.6 (–)	5.8 (–)	5.88 (–)	18.28 (–)	0.16 (–)
3.	*Calotropis gigantea*	– (21.4)	– (31.5)	– (31.3)	– (84.2)	– (0.36)
4.	*Cassia occidentalis*	6.6 (7.1)	5.8 (5.2)	14.37 (6.0)	26.77 (18.3)	0.16 (0.15)
5.	*Cestrum nocturnum*	13.3 (–)	17.6 (–)	24.18 (–)	55.08 (–)	0.30 (–)
6.	*Duranta repens*	13.3 (14.2)	11.7 (15.7)	12.41 (18.0)	37.41 (47.9)	0.25 (0.28)
7.	*Hibiscus cannabinus*	– (7.1)	– (5.2)	– (4.6)	– (16.9)	– (0.15)
8.	*Jasminum amplexicaule*	– (7.1)	– (5.2)	– (5.3)	– (17.6)	– (0.15)
9.	*Mussaenda frondosa*	20 (21.4)	17.6 (21.05)	21.56 (19.3)	59.16 (61.75)	0.30 (0.32)
10.	*Mussaenda roxburghii*	6.6 (–)	11.7 (–)	3.92 (–)	22.22 (–)	0.25 (–)
11.	*Nyctanthes arbortristis*	6.6 (–)	5.8 (-)	3.26 (–)	15.66 (–)	0.16 (–)
12.	*Pogostemon benghalensis*	6.6 (–)	5.8 (–)	3.26 (–)	15.66 (–)	0.16 (–)
13.	*Vitex negundo*	13.3 (14.2)	11.7 (10.5)	9.80 (14.0)	34.8 (38.7)	0.25 (0.23)

Values in parentheses are for study site II

From site II, the highest basal area was exhibited by *Tectona grandis* (5.51m^2 ha^{-1}), whereas the lowest basal area was observed in two species, viz., *Parkia roxburghii* and *Ficus hispida* (0.19 m^2 ha^{-1}). In site II, the highest basal area for shrubs was exhibited by *Calotropis gigantea* (0.47 m^2 ha^{-1}), whereas the lowest basal area was observed in *Adhatoda vasica* (0.024 m^2 ha^{-1}).

Distribution patterns exhibited by trees and shrubs at both the study sites of Teak—*Dipterocarpus* forests are represented in Table 14.5. Majority of the tree species were found to be randomly distributed in site I. Some species like *Aquilaria agallocha, Dipterocarpus turbinatus, Xylosma longifolium, Cassia fistula, Bauhinia variegata, Mangifera indica* exhibited random distribution as their abundance to frequency (A/F) ratio was between 0.025 - 0.05, while *Tectona grandis* and *Dipterocarpus tuberculatus* were found to be regularly distributed as their A/F ratio was below 0.025 and *Ficus elastica* and *Salix tetrasperma* exhibited clumped distribution as their A/F ratio was > 0.05 (Table 14.1).

Table 14.5 Distributional pattern (%) exhibited by trees, shrubs and herbs at two study sites of Tengnoupal subdivision of Manipur

Trees			
Sites	*Distributional patterns*		
	Regular	*Clumped*	*Random*
Site I	13.3	13.3	73.3
Site II	11.7	5.8	82.3
Shrubs			
Sites	*Distributional patterns*		
	Regular	*Clumped*	*Random*
Site I	10.0	10.0	80.0
Site II	12.5	–	87.5

It was found that the majority of tree species from site II were randomly distributed. One species (*Aquilaria agallocha*) was found to be clumped in distribution as its A/F ratio was >0.05, while *Tectona grandis, Dipterocarpus tuberculatus* and *Gmelina arborea* were found to be regularly distributed as their A/F ratio was below 0.025. The remaining species exhibited random distribution as their A/F ratio was between 0.025 – 0.05 (Table 14.1).

Majority of the shrub species were found to be random in distribution. *Bougainvillea spectabilis, Adhatoda vasica, Cassia occidentalis, Duranta repens, Nyctanthes arbortristis, Vitex negundo* exhibited random distribution as their abundance to frequency (A/F) ratio was between 0.025 to 0.05 while *Mussaenda roxburghii* showed clumped distribution as its A/F ratio was >0.05 and *Mussaenda frondosa* was found to be regularly distributed as its A/F ratio was below 0.025 (Table 14.2).

The shrub species from site II did not exhibit clumped distribution at all. Only one species, viz., *Mussaenda frondosa* exhibited regular distribution as its A/F ratio was below 0.025. The other remaining 7 species exhibited random distribution as their A/F ratio was between 0.025 – 0.05 (Table 14.2).

H^1 varied as 2.36 and 2.37 for trees in site I and site II respectively. Index of co-dominance, λ varied as 0.053 and 0.064 for trees in both the sites. For shrub, H^1 varied as 2.15 and 1.79 in site I and II respectively. Index of co-dominance also varied as 0.129 and 0.122 in the two study sites. Other values for hill diversity numbers and evenness indices for trees and shrubs are given in Table 14.6.

Table 14.6 Diversity indices, co-dominance indices, Hill diversity numbers and evenness indices of different plant species in two study sites of Tengnoupal subdivision of Manipur

Trees			
Diversity parameters		*Study site I*	*Study siteII*
Diversity Index (H')		2.36	2.37
Index of Co-dominance (λ)		0.053	0.064
Hill diversity number	N_1	10.59	10.69
	N_2	18.86	15.62
Evenness Indices	*E*	1.78	1.46
	E'	1.86	1.50

Shrubs			
Diversity parameters		*Study site I*	*Study site II*
Diversity Index (H')		2.15	1.79
Index of Co-dominance (λ)		0.129	0.122
Hill diversity number	N_1	8.58	5.98
	N_2	7.75	8.33
Evenness Indices	*E*	0.90	1.39
	E'	0.89	1.47

Species diversity is an important attribute of a natural community that influences the functioning of an ecosystem (Hengeveld, 1996). The number of tree species was found to be lower than the number of species reported by several workers in other tropical forests (Chowdhury et al., 2000; 85 species; Fox et al., 1997, 94 species; Kadavul and Parthasarathy, 1999, 89 species; Khera et al., 2001, 92 species). The mean stand density of 860 stems ha^{-1} and 760 stems ha^{-1} in the Teak-Dipterocarpus forest of study area is well within the range of 276-905 stems ha^{-1} reported in the tropics (Ghate et al., 1998; Sundarapandian and Swamy, 1997; Murali et al., 1996). The densities reported in other tropical sites across the world also show similar trend as in Costa Rica-448 to 617 ha^{-1} (Heaney and Proctor, 1990); Brazil 420 to 777 ha^{-1} (Campbell et al., 1992) and Malaysia 250 to 500 ha^{-1} (Primack and Hall, 1992).

Importance value index (IVI) of plant species indicates the richness of species in a community. Higher IVI indicates high ecological significance and it is a reflection of better natural regeneration status of plant species. On the basis of IVI, *Dipterocarpus tuberculatus* (54.79) in site I and *Tectona grandis* (65.78) in site II were recorded as dominant species. *Tectona grandis* (51.63), *Aquilaria agallocha* (42.19) in site I and *Dipterocarpus tuberculatus* (47.88) and *Gmelina arborea* (33.33) in site II were co-dominant species. Higher values of IVI of these species indicated their good natural regeneration power, wide ecological amplitude and more adaptability to respective study sites. Monk (1967) reported that diversity tended to reach a maximum level after a community is composed of more than 12-15 species.

Among the shrubs, *Mussaenda frondosa* (59.16) in study site I and *Calotropis gingantea* (84.2) in study site II were observed as dominant species. Two co-dominant species were *Cestrum nocturnum* (55.08) and *Duranta repens* (37.41) in study site I, whereas in study site II they were *Mussaenda frondosa* (61.75) and *Duranta repens* (47.9).

Dominance-diversity curves (d-d curves) for tree and shrubs layers have been drawn on the basis of IVI to interpret the community organization in terms of resource share and niche space (Figs. 14.3 & 14.4), as resource apportionments in a community have been

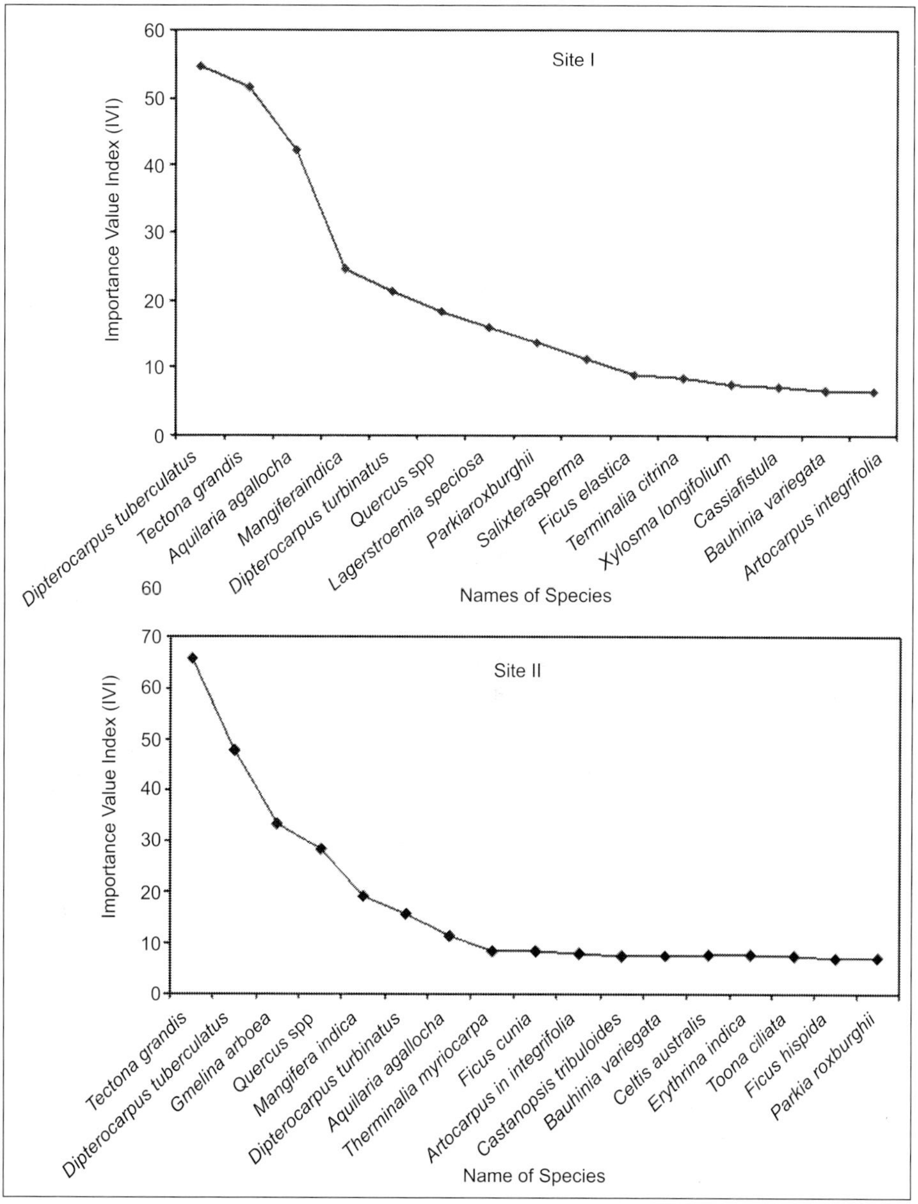

Fig. 14.3 IVI of various tree species in two study sites of Tengnoupal sub-division of Manipur

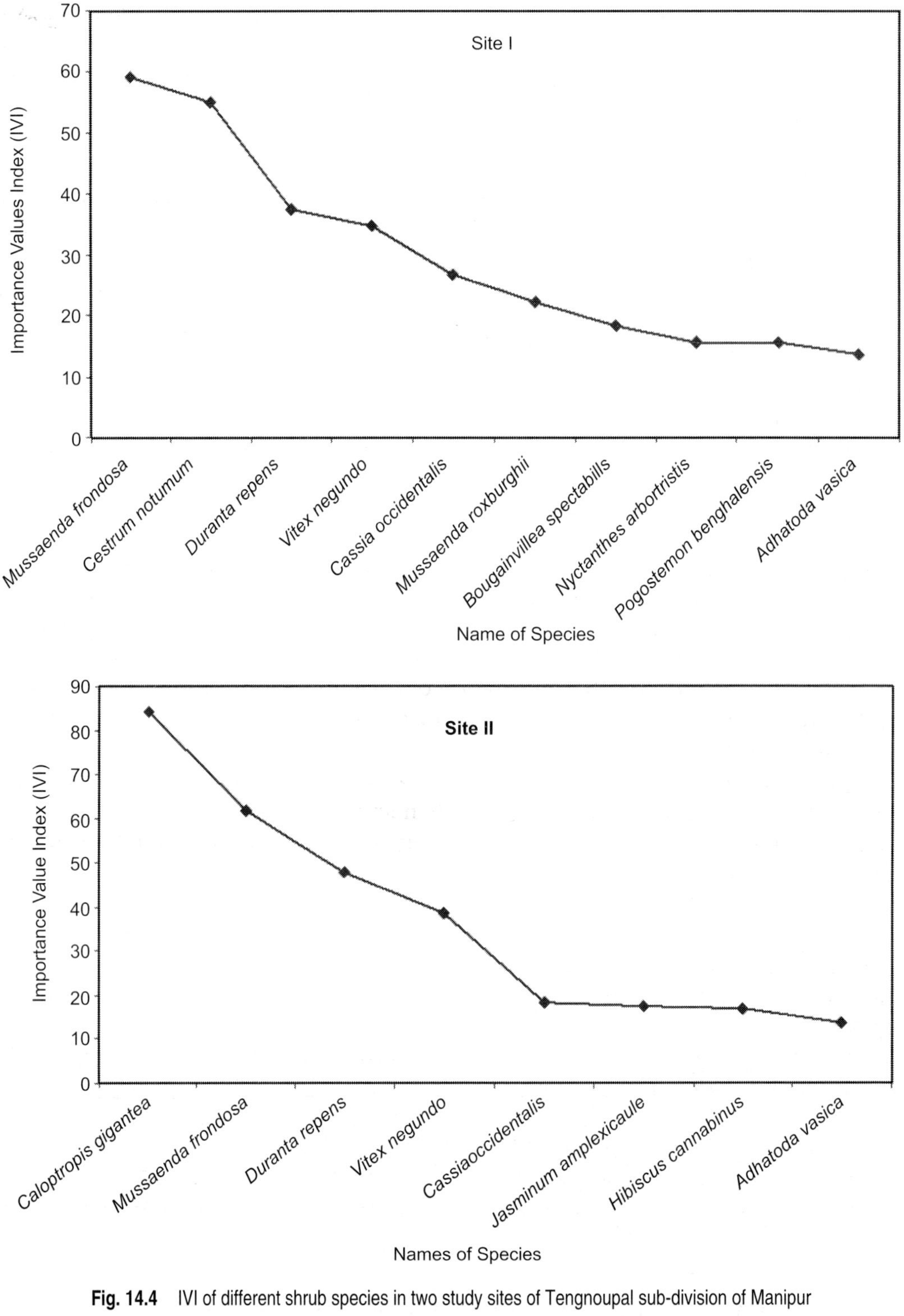

Fig. 14.4 IVI of different shrub species in two study sites of Tengnoupal sub-division of Manipur

interpreted from them (Whittaker, 1969). The d-d curves for tree and shrub approached to log normal distribution model, with less number of species in higher IVI range.

In the present study, basal area ranged from 19.69 m^2ha^{-1} (site II) to 50.88 m^2 ha^{-1} (site I), and the values are comparable with the ranges reported from dry and wet forests of the world, 17 to 40 and 20 to 75 m^2 ha^{-1} respectively (Murphy and Lugo, 1986).

Majority of the tree species were characterized by random pattern of distribution in the study sites, 73.3% in study site I and 82.3% in study site II. Species like *Tectona grandis* and *Dipterocarpus turberculatus* showed regular pattern of distribution in both the study sites. *Gmelina arborea* in study site II also exhibited regular pattern. Trees exhibited regular distribution pattern as 13.3% and 11.7% at study sites I and II respectively. In study site I, *Ficus elastica* and *Salix tetrasperma*, and in study site II, only *Aquilaria agallocha* showed clumped distribution by 13.3% and 5.8% plants respectively in both seasons (Table 14.5).

Majority of shrub species was characterized by random pattern of distribution in both the sites (80% in study site I and 87.5% in study site II). Only one species, i.e., *Mussaenda frondosa* showed regular pattern of distribution in both study sites (10% in study site I and 12.5% in study site II). However, clumped distribution was completely absent in site II whereas one species showed clumped distribution in study site I (Table 14.5).

The pattern of plant distribution depends both on physicochemical nature of the environment as well as on the biological peculiarities of the organism themselves. Majority of trees and shrubs reveal random type of distribution exhibiting non-selective behaviour pattern. Most of the herb species showed clumped pattern of distribution except a few that showed random pattern. Regular distribution of plants was found to be totally absent. This confirms the aggregation of individuals in favourable part of habitat which may be attributed to reproductive mode of species. Similar pattern is observed by Rastogi and Rastogi (2007).

Variation in the distribution pattern across the slopes and vegetation strata seems to be associated with multitude of factors, specially micro-environment and biotic. According to Odum (1971), contagious distribution is common in nature, random distribution is found only in uniform environment only and regular distribution occurs where severe competition exists between individuals.

In the present study, the diversity index ($H^/$) valued for both the study sites ranged from 1.79 to 2.37. Pascal (1988) reported that this value ranged between 2.1 and 4.3 in different forest ecosystems in Western Ghats. Species diversity varied from 2.26 to 3.58 in tropical forests of Andaman (Tripathi et al., 2006). Thakur and Khare (2008) observed the diversity index value ranging from 2.22 to 3.66 in the forest vegetation of Sagar district in Central India. While comparing the present values with the above, they are found to be low which might be due to severity in the environment (Table 14.6).

The Shannon-Wiener values for tree species diversity in the present study at site I as 2.36 and at site II as 2.37 are within the range compared to 2.20-2.65 reported for the tropical forests of Kodayar in the Western Ghats of southern India (Sundarapandian and Swamy, 2000).

The index of co-dominance values ranged from 0.053 to 0.122 in above ground flora in the present study, which are similar to those for forest vegetation in Sagar districts (0.052

to 0.141) indicating that dominance was shared by more than one species. Knight (1975) reported an average value of co-dominance as 0.06 for a tropical forest. Our values are comparable with the H′ values in dry tropical forest in Vindhyan Hill (1.93 to 2.82) and in dry tropical forest located along the disturbance gradient (1.398 to 2.629) as revealed in Table 14.6.

The concentration of dominance (Simpson's index) in the present study sites is very much lower as 0.053 and 0.064 at sites I and II respectively than the reported range (0.21 to 0.92; 0.10 to 0.99) for tropical forests by several workers (Knight, 1975; Visalakshi, 1995).

CONCLUSION

A sound knowledge of the floristic composition is required for appropriate conservation and restoration of the biological diversity. Floristic inventory can help to decide how to stratify sampling efforts for monitoring forest processes, or to interpret the ecosystem significance of results from individual species–level experimental manipulations. On a wider-scale, inventories are also critical for protected areas and development planning, which require geographically referenced, replicated and comparable samples to support decisions on where to focus conservation resources or development activities.

Information on distribution and abundance of tree species and biodiversity is highly needed for the conservation and management of tropical forest. Vegetation analysis in Teak-Dipterocarpus forest from Indo-Burma hotspot is an effort to record changes in vegetative parameters and various diversity measures in two study sites. The study is of much significance as Indo-Burma hotspot has a major contribution to the mega-biodiversity of North-East India, and people here rely on jhumming or shifting cultivation depending upon forest productivity to maintain livestock and soil fertility.

With proper management, there is a hope that these forests which are harbouring important germplasm, however, at present and are much threatened, could be preserved for posterity. Sustainable management of valuable forest resources can only ensure the safeguard and maintenance of the valuable biodiversity present in this hotspot.

REFERENCES

1. Bawa, K.S. Kress, W.J. and Nadkani, N.M. (2004). Beyond paradise: Meeting the challenges in tropical biology in the 21st Century. *Biotropica* 36: 276-284.
2. Campbell, D.G. Daly, D.C. Prance, G.T. and Maciel, U.N. (1992). A comparison of the phytosociology and dynamics of three floodplain (Varzea. Forest of known ages, Rio Jurna, Western Brazilian Amazon). *Botan. J. Linn. Soc.* 108: 213-237.
3. Chowdhury, M.A.M., Auda, M.K. and Iseam, A.S.M.T. (2000). Phytodiversity of *Dipterocarpus turbinatus* Gaertn. F. (Garjan) undergrowths at Dulahazara Garjan forest. Cox's Bazar, Bangladesh. Indian For. 126: 674-684.
4. Curtis, J.T. and Cottam, G. (1950). The use of distance measures in phytosociological sampling, *Ecology*. 37: 451-460.
5. Fox, B.J. Jennifer, E.T. Marelyn, D.F. and Williams, C. (1997). Vegetation changes across edges of rainforest remnants. *Biol. Conser.* 82: 1-13.

6. Ghate, U. Joshi, N.V. and Gadgil, M. (1998). On the patterns of tree diversity in the Western Ghats of India. *Curr. Sci.* 75: 594-603.
7. Heaney, A. and Proctor, J. (1990). Preliminary studies on forest structure and floristics on volcan Barva, Costa Rica. *J. Trop. Ecol.* 6: 307-320.
8. Hengeveld, R. (1996). Measuring ecological biodiversity. *Biodiversity Letters* 3: 58-65.
9. Kadavul, K. and Parthasarathy, N. (1999). Structure and composition of woody species in tropical semievergreen forest of Kalayan Hills, Eastern Ghats, India. *Trop. Eco.* 40: 247-260.
10. Khera, N. Kumar, A. Ram, J. and Tewari, A. (2001). Plant biodiversity assessment in relation to disturbance in mid elevation forest of Central Himalaya, India. *Trop. Eco.* 42: 83-95.
11. Knight, D.H. (1975). A phytosociological analysis of species rich tropical forest on Barro Colorado Island, Panama. *Ecol. Monogr.* 45: 259-289.
12. Kumar, A. Gupta, A.K. Marcot, B.G. Saxena, A. Singh, S.P. and Marak, T.T.C. (2006). Management of forests in India for biological diversity and forest productivity – a new perspective. Volume IV: Garo Hills Conservation Area (GCA). Wildlife Institute of India, USDA Forest Service Collaborative Project Report, Wildlife Institute of India, Dehradun.
13. May, R.M. and Stumpf, M.P.H. (2002). Species area relations in tropical forest. *Science* 290: 2084-2086.
14. Monk, C.D. (1967). Tree species diversity in the eastern deciduous forest with particular reference to north central Florida, Amer. *Nature* 101: 173-187.
15. Murali, K.S. Uma, S. Shaanker, U. Ganeshaiah, K.N. and Bawa, K.S. (1996). Extraction of forest products in the forests of Biligirirangan Hills, India and impact of NTFP extraction on regeneration, population structure and species composition. *Eco. Bot.* 50: 252-269.
16. Murphy, P.G. and Lugo, A.E. (1986). Ecology of tropical dry forests. *Annu. Rev. Ecol. Syst.* 17: 67-88.
17. Nirmal Kumar, J.I. Kumar, R.N. and Joseph, S. (2001). Tree species diversity of Khatana forest of south Gujarat, *J. Nat. Conserv.* 13: 149-166.
18. Nirmal Kumar, J.I., Kumar, R.N. and Joseph, S. (2002). Tree species diversity of Waghai forest of Northern part of Western Ghats. *Int. J. Ecol. Environ. Conserv.* 8: 235-248.
19. Nirmal Kumar, J.I. Kumar, R.N. Hiren Kumar, B.S. and Amit N. Gohil (1999). Biodiversity and socio-cultural status of Khatana and Waghai forests of North Western Ghats, South Gujarat. *International Journal of Ecology, Environment and Conservation*. 5: 171-176.
20. Nirmal Kumar, J.I. Kumar, R.N. Hiren Kumar, B.S. and Gohil Amit, N. (2000). Preliminary investigations of plant diversity of Khatana and Waghai forests of North Western Ghats, South Gujarat. *International Journal of Ecology, Environment and Conservation*. 6: 87-92.

21. Odum, E.P. (1971). Fundamental of Ecology 3rd Edition. W.B. Saunder and Co- Philadelphia. 527.
22. Pascal, J.P. (1988). Wet Evergreen Forests of The Western Ghats of India: Ecology, Structure, Floristic Composition and Succession. Pondicherry Institute, Français Pondichéry, pp. 345.
23. Pielou, E.C. (1966). The measurement of diversity in different type of Biological Collection. *J. Theor, Biol.* 13: 131-144.
24. Pitman, N.C.A. Terborgh, J.W. Silman, M.R. Percy, N.V. Neill, D.A. Ceron, C.E. Palacios, W.A. and Aulestia, M. (2002). A comparison of tree speciies diversity in two upper Amazonian forests. *Ecol.* 83: 3210-3224.
25. Primack, R.B. and Hall, P. (1992). Biodiversity and forest change in Malaysian Borneo. *Bio. Science.* 42: 829-837.
26. Raghubanshi, A.S. and Tripathi, A. (2009). Effect of disturbance, habitat fragmentation and a line invasive plants on floral diversity in dry tropical forest of Vindhyan highlands: a review. *Trop. Ecol.* 50-69.
27. Ramakrishnan, P.S. (2003). Linking natural resource management with sustainable development of traditional mountain societies. *Trop. Ecol.* 44: 43-54.
28. Rastogi, N. and Rastogi, A. (2007). Phytosociological Analysis of the Restored Sal (*Shorea robusta*) Plantations and Natural Sal Forest of Tripura. *Indian J. For.* 30(4): 377-385.
29. Simpson, E.H. (1949). Measurement of diversity. *Nature* 163: 688.
30. Sundarapandian, S.M. and Swamy, P.S. (1997). Plant biodiversity at low elevation evergreen and moist deciduous forests at Kodayar (Western Ghats, India). *Int. J. Ecol. Environ. Sci.* 23: 363-379.
31. Sundarapandian, S.M. and Swamy, P.S. (2000). Forest ecosystem structure and composition along an altitudinal gradient in the Western Ghats, South India. *J. Trop. For. Sci.* 12: 104-123.
32. Thakur, A.S. and Khare, P.K. (2008). Species diversity and composition of forest vegetation of Sagar district in Central India. *Indian For.* 134(6): 801-813.
33. Tripathi, K.P. Mehrotra, S. and Pushpangadan, P. (2006). Vegetation characteristics of tropical forest of Andaman Island. *Indian For.* 132(2): 165-180.
34. Visalakshi, N. (1995). Vegetation analysis of two tropical dry evergreen forests in Southern India. *Trop. Ecol.* 36: 117-127.
35. Whittaker, R.H. (1969). Evolution of diversity in plant communities. *Brook Haven Symp. Biol.* 22: 178-196.
36. Yadav, A.S. and Yadav, R.K. (2005). Plant community structure of Bala-fort forest in Alwar, Rajasthan. *Int. J. Ecol. Environ. Sci.*, 31-109-117.

15

Strategy for Forest Conservation to Facilitate Ecosystem Services in the Context of Climate Change

S.B. Roy and Pinaki Deb

Indian Institute of Bio-Social Research and Development (IBRAD)

Prafulla Kanan, Kestopur (JL-17), VIP Road, Kolkata – 700 101

E-mail: sbroy_chairman@yahoo.com

ABSTRACT

The forest ecosystem covers the flora, fauna and physical environment. The strategy for forest conservation in the context of climate change is required to improve the biodiversity and system as a whole to improve ecosystem services. Forest Ecosystem Services (ES) involve management of a complex system where institutional, ecological and economic elements form a web of human and environment interaction. Ecosystem management recognizes that the interconnectivity of ecological, socio-cultural, economic and institutional systems is fundamental to our understanding of the factors which influence environmental objectives and outcomes in a given situation. Lack of accounting of benefits is mostly due to the fact that ecosystem services are regarded as *positive externalities* or *public goods*. If finance mechanisms can be established well, it could provide a cost-effective solution to further internalize the value of nature into the (global) market by seeing it as a good or services that can be traded between suppliers and users of these ES. This paper argues how the ES depends on the effective functioning of ecosystem. Since human intervention influences the natural ecosystem, the problem of ES or socio-ecological processes cannot be understood in isolation. These are systemic problems, which means they are interconnected and interdependent.

INTRODUCTION

Strategy is meant for a long-term programme in view of the challenges. Strategy is based upon the strength of the area. It may be the number of natural assets that represent growth. This includes the primary sector, both agriculture and forestry, agro-food industry and a number of tourism sites. The weaknesses of the area may consist of an outdated and insufficiently maintained infrastructure system, which is largely inadequate to provide good accessibility to markets and services and to act as a catalyst for investment. The opportunities may be in the form of agro-processing industry, while the threat may be in the form of substantial migration of skilled labour force.

Here the strategy for forest conservation in the context of climate change is required to improve the biodiversity and system as a whole to improve ecosystem services.

It is crucial to recognize the influence and potential of forests and other natural ecosystems on climate adaptation/mitigation, environment and livelihood security of forest dwellers specifically and the nation at large, in the context of climate change. The forests would be vulnerable on account of the altitudinal and latitudinal shift of the species of the forest ecosystems, and also on account of increased occurrences of fire, pests, invasive species, change in species and loss of biodiversity. All this may severely affect forest-dependent livelihoods of the forest dwellers, especially of women who depend on fuel, fodder and food from forests, thus enhancing vulnerability of forest dependent communities.

The forest ecosystem covers the flora, fauna and ground conditions within the parameters of a forest. The forest ecosystem is dependent on the major resources available from the climatic conditions to the members and relationships in the food chain. The amount and proportion of flora, including the varieties of trees, grasses, fungi and flowers will affect the way in which fauna exist in forest ecosystem. The smallest creatures and plants are still important to the structure of the environment. The relationship between the food chain is vital to the balance of the ecosystem right from the smallest gnat to the largest predator.

Ecosystem services are the benefits people obtain from ecosystems. These include provisioning services such as food and water; regulating services such as flood and disease control; cultural services such as spiritual, recreational, and cultural benefits; and supporting services, such as nutrient cycling, that maintain the conditions for life on the earth.

Ecosystem services (ES) involves management of a complex system where institutional, ecological and economic elements form a web of human and environment interaction. This system as a web provides checks and balances of cultural values with ecological and economical implications governed by state bureaucratic and community institutions.

In nature the interaction between the entire complex of social systems may be called eco-governance, which has direct impact on biogeophysical systems. To understand ES and sustainability, therefore, requires some understanding of the behaviour of systems, in general, and of social and ecological systems, in particular.

HYPOTHESIS

The ES depends on the effective functioning of ecosystem. Since human intervention influences the natural ecosystem, the problem of ES or socio-ecological processes, cannot be understood in isolation (Capra, 1996). They are systemic problems, which means they are interconnected and interdependent.

Unless there is clarity on the intended purpose of long term forest land management, and defined roles and responsibilities related to gender and equity are socially accepted and implemented, conflict is inevitable causing disruption of ES.

Since we cannot imagine a world without government, formal or informal, there has to be unanimous agreement amongst the members of community and government functionary on the issue of natural resource management on sustainable basis for ESPA. Both the institutions, community and government, have their own perception, belief, norms, procedures and behaviours, which need to be matched for the purpose of ESPA, and may be termed bilateral matching institution.

Participatory monitoring supports the process of effective decision making. This provides feedback on ongoing programme effectiveness, allowing for adaptation during implementation. The perspectives and insights of all stakeholders (beneficiaries as well as project implementers) are considered.

The information generated through participatory vegetation monitoring and participatory monitoring of social institutions may certainly help both the community and forest department in decision making for improving productivity on a sustainable basis.

APPROACH

Ecosystem management recognizes that the interconnectivity of ecological, socio-cultural, economic and institutional systems is fundamental to our understanding of the factors, which influence environmental objectives and outcomes. It is a holistic, multidisciplinary and integrated approach, which requires a substantial shift in the way we perceive and approach the management of both our natural and modified environments.

The complexity of the earth's ecosystems poses a challenge for scientists as they try to understand how relationships are interwoven among organisms, processes and their surroundings. As it relates to human ecology, a suggested research agenda for the study of ecosystem services includes the following steps:

1. identification of ecosystem service providers (ESPs) - species or populations that provide specific ecosystem services - and characterization of their functional roles and relationships;
2. determination of community structure aspects that influence how ESPs function in their natural landscape, such as compensatory responses that stabilize function and non-random extinction sequences which can erode it;
3. assessment of key environmental (abiotic) factors influencing the provision of services; and
4. measurement of the spatial and temporal scales ESPs and their services operate on.

Forests need to be managed in large units, particularly where they are being managed not only for products that can be removed, but also for their environmental protection value, both at micro and macro levels.

What will be Forest Management Unit (FMU)?

Landscapes are usually large areas encompassing several ecosystems or habitats, consisting of a mosaic of forests, grasslands and agricultural areas, and waterbodies such as lakes and human settlements. Although, from an ecological perspective, the delineation and management of FMUs should coincide with landscape boundaries, this arrangement is rarely the case as FMU boundaries are determined by political, administrative or market factors. FMUs will usually operate at scales smaller than landscapes.

Therefore, while the assessment of sustainability should most appropriately use a landscape scale, this approach may not be feasible, and some indicators and verifiers may only be incompletely applied. The selection of site will be based on the information where there is imbalance between the ecological and social processes due to several factors causing conflict and imbalance.

METHODOLOGY

Methodology for Studying Governance

Eco-Governance: Institutional Arrangements

The local community institution has to be compatible with the ecosystems. It should operate within the framework of open, transparent, participatory and partnership relationships between community institutions and local government. The local governance practices and regulations for sustainable forest ecosystem processes and methodologies such as social assessment, participatory resource assessments have to be designed in case it is not in place. These include strengthening and localization of resource assessment, monitoring and evaluation into the local planning system; integration of ecosystem-based/environmental planning, indicators and benchmarks into local decision-making and policy-formulation processes; regular budgetary support and allocation for eco-governance activities; and creating enabling local and national policy environments for sustaining eco-governance initiatives.

For the community level, one has to study and intervene at the social system following sequential steps such as identifying:

- Perception of the community related to forest conservation, identity and belongingness of the community towards public forest and other natural resources around their village and how they foresee the benefit.
- What kind of communication mechanism exists within Forest Protection Committee(FPC) and between FC and FD?
- What is the process of community group formation and dynamics?
- What kind of social institutions favour forest conservation?
- Whether there is clarity in roles and responsibilities related to gender and equity.
- What kind of leadership exists to facilitate NRM?
- What are the values and norms related to NRM through which community actions are integrated?
- What are the social symbols and rituals practiced by the community relating to natural resource management?
- Whether the socialisation process of children help to see how the parents interact and value natural resources and social capital through formal and informal orientation.

Bilateral Matching Institutions

Any kind of intervention through physical activity (such as plantation, soil work, pit making, thinning, pruning, harvesting) to improve forest conditions will not yield the required result unless the desired behaviour of individuals in society is monitored and controlled as social norms and institutions.

In the same way, any kind of new behaviour by an individual forester will not have sustained action unless organisation as a whole changes its philosophy, policy and operational plan. Both the institutions have their own perceptions, beliefs, norms, procedures and behaviours, which need to be matched. The members of the community could then also set norms of harvesting non-timber forest produces (NTFP) and timber, based on sustainability. Any kind of economic activity such as harvesting of forest produces, value addition, distribution, marketing will be successful only when:

(1) bilateral matching institutions continue; and
(2) ecological activities such as forest protection, enrichment assisted natural regeneration (ANR), thinning, pruning, harvesting, etc., are carried out following the rules of sustainability.

Community and foresters as government institutions try to achieve the common goal of forest conservation, and both the institutions, FD and FC have their own perception, belief, norms, procedures and behaviours, which need to be complimentary. Participatory monitoring helps in getting diagnostic information based on common understanding, which further helps in evolving control mechanism to check indiscriminate use of the resources as well as meet common goal of FD and FC for resource development to facilitate ES.

A number of scientific research methods of monitoring the forest and human resources exist, but the challenge is to develop such criteria and indicators for monitoring which can be understood and adopted by the local people themselves so that people and the forest department can also make use of the findings to support their decision-making process.

How the monitored information is being used by the forest department and local people for decision making to support sustainable forest management? *Coordination between the forest community and forest department as an organisation.* As in JFM both FC and FD work as partners and have developed common goal and agreed plan of action, it is expected that they would follow certain set norms for regular interaction. The communication mechanism between these two institutions are to be taken into account.

Authors had experience of working in three state in India and have found that monitoring the social institutions by the committee members themselves in a cluster have helped them immensely to take corrective measures and procedures for conducting meetings, writing minutes of the meeting, etc. Inefficient leaders are replaced by efficient leaders.

Utilization by the community members

In case of vegetation monitoring, information has influenced the community to debate, analyse, interpret data relating to the forest condition. For example, people of Ghugimura village used to bark tree for rope making, which resulted in loss of several trees in course of time. After monitoring, people got the facts, discussed the alternative options and banned barking. In another village, loss of biodiversity due to unchecked harvesting has been restricted by villagers. Participatory NTFPs monitoring at Tungaduha village in Midnapore district is one of the examples where one can validate its impact on decision making. It has effected a change in perception and attitude of people towards justifiable use of NTFPs and hence people have decided to go for sustainable harvesting of NTFPs.

Now they keep records of daily NTFP collection, analyze data on monthly basis and then decide on the actions to be taken for collection of particular NTFPs.

Monitoring is one of the basis of enhancing community capability to think for better options and opportunities. Therefore, vegetation / NTFPs monitoring has not only influenced forestry-related decisions, but also encouraged the community to decide on taking up income generation / village development activities based on local resources and expertise, such as pulse cake making, puffed rice selling, tailoring, cleaning of village, FPC office construction, conducting quiz, competition, awareness rally, and so on.

Enhance community voice with respect to other stakeholders.

Utilization by the forest department

The information generated through participatory vegetation monitoring and participatory monitoring of social institutions would help forest department in the decision-making process. As in Andhra Pradesh the committees are being classified based on the information, and appropriate actions are taken to improve the capacity of the weak FPCs.

The data from vegetation monitoring may help the forest to take appropriate action for harvesting timber and non-timber forest products, plantation, etc. This information may be incorporated in their macro level working plans.

Indian Institute of Bio-Social Research and Development (IBRAD) has been assigned by the Ministry of Environment and Forests, Government of India, to develop a format for monitoring the JFM system in the country. This has been developed and tested in different states of India. Now, a systematic monitoring mechanism for JFM is being planned.

Human well-being, food security and sustainable livelihoods are intimately linked to the future health of our diverse ecosystems. Ecosystem management demands an understanding of the functions of ecosystems in supporting and regulating the processes which underpin life on the earth. Furthermore, the approach recognises that ecosystems provide diverse goods and services which are directly or indirectly valued by society in ecological, economic and socio-cultural terms.

Many conventional approaches to resource management have usually been single-purpose and limited in space (e.g., local, regional, national and trans boundary) and time (e.g., short-, medium- and long-term). The resulting decision-making processes have therefore generally failed to address the human-induced pressures on the environment in terms of understanding their effects on essential ecosystem services.

Ecosystem management, in recognising the critical role humans have as managers of biodiversity, aims to avoid these shortcomings by identifying and communicating the benefits and values of healthy ecosystems in ensuring both species survival and human well-being.

Forests provide several ecosystem services for human well-being. These include provisioning services such as food, water and fiber; regulatory services such as climate, floods, disease, wastes and water quality; cultural services such as recreation, aesthetic enjoyment; supporting services such as soil formation, photosynthesis and nutrient cycling.

- Crop production quantity and quality would be directly affected by climate, and also indirectly through diseases, insects, and weed effects.
- Greater frequency and intensity of rainfall, together with increased use of nitrogen, mineral elements and pesticides would lead to leaching or run-off of these substances.
- Rural environment would be abused, even to the stage of land abandonment due to climatic change impacts on certain species or the introduction of new ones, the modification of land-use especially with the development of irrigation projects.

How Finance Mechanism is Linked with Ecosystem Services?

A "financial mechanism" or "payment mechanism" can be defined as a mechanism to calculate the unitary charge of a service provider to a service buyer in return for the service. Related to ecosystem services, it can be defined as a tool to capture the value of an ecosystem good or service in monetary terms. A set of rules and procedures is usually used to effect the movement of money or a payment instruction from a service buyer to a service seller.

Types of financing mechanisms

Finance mechanisms will vary according to their ecological, social and political context. Four broad types of mechanisms can be distinguished, in declining level of government involvement (Scherr et al., 2004; Powell et al., 2002):

1. Public payment schemes. The government provides the institutional foundation for a programme and directly invests in it as well. Examples include:
    - permanent conservation easements;
    - contract farmland set aside for conservation;
    - programmes to co-finance investments in afforestation or sustainable forest management; and
    - payments for the confirmed presence of endangered wildlife species.
2. Open trading under a regulatory cap or floor. The government defines a mandatory level of a specific ecosystem service to be provided. But to achieve this level the

regulated party can choose either to comply directly with the requirement or to pay others, who are in the position to supply the service more cheaply, to do so. Examples include:
- wetland mitigation banking in the USA;
- tradable development rights;
- salinity credit trading in New South Wales, Australia; and
- carbon emission offset trading.

3. Self-organized private deals. Direct, usually closed, transactions between offsite beneficiaries of forest services and forest landholders responsible for the services. Examples include:
   - hydropower companies paying land owners upstream for managing forest cover in the designated watershed;
   - a non-governmental organization (NGO) partners with local people to restore degraded land; and
   - a pharmaceutical company paying for access to a certain area or for the collecting work by local people of potentially valuable organisms.
4. Ecolabeling of forest or farm products. Also this process is also handled by private actors, payments for ecosystem services are embedded in a traded product. Examples include:
   - FSC (Forest Stewardship Council) timber and non-timber certification
   - "Salmon-safe" labelled products from farmers in the northwerst US.

Why are Finance Mechanisms Needed?

While scientists can determine the (monetary) value of "nature" pretty good, its value is for the most part not accounted in the world market. Only the most obvious benefits, such as the production functions of nature (timber, fish, etc) and just recently carbon sequestration, are being accounted for. This lack of accounting of benefits is mostly due to the fact that ecosystem services are regarded as *positive externalities* or *public goods*. A key characteristic of these type of public goods is that they are *non-rival* and *non-excludable*, meaning that nobody can be refused to use the public good even if they do not pay for it, while at the same time the use of the good does not reduce its availability to others (Landell-Mills and Porras, 2002). These two characteristics undermine the creation of a market for these goods, as nobody is prepared to pay for it.

A result of this lack of payment has repercussions on welfare, since there is systematic under-investment in the protection and management of these services. Consequently, the depletion of ecosystem services is far greater than it would have been without market failure. However, everything has a price: sooner or later we are going to face the consequences of our current behaviour. There are already cases known in the world where the (threat to) deterioration of the environment causes individuals, communities, government agencies and the private sector to look for innovative solutions to cope with. Finance mechanisms can, if well established, provide a cost-effective solution to further internalize the value of nature into the (global) market by seeing it as a good or services that can be traded between suppliers and users of these ES.

Incentives to become involved in markets for ecosystem services

Forced to:

- Regulatory compliance. Examples include:
  - law
  - cap-and-trade; and
  - international convention.

Clear business benefits:

- Business opportunity. Examples include:
  - ability to earn money through carbon offsets or the water market (financial intermediaries);
  - reduce environmental risk (insurance firms); and
  - eco-enterprise operations.
- Secure, sustain or reduce costs of key natural resource inputs required for business operations. Examples include:
  - uncontaminated water needed for bottling plant;
  - "charismatic" macrofauna for ecotourism; and
  - genetic resources for pharmaceutical companies (i.e., bioprospecting); and
  - conservation of watershed to secure water flow regulation/high quality drinking water.
- Securing licence to operate: Reputational risk management/managing potentially difficult relationships:
  - Better relations with regulators, supporting formal licence to operate in the future.
  - Better relations with local communities, supporting informal license to operate, avoiding disruption/losses from protests.
- Enhance or maintain the financial value of land, forest or other assets belonging to the company.

Indirect business benefits

- Enable strong "green" branding by the company (for marketing to consumers, investors or others committed to green products.
- Improved staff pride and morale, enhanced recruitment and retention of superior staff.
- Reflect/consistent with broader business values of the corporation (commitment of CEO to "good husbandry").

Not-business related

- Philanthropy/charity.

Methodology for studying forest ecosystem services:

Understanding of interdependence—whereby all species are, to some extent, dependent on one another— is a key characteristic of the forest ecosystem.

- *Identification of species relying on other species for pollination and seed dispersal to predator-prey relationships to symbiotic relationships.*
- *Identify how competitive species have developed complex symbiotic relationships with other species in order to survive.*

What are the symbiotic relationships? As we know, certain caterpillar species produce sweet chemicals from "dew patches" on their backs, upon which a certain ant species will feed. In return, the ants vigorously protect the caterpillar and have even been observed carrying the caterpillar to the nest at night for safety. This relationship appears to be species-specific in that only one caterpillar species will cater to a particular ant species.

KEYSTONE SPECIES

Understanding the nexus of keystone species that disappear from the ecosystem may weaken the survival chances of another, while the loss of a keystone species—an organism that links many other species together, much like the keystone of an arch—could cause a significant disruption in the functioning of an entire system.

Development of communication strategy which will articulate, explain and promote a vision and a set of well-defined goals of managing ecosystem to get optimum ES for PA. It creates a consistent, unified "voice" that links diverse activities and goals in a way that appeals to stakeholders.

This will include the concrete steps and techniques used in implementation of communication goals.

REFERENCES

1. Capra Fritjof, (1996). The web of life: A new scientific understanding of living systems, Anchor Books.
2. Barbier, E.B. (1993). Valuing tropical wetland benefits: Economic methodologies and applications. *Geographical Journal*, Part 1, 59: 22-32
3. Clark, C. (1976). Mathematical Bioeconomics. New York: John Wiley and Sons.
4. Ellis, G.M., Fisher, A.C., (1987). Valuing the environment as input. *J. Environ. Manag*. 25: 149-156.
5. Landell-Mills, N. and Porras, I. (2002). Markets for Forest Environmental Services: Silver Bullet or Fool's Gold? Markets for Forest Environmental Services and the Poor, Emerging Issues. International Institute for Environment and Development, London, UK.
6. Mishra, T.K. Sahoo, T.K. A. Bhuian A. and Roy, S.B. (1997). Assessing Impact of extraction on sustainability of forests under joint forest management: A case Study published in the proceedings of XI World Forestry Congress.
7. Mukhopadhyay, D. Tewari, H.R. and Roy, S.B. (2007). Role of Community Institutions in Joint Forest Management. *J. Hum. Ecol.*, Vol. 21, No. 1, 37-42.
8. Freeman, A.M., III. (1991). Valuing environmental resources under alternative management regimes. Ecological Economics.

9. Porter, M.E. (1996). (November-December). What is strategy: Harvard Business Review.
10. Powell, I., White, A. and Landell-Mills, N. (2001). Developing Markets for Ecosystem Services of Forests. Forest Trends, Washington, DC, USA.
11. Roy, S.B. (1992). Bilateral matching institutions: An illustration in forest conservation. *J. Ind. Anthropol. Soci.* 27: 253-262.
12. Roy, S.B. (1996b). From conflict to collaboration: Institutional issues in community management. Paper published in Village Voices, Forest Choices. Forest Management in India. Edited by Poffenberger M and B. Mcgean. Oxford University Press, New Delhi.
13. Roy, S.B. et al. (2000). Forest Protection and Management by Community in Midnapore district, West Bengal; published in book 'Joint Forest Management and Community Forestry in India - An Ecological and Institutional Assessment'; Oxford and IBH Publishing Co. Pvt. Ltd.
14. Roy, S.B. Mukhopadhyay, D. and Das, S. (2001). Strengthening Institutions in Joint Forest Management: Systemic Approach to Forest Conservation. *J. Soc. Sci.*: Interdisciplinary Reflection of Contemporary Society Vol. V, No. 1 & 2, Jan- April.
15. Roy, S.B. Mukherjee, R. and Mukhopadhyay, S. (2003). Conceptual framework for Criteria & Indicator for Assessment of Sustainable Development: An illustration from Joint Forest Management. *Ind. J. Land. Sys. Ecol. Stu.* Vol. 26, No. 2.
16. Roy, S.B. (2004). Participatory Vegetation Monitoring: Examples from West Bengal; Root to Canopy, by Winrock International India, New Delhi.
17. Roy, S.B. Mukhopadhyay, D. and Mukhopadhyay, R. (2011). Converging Sectoral Land Management Plans and Sustainable Livelihood; *Ind. J. Land. Sys. Ecol. Stud*, Kolkata, Vol. 34, No. 2.
18. Scherr, S. White, A. and Kaimowitz, D. (2001). Making Markets Work for Forest Communities. Forest Trends, Washington, DC, USA.

16

Afforestation: An Effective Intervention for Sustainable Development of Fragile Desert Ecosystem

Pradeep Chaudhry and V.P. Tewari

Arid Forest Research Institute

P.O. Krishi Mandi, New Pali Road, Jodhpur – 342 005

E-mail: pradeepifs@yahoo.com and vptewari@yahoo.com

ABSTRACT

Deserts are a special kind of ecosystem bearing little vegetation. The edapho-climatic conditions do not favour a great variety of plants and animals. Special efforts are required to improve the vegetational status of such places which, in turn, improve the environmental conditions of the area.

With the example of Thar desert of Rajasthan, a brief description of afforestation methodology has been given in the present paper which suggests that the selection of suitable plant species is the first and foremost point to be decided wisely. Social acceptability of those species is the second factor responsible for the success of all plantation programmes. If the tree species produce some useful minor forest products (MFPs) for the people or their cattle, they are liked by the rural masses and their protection can be done by them. Urban afforestation must improve the aesthetic beauty of the area. Appropriate plantation technology for saline-alkaline lands has also been suggested in the present paper.

DESERTS AND DESERTIFICATION

As per an estimate of UNEP, about 36.3 per cent of earth's surface is extremely arid or semi-arid which may be put under the general headings of "deserts", yet a world survey of land conditions, based on soil and vegetation data, indicates that 43 per cent of the earth's surface falls within these categories. The difference of 6.7 per cent is accounted for by the estimated extent of manmade deserts.

Desertification is one of the world's most alarming processes of environmental degradation. The issue is often obscured, however, by a common misperception that it is a natural problem of advancing deserts in faraway developing countries. In fact, desertification is about land degradation: the loss of the land's biological productivity, caused by human-induced factors and climate change. It affects one-third of the earth's surface and over a billion people. Moreover, it has potentially devastating consequences in terms of social and economic costs.

Desertification commonly appears as degradation of plant, animal, soil and water resources and general loss of biological productivity in areas under ecological stress. In fragile ecosystems, such as those on desert margins, this degradation can easily become irreversible, and permanently reduce the capacity of the area affected to support human life.

With the adoption in 1994 of the United Nations Convention to Combat Desertification (UNCCD), the issue was given proper recognition. Desertification as a global challenge, together with climate change and biodiversity, now enjoys the support of a strong coalition of partners (Nation Action Programme (NAP), 2001). But public awareness has not kept pace. In relation to the true scope and magnitude of the problem, desertification still receives too little attention and is little understood by the public at large.

At its 58th ordinary session, the U.N. General Assembly declared 2006 the International Year of Deserts and Desertification (IYDD). In doing so, the General Assembly underlined its deep concern for the exacerbation of desertification, particularly in Africa, and noted its far-reaching implications for the implementation of the Millennium Development Goals (MDGs), which must be met by 2015. At the 2002 World Summit on Sustainable Development, the Convention was singled out as a key instrument for poverty eradication in dryland rural areas.

THAR DESERT

Indian hot arid zone is spread out in 0.32 million square km of which 0.29 million square km are in northern-western portions of the country, mainly confined to the states of Rajasthan, Gujarat and Haryana. The Thar Desert in Rajasthan occupies an area of 196,150 square km in Rajasthan involving 12 districts located in the west and north-west parts of the state. These account for 67.5% of the Thar and 61% of the total hot arid zone of India. Substrates and soils are sandy, rocky, gravelly or saline. Sand dunes are a dominant formation of the Thar, accounting for over 58% of the land area of the region, with various densities of occurrence from place to place (Desertification & Land Degradation Atlas of India, 2007). Vegetations in arid zones are sparse with scattered thorny trees, shrubs and grasses, and are ineffective from soil and water conservation point of view.

PROBLEMS AND CHALLENGES

The production and life support systems in this region are constrained by bioclimatic and environmental limitations:

- Low precipitation (100-400 mm MAR).
- High pre-monsoon temperatures (mean maximum of the hottest month: 45-47°C).
- High wind-speed: average annual 8-10 km h^{-1} with figures upto 30-40 kmh^{-1} occurring in summers.
- High potential evapo-transpiration with an annual total of 1,500 to 2,000 mm.

These conditions are most reminiscent of the Sahel of Africa. In the past, agriculture and urban expansion has been at the expense of forests, woodland and grassland ecosystems. The accumulated effects of these changes are reflected in serious soil degradation, desertification and loss of biodiversity, leading to food insecurity and extreme poverty in the most vulnerable areas.

Tree planting and management is one of the solutions for addressing these challenges. Trees have the potential to restore degraded ecosystems, to provide food, medicines and other goods, and render environmental services alongwith socio-economic benefits.

DEVELOPING AFFORESTATION TECHNOLOGY FOR STRESS SITES IN ARID REGIONS

Silvicultural problems demand (a) a careful study regarding selection of tree species; (b) production of quality planting stock in nurseries; (c) silvicultural characteristics of species selected; (d) soil working, weeding and cultural operations; and (e) protection against diseases and animals.

The local tree species are the best indicators for selection of species for planting work. The local species of the arid zone are not only a few but are also extremely slow growing. Therefore need arises to introduce tree species of higher growth rate and of better resistance to drought and cold conditions. Many species of *Acacia* and *Eucalyptus* were tried. The best suited was *Acacia tortilis* from Israel. Next to this was *E. camaldulensis*. Local species like *Acacia nilotica, Prosopis cineraria* and *Acacia senegal* were also found suitable. Arid Forest Research Institute (AFRI), Jodhpur, successfully demonstrated a sand dune stabilization technique by introducing indigenous species like *Calligonum polygonoides* associated with under shrub *Cassia angustifolia* and grass *Cenchrus ciliaris,* which provided beneficial effects in controlling sand reactivation and drift. In another study conducted by AFRI in arid salty land, the indigenous multipurpose halophytic tree, *Salvadora persica,* maintained 67% to 85% survival after eight years with gypsum and 9g N treatment giving best biomass and growth results. Exotic shrubs of genus *Atriplex,* a fodder species, performed well with FYM and nitrogen. These shrubs produce nitrogen rich fodder used for sheep and goat. *Acacia ampliceps* (exotic tree species) performed very well on deep alkali soils and yielded twofold biomass (12 kg/tree to 5.5 kg/tree for gypsum treated and 8.1 kg/tree to 3.9 kg/tree for untreated trees) at the age of five years.

Agroforestry land use systems hold significant social and economic implications in the fragile ecosystem of arid areas. Models of *Prosopis cineraria* (tree legume) and *Tecomella undulata* (Marwar Teak) based agroforestry models were developed at AFRI, Jodhpur to maximize total production. Optimum tree density at which crop productivity is almost similar to sole agricultural crop is 833 at 2-3 years, 417 at 4-6 years, 278 at 7-8 years and 208 tree per ha at 9-11 years of tree age. The best suitable agroforestry species, as per AFRI study, is *P. cineraria,* which provides utilizable biomass of 21-32 tonnes per ha (2 × 3 m spacing) at 8 years and 19.96 tonnes per ha including leaf fodder of 0.85 tonnes per ha at 12 years age (208 tree per ha). In case of *T. undulata,* the crop yield is 10-20% lesser under than *P. cineraria. Colophospermum mopane* was found more suitable for rehabilitation of degraded lands rather than its integration with agricultural field.

DESERT DEVELOPMENT PROGRAMME (DDP)

The degraded environment from ecological point of view and the hardships of the people living in these areas necessitated state intervention in introducing Desert Development

Programme (DPP) which was launched as a centrally sponsored scheme in 1977-78 with the prime objective of controlling desertification and restoration of ecological balance. This objective was sought to be achieved through the activities like afforestation (with special emphasis on sand dune stabilization), shelterbelt plantations, grassland developments, soil moisture conservation and water resources development. Initially, it was implemented in hot desert areas of Rajasthan, Gujarat and Haryana and cold desert areas of Jammu & Kashmir and Himachal Pradesh. From 1995-96, the scheme was extended to Andhra Pradesh and Karnataka as well. Since April 1995, the programme is implemented on watershed basis.

The DDP aims at integrated development of the desert area by increasing the productivity, income level and employment opportunities for its inhabitants through optimum utilisation of land, water, livestock and human resources. The main thrust has been on arresting desertification through activities, which would restore ecological balance and stabilise sand dunes. For arresting desertification and ingress salinity of the soil, afforestation activity is of utmost importance.

The various schemes taken up in Rajasthan under DDP are shelterbelt plantations, rehabilitation of degraded forests, village fuel wood and fodder plantation, farm forestry, sand dune stabilization, pasture development and establishment of nurseries (Sinha et al., 1997).

The development of micro-watershed as a unit of planning is considered to be the most scientific approach for optimum utilisation of land, water, human and livestock resources leading to a balanced development of the area through increased production and productivity. In a watershed area the available resources have to be properly assessed and development plan prepared in an integrated manner.

In Rajasthan, where all districts are covered under DDP, it was envisaged that wherever watersheds are feasible, planning has to be done on watershed basis only. The watershed development programme is based on the area development approach in which a patch of 5 to 20 thousand hectares of the watershed area is identified/selected. It is called macro-watershed. This project area is divided into smaller areas called micro-watershed, development of which in an integrated manner could be managed by a single agency. The area of a micro-watershed is generally below 500 hectares.

IGNP Area Plantations

The idea of Indira Gandhi Nahar Pariyojana (IGNP) was thought of against the background of poor agricultural growth and problem of desertification in the drought-prone arid parts of Rajasthan. The IGNP canal system is divided into two stages: Stage-I covering the northeastern parts and Stage-II covering southwestern parts of the command area. About one lakh ha. area has been covered in Stage-I under various works.

In spite of the prevailing adverse conditions, afforestation work undertaken, taking advantage of the availability of water, was very successful in the Stage-I area. Most of the irrigated plantations have done very well and what used to be a barren sandy tract 15 years ago is now giving a look of thickly vegetated tract. The impact of this massive afforestation is well visualized. As a result of afforestation, the barren desert areas have been transformed into long green belts and the microclimate of the area has changed markedly (ISFR, 2009).

In Stage-II of the main canal up to 620 R.D., all the canal systems, almost all roads, fuelwood Abadi plantation have been planted and about 80,000 hectare area has been treated under pasture development. The impact of this massive afforestation is being felt in the form of green belts providing protection to the canal system and agricultural fields. The plantation activities in IGNP Stage -II have twin objectives of protection and production forestry. Nonetheless, the former remain the overriding consideration to protect the canals from wind-borne sand deposition and shifting sand dunes, a spectacular phenomenon in the region. The plantations have been raised under various activities like canal side, block, sand dune stabilisation, pasture development, roadside and environmental improvement.

The canal system proved a boon for the local inhabitants as it facilitated growing of trees for improving microclimate of the region and made available wood for local needs, provided water for drinking purposes, irrigation and electricity generation and thus enhanced crop (grain & fodder) productivity for life sustenance, however, injudicious use of water has created problem of waterlogging and secondary salinity in many areas. It has also affected biodiversity of the region to some extent.

BIODIVERSITY CONSERVATION AND TRADITIONAL KNOWLEDGE

Traditional knowledge is vital for sustainability of natural resources including forests, water, and agro-ecosystems across landscape continuum spanning from households through farms, village, commons and wilderness.

Management of natural resources cannot afford to be the subject of just any single body knowledge, but it has to take into consideration the plurality of knowledge systems. Both application of scientific research and local knowledge contribute to the equity, opportunity, security and empowerment of local communities, as well as the sustainability of the natural resources. Local knowledge helps in scenario analysis, data collection, management planning, designing of the adaptive strategies to learn and get feedback, and institutional support to put policies into practice. Local knowledge systems have been found to contribute to sustainability in diverse fields such as biodiversity conservation and maintenance of ecosystems services, tropical ecological and biocultural restoration, sustainable water management, genetic resource conservation and management of other natural resources. Local knowledge has also been found useful for ecosystem restoration and often has ingredients of adaptive management.

Over thousands of years local people have developed a variety of vegetation management. People also follow ethics that often help them regulate interactions with their natural environment. Such systems are often integrated with traditional rainwater harvesting that promotes landscape heterogeneity through augmented growth of trees and other vegetation, which, in turn, support a variety of fauna.

The traditions are also reflected in a variety of practices regarding the use and management of trees, forests and water. These include:

- collection and management of wood and non-wood forest products;
- traditional ethics, norms and practices for restraint use of forests, water and other natural resources;
- traditional practices on protection, production and regeneration of forests;
- cultivation of useful trees in cultural landscapes and agroforestry systems; and
- creation and maintenance of traditional water harvesting systems such as tanks along with plantation of the tree groves in the proximity.

Local knowledge has proved useful for forest restoration and protected area management in Rajasthan — one of the driest regions of India with scanty rainfall. Cultural landscapes in rural and urban areas and agro-ecosystems, created by the application of scientific and local knowledge, also support a variety of trees, birds and other species, and provide opportunity of integration of nature and society.

AFRI for Managing and Improving Desert Ecosystem

Arid Forest Research Institute (AFRI) was established by the MoEF, Govt. of India to cater to the forestry research needs of the states of Rajasthan and Gujarat. The institute is assigned the task of developing suitable technology for improving the environment in harsh climatic conditions of arid region of Rajasthan. It has made significant strides in the fields of soil and water management, sand dune stabilization, selection of suitable species and planting practices in salt affected areas, use of industrial effluents in raising plantations, site-specific agroforestry and silvi-pasture models, tree improvement, biofertilisers and their use in improving growth and productivity, non-timber forest products and value additions, management of plantations and integrated pest management.

Urban afforestation work has been done in Jodhpur city, which not only added to the aesthetic beauty but also contributed in improving microclimate. Few species like *Cassia siamea, Dalbergia sissoo, Azadirachta indica, Alastonia scholaris and Pongamia pinnata* performed very well in comparison to others while using sewage water. A status report has been prepared on the pasture lands (Orans & Gauchers) of Jodhpur district, and models have been developed on Ostran and Tulesar villages for improving productivity in pasture lands (Parihar G.R. 2015). Work has been done on identifying the Khejri mortality in western Rajasthan, and measures have been prescribed to control it. Growth and yield models have been developed for managing shisham and eucalyptus plantations in IGNP area and neem plantations in Gujarat on sustained basis. Also, cultivation and silvicultural management practices have been developed for some important medicinal plants of arid zone. Presently, the institute is engaged in many important research activities such as propagating bamboo plantations in the mandated states, integrated management of Rohida plantations, biofuels, assessing potential of non-timber forest products in southern Rajasthan and their impact on rural economy, demonstrating water conservation techniques in Arawali region, developing appropriate technology for afforestation in saline-alkaline lands in Kutch, exploring effective extension technology, producing salt-resistant plants through gene transfer technique, ensuring peoples' participation in forestry related activities and identifying criteria and indicators for successful joint forest management and bio-drainage for solving the problem of waterlogging in IGNP area, a problem which has been developed due to injudicious use of water.

Suggestions for Development of Fragile Desert Ecosystem

- Uncontrolled grazing has to be curbed with the active cooperation of the people and pasture developed on scientific lines with increased carrying capacity. The introduction of silvi-pastoral system with supply of seeds of improved variety of grasses, legume, fodder crops, etc., should be encouraged.
- Pastures are the backbone of the desert economy and as such their identification through proper surveys to find out the extent of village-wise area is necessary. For getting this work done, help of Revenue, Agriculture, Animal Husbandry and other like departments may be taken. Conservation, management and improvement of Orans and Gochers of the area should be given high priority.
- There should be no grazing on the stabilized sand dunes during the first two years and grasses had to be harvested manually to be stall fed to animals. This generally did not happen and animals were left to graze which adversely affected the stabilization process.
- The period of maintenance of plantations for the desert areas be increased and more village people be involved in farm forestry. More tubewells may be provided with power connections as additional irrigation facilities are likely to bring prosperity to the area, but a cautious approach should be adopted as more and more areas are coming under "dark zone". Groundwater should be monitored regularly. Efficient and effective water conservation techniques need to be adopted to solve water scarcity and recharge of groundwater.
- It is suggested that the Block Agency, which has a team of extension workers and generally have a support with the village people should be effectively involved especially in the schemes under agriculture, forestry and pasture development. The programme of afforestation can make a big dent if uncultivable land owned by individuals could be covered with proper incentives to them and keeping their rights intact.
- The marginal lands should be removed from cropping by acquisition, by financial inducements or by the establishment of forest, grazing or water catchment reserves. When such measures involve the disruption of traditional livelihood systems, they are unlikely to succeed unless they form part of larger schemes of rural reconstruction involving appropriate changes in land tenure, the consolidation of holdings or resettlement programmes offering alternative livelihoods.
- Wind erosion, which blows soil away from rain-fed crop land and which causes sand drift and dune encroachment, can be countered by planting shrubs and trees in shelter belts. Fences can be constructed or lines of resistant shrubs and trees planted as barriers against oncoming sand, upwind of threatened areas. Bare sand can be covered with mattering, bituminous coating or mulches of vegetation litter.
- It is now well established that symbiotic root microorganisms (*Rhizobium*, *Frankia* and mycorrhizal fungi) can effectively contribute to tree productivity in

marginal climactic and edaphic conditions. Since significant advances have been made recently in the manipulation of these microorganisms, it is now possible to contemplate their use in the field.

- The main aim of afforestation must be the control of desertification and production of wood for local need, whereas possibilities for commercial timber production are limited. It should be pointed out, however, that some drought-resistant tree species, which are used for shelter belts and desert reclamations, also might produce useful by-products with high commercial value. Furthermore, tree plantations may produce valuable forage for cattle, even fruits for human consumption, and food and shelter for game animals.
- People's participation and cooperation of the rural and urban people is must for protecting environment and managing fragile ecosystem. It is the need of hour that both, the public and government research institutions, should join hands together for environment and wildlife protection and maintaining ecological balance for making our future life safe.

REFERENCES

1. Desertification and Land Degradation Atlas of India (2007). Space Applications Centre, Indian Space Research Organisation (ISRO), Department of Space, Government of India.
2. Indian State of Forest Report (2009). Forest Survey of India, Ministry of Environment and Forest, India.
3. Nation Action Programme (NAP) to Combat Desertification in the context of United Nations Convention to Combat Desertification (UNCCD). (2001) Ministry of Environment & Forests Government of India, New Delhi.
4. Parihar, G.R. Faunal Diversity Richness of Ecological Institute in the Village Landscape of the Thar Eco-Region Rajasthan, India (2015).
5. Sinha, R.K. Bhatia, S., and Vishnoi, R. (1997). Desertification control and rangeland management in the Thar desert of India, RALA REPORT NO. 200. Indira Gandhi Centre for Human Ecology, Environment and Population Studies, Rajasthan University, Jaipur, India.

17

Economic Management of Urban Forestry Amenities of Chandigarh City (India)

Pradeep Chaudhry

Silviculture Division, Arid Forest Research Institute, Jodhpur-342 005, India.

E-mail: pradeepifs@yahoo.com

ABSTRACT

Urban forests or urban greens are responsible for various environmental, social and educational benefits to the human society. These benefits are called non-market benefits (NMBs) or intangible benefits as these are not traded in markets. Pressure of ever-increasing human and traffic population is not only telling upon cities' infrastructure but also on green belts of these big cities. Certain unscrupulous elements in the society always try to grab such green open spaces, especially in developing countries, for creation of concrete jungles. This happens because NMBs of such areas are not correctly valued and not fully incorporated into cost-benefit analysis of different developmental projects and into decision making on resource allocation. Chandigarh is one of the planned cities of India, established after the country got independence in 1947, which is known for its magnificent urban greenery (including its reserved forests and one wildlife sanctuary). In India, very little work has been done on quantification of recreational use value of forest resources. Basically, these types of studies have been restricted so far to only few National Parks and Sanctuaries. No study has been undertaken so far for finding out recreational use value of urban forestry of a big Indian city. Therefore, Chandigarh city was selected for this purpose. The non-market value of recreational benefits provided by the urban forestry of the city from the point of view of residents was estimated. The mean willingness to pay (WTP) for the betterment of existing green landscape features and for creating new parks/gardens on the part of each reasonably earning family residing in the city was found to be Rs. 153/- per year for a period of five years, which converts to an annual recreational use value of city's urban forestry assets to Rs. 27.50 million at 2002-03 prices.

Keywords: Urban forestry, CVM, Chandigarh, use value, parks/gardens.

INTRODUCTION

The various benefits – both market and non-market – of urban forests are often ignored or grossly underestimated. On the other hand, various developmental options for the area may be attractive to politicians and bureaucrats due to immediate returns from these projects. It has been observed that urban parks and open spaces left for green belt formation in future are subject to development pressure, developing countries (Dwyer, 1994). One possible reason for this is that the planners and researchers have been unable to articulate their value in economic terms (More, TA et al., 1988). If benefits from the urban forest areas are not properly assessed, the authorities' decision may be biased to implement development activities such as construction of schools, shopping complex, housing flats, etc., in open and green spaces. Once the monetary valuation of recreational and other non-market benefits is done, it can be introduced into public decision making and cost-benefit analysis of projects. Moreover urban development projects often decrease the amenity values of green spaces, which should be taken into consideration in planning. Therefore, quantitative information on residents' valuations attached to urban forestry is needed for assessing urban land-use. Keeping these facts in view, an attempt was made to quantify the recreational use benefits of urban forestry of Chandigarh city of India, which was created after country's independence in 1947 and is known for its urban forestry (Mohd. Shahwahid et al., 1998).

The present study was designed and conducted during 2002-03, mainly to find out use value of urban forests/parks of the Chandigarh city of India. In other words, to assess the 'public good' value of the urban forestry of the city in terms of recreational benefits from the point of view of city's residents and tourists visiting the city. Although there are a number of studies on the topic of assessing recreational benefits, (Cole D, 1996), majority of them have been carried out on a specific wildlife national park, protected areas, city forests, and places of scenic beauty and cultural heritage in developed countries of North America, Europe and Australia. No such study on valuation of recreational aspects of urban forestry has been carried out in any city of India so far. Outside India, a limited number of references have been found when recreational benefits of urban forests/ wooded areas/ parks have been quantified in monetary (Willis, 1989; Willis et al., 1991). During the last three decades a number of valuation methods have been developed to derive monetary measures of the value of the changes in the quality or quantity of unpriced goods or services or to place monetary value on their recreational benefits. Two methods, i.e., Travel Cost Method (TCM) and Contingent Valuation Method (CVM) were used in the present study as these are universally regarded as proper techniques for estimating recreational value of various non-market goods/services, and sufficient literature is available on them in developed countries (Harrison, 1993; Hausmann, 1993). Zonal travel cost method was used on the tourists visiting the city, and Contingent valuation method (open-ended) was used on the residents/working people of the city. However, in the present paper, only the CVM analysis has been presented.

STUDY SITE

Le Corbusier, an eminent French architect and urban theorist, is known to be the main architect of Chandigarh city's planning and master plan. Tree plantation and landscaping has been an integral part of the city's master plan. The most fascinating feature of the city's landscaping is perhaps the tree plantation along avenues, open spaces, green belts and around building complexes. A number of beautiful avenues with conspicuous tree species, well wooded forests along the periphery of city, Sukhna Lake against the backdrop of lake reserve forest and green belts running across the length and breadth of the city enhance recreational and aesthetic value of the whole city. The total forest cover in the U.T. is 32.42 km^2, which forms 23.50% of the overall geographical area. The green spaces like 1900 numbers of small and big parks/gardens maintained by Municipal Corporation, Chandigarh; green belts, road-avenues, etc. are in addition to this forest cover of 23.50%. Thus, the green cover of the city is more than 33% of its area (Action Plan-2002-03). In spring, it is a symphony of changing colours as different trees bloom with radiant blossoms. With the onset of summer, atmosphere starts rippling and sparkling with different shades of violet colours of *Millettia ovalifolia* (Rosewood), *Jacaranda mimosifolia* (Nili gulmohar) and *Lagerstroemia thorelli* (Queen's flower). Pinkish flowers of *Cassia javanica* (Java ki Rani) and *Lagerstroemia rosea* (Pride of India) on one hand, and red flowers of *Delonix regia* (Gulmohar) on the other indicate peak period of summer season. During winter season, groves of *Koelreuteria apiculata* in different parks like Rose Garden, Garden of Fragrance and along roads produce yellow flowers in October-November which turn bronze, dull crimson, red-coloured pods and look very attractive. A few other majestic ornamental trees, which have the capacity of attracting an environmentally conscious tourist's attention because of their impressive foliage in the city, include *Chukrasia tabularis* (Chukrasia), *Ficus benjamina* (Golden fig), *Magnolia grandiflora* (Bara champa), *Pterospermum acerifolium* (Kanak champa), *Mimusops elengi* (Moulsari), *Polyalthia longifolia* (Ashok), *Putranjiva roxburghii* (Jiva putra), *Schleichera oleosa* (Kusum), *Swietenia mahagoni* (Mahagony), *Terminalia arjuna* (Arjun) and *Barringtonia acutangula* (Samunder phall). The meandering Leisure Valley with its myriad colours and texture of its trees and flowers, attracts not only city-dwellers but tourists and other city-planners also. On the whole, the city gives the impression of a luxuriant garden.

METHODS USED

The CVM is used to estimate economic values for all kinds of eco system and environmental services. This method involves directly asking people, in survey, how much they would be willing to pay for a specific environmental service. The CVM is referred to as a "stated preference" method, because it asks people to directly state their value, rather than inferring values from actual choices, as the "revealed preference" methods do. The fact that CVM is based on what people say they would do, as opposed to what people are observed to do, is the source of its greatest strengths and its greatest weaknesses. It is also the most controversial of the non-market valuation methods.

Primary data was collected using structured and unstructured interview schedule with the respondents who were above twenty years in age and working somewhere in the city. In addition, housewives were also included in the survey as they constitute nearly 44% of city's population and have a definite role in managing household finances. Moreover, most of the women (literacy rate: 76.65%) also earn something and add to family's income by way of offering tuitions to children or working part time somewhere in the city.

The first few warm-up questions were related to the socio-economic status of the respondents like occupation, net household monthly income, household size, educational qualifications, etc. Chandigarh city has substantial number of people belonging to salaried class, either in government or in private sector, which is one of the main reasons for this U.T being first in India in high per capita monthly income (Tribune News Service, 2002). Therefore, it was decided, in the beginning, to accommodate maximum number of people from these two categories in the sample followed by business class, professionals and housewife category.

The valuation question was based on an open-ended CVM format, in which respondents were asked to indicate the maximum sum that they would be willing to contribute per family, once in a year, in the "Environment Fund" if created by the U.T. administration for the maintenance of existing and for creating new parks/gardens in the city. This simple approach was adopted, in preference to the dichotomous choice question format recommended in the NOAA (Arrow et al., 1993), because the use value of the city's urban forestry was being estimated rather than its non-use value (Garrod and Willis, 1997). Dichotomous choice questions can offer certain methodological advantages over open-ended CVM formats (Willis, 1995); however, WTP estimates based on dichotomous choice data are typically several times greater than those derived using data from open-ended CVM questions (Schulze et al., 1996). It is possible that dichotomous choice formats encourage respondents to overestimate their WTP through a phenomenon known as "yea" saying, and these formats may produce less accurate results than carefully constructed OE questions, particularly when respondents are familiar with the good and are clear about the increase or decrease in amenity that they are being asked to value (Fisher, 1996).

Currently, a number of questions are being asked regarding the ability of CVM to produce reliable and robust estimates of values for environmental and private goods. These arise from concerns about strategic behavior, hypothetical bias, information effects, the sequencing of questions and the resultant problems of embedding and scope effects (Hanemann, 1996). There are no definitive tests for strategic behaviour, but the modest WTP amounts in relation to socio-economic status of respondents suggest that such behaviour was either absent in the sample or led to an underestimation of the value of the environmental good/service. Hypothetical CVM questions can produce unrealistic answers, but the city's urban forestry is a well-defined good with which respondents are familiar, and for which it would be feasible to levy a monthly or yearly maintenance charge.

A sample size of 2400 was selected based on the criteria adopted earlier during quesitionaire survey for tourists and adopted by Mitchell and Carson (1989). The actual survey produced 2358 number of responses, which were complete in almost all the aspects and covered residents of different occupations, spread over different sectors of the city.

In order to feel comfortable with the questionnaire survey, the respondents were asked a few environmental related questions. The "Green Cover" in the form of tree avenues, boulevards, parks, gardens etc. in the city during last 15-20 years has improved or deteriorated, was the first question related to urban forestry of the city. Majority of residents (61.50%; 1456/2358) were of the view that urban forestry/green cover of the city had improved during this period, while 35.37% (834/2358) differed from this view and voted for "deterioration" clause because of reasons like road-widening, encroachments, termite-related problems on the trees, etc. Only 3.13% people did not reply the question as they could not differentiate between the status of urban forestry at present and 15-20 years ago.

The city has one wildlife sanctuary, called Sukhana Wildlife Sanctuary covering an area of just behind lake in the foothills of Shivalik hills. The residents were asked, whether they had visited the sanctuary? If not, then what was the reason? Only 43.25% of residents (1020/2358) had visited the sanctuary. Out of the rest, nearly one-third (i.e., 34.18%; 806/2358) were not aware of any such sanctuary in the city, and (17.47%; 412/2358 cited the reason of not getting time to visit it, though they were aware of its existence. A meagre 2.24% (53/2358) cited the reason of difficulty in getting permit for such areas. 2.83% (57/2358) gave no reply on this subject, as they were not interested in it.

The residents were asked to specify the tourist site within the city, which their family likes most. Maximum number of response (1657/2358, 70.30%) was received in favour of Sukhana Lake, followed by Rock Garden (426/2358, 18.0%) and Rose Garden (210/2358, 9.0%). Rest of the spots like Sector-17 Plaza, Leisure Valley, Shanti Kunj, Garden of Fragrance, etc., got rest of the responses. In case of tourists, it was the Rock Garden, which received maximum response. The outcome of this question is not surprising as all categories of residents, right from elders to youngers, prefer lake view for different reasons like walking, jogging, boating, eating out, etc., in open space restaurants.

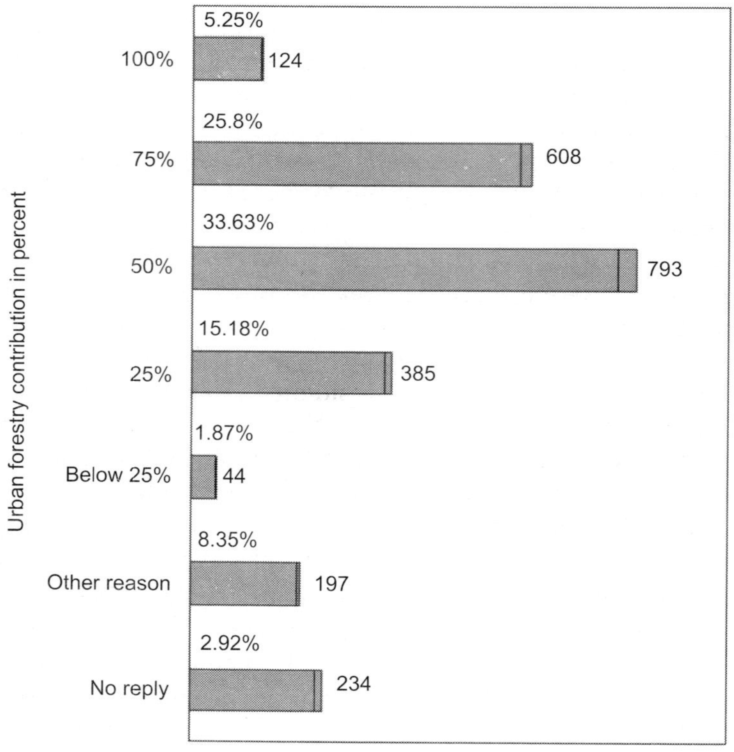

Fig. 17.1 Percent contribution of greenery vis-à-vis place of residence/working

Respondents were asked specifically to mention the name of the city in India, which they think was ahead of Chandigarh in the field of urban forestry. During survey, it was found that 36.85% (869/2358) of residents, mainly below 35 years of age, had not got much opportunity to move out to other cities in the country. Two hundred and thirty-five people did not reply the question, while 53.20% (1254/2358) replied the question in an appropriate manner. Out of the 1254 answers, 389 voted for Bengaluru and 424 for Chandigarh, i.e., very close race was observed between the two cities to grab top position in urban forestry. New Delhi, Pune, Gandhi Nagar, Bhubaneshwar, Jamshedpur, Hyderabad, Bhopal, Mysore, Dehradun, Panchkula and Mohali, etc.,were few other cities which also got some response in this respect. Nevertheless, during survey, the researcher came across a few residents, who declared confidently that Bangalore was not ahead of Chandigarh in terms of urban forestry. Sharma (2003) has mentioned that Bangalore, which is called the "city of gardens", and the coastal areas, which are lined with endless palm trees, can only be dubbed poor cousins of Chandigarh as far as parks/gardens are concerned.

Due to comparatively clean and green environment of the city, many people are being attracted to this city to permanently settle down here. Respondents were asked that what was the percentage contribution of green cover of the city, prompting them to select Chandigarh as a place of working or as permanent residence? While deciding about this percentage contribution (equivalent to marks out of 100), they were to consider employment opportunities, educational facilities and other infrastructural and entertainment facilities available in the city, competing with the green cover of the city. City's parks, gardens, ornamental trees and green avenues were found responsible to the extent of 55.65%, in attracting people to reside or work permanently in the city (Fig. 17.1). The other factors like employment opportunities, educational facilities, etc., together account for the rest of 44.35 % marks given by the respondents.

During the last few years, particularly due to large-scale influx of migrant labourers from the states of UP and Bihar, the civic facilities of the city are under strain (Banerjee, 2003). The respondents were asked to give their opinion whether Chandigarh Administration was taking adequate steps and doing good work to improve overall environmental quality? 68.78% residents (1622/2358) seemed happy with the Administration on this aspect. Rest 28.58% were worried on different accounts of environment degradation, i.e., unauthorized colonies, slum-development, poor response of municipal corporation in lifting garbage, unhygienic sanitary conditions in some parts of the city, termite and pest attack on roadside trees, non-development of parks in southern sectors of the city and problem of stray cattle on the roads. The latter category of people, in general, were older residents of the city who had seen the city growing and developing during past 3-4 decades, and they were in a better position to compare earlier Chandigarh with the present one.

After some more questions, the respondents were asked the C.V. question, who's actual wording was:

"For the proper maintenance of city environment, mainly for creation of more green areas like parks and gardens for the enjoyment and recreation of your family, for better maintenance of present parks and for more effective pollution control measures, suppose Chandigarh Administration creates an "Environment Fund", what amount would you be willing to contribute per family, per year, for a period of five years in this fund?"

Please bear the following points in your mind while answering the above question:

- Chandigarh is a well-planned and green city in India.
- Your income is limited and has important alternative uses.
- There are other cities also in India. At present, kindly concentrate upon the Chandigarh city only for valuation of recreational aspects of urban forestry.

Weighted mean WTP on the part of the residents of the city in the "Environment Fund" was calculated separately for five different categories, while taking into account number of people quoting different amounts per family per year, ranging from Rs. 10 to Rs. 2000. A few respondents belonging to highest income band of above Rs. 15000 per month net household income, quoted for Rs. 5000 and above as WTP amount. Such respondents quoting amounts greater than Rs. 2000 per year per family were not taken in to account for calculating MWTP so as to arrive at a genuine MWTP, which should not be deceivingly too high due to a few individuals reporting very high WTP amounts (Carson, 1991). One or two such outliers can increase average WTP dramatically and can provide a false indication of the worth of a public good. Majority of respondents lying above and below the average income (which was Rs. 10216) quoted WTP amounts ranging from Rs. 50 to Rs. 500 per year per family. By and large, the overall impression in respect of residents of the city was that people who were willing to participate in the interview did their best to give an honest answer to the CV questions, and uninterested persons due to lack of time or other reasons like fear of giving something in writing or avoiding any government agency, etc., were separated out at the time of primary data collection. To avoid making decisions on the truthfulness or falsity of individual observations in a data set, a common practice is to simply trim the top and bottom 5% or 10% of the distribution of WTP observations. Clearly this is not strictly legitimate, some of these observations may be true estimates (Garrod and Willis, 1999). Therefore, WTP responses with zero bids have not been removed in the analysis for calculating MWTP, which otherwise would have increased its value (i.e., of MWTP).

Overall sixty four percent of respondents, which include people either working or residing in Chandigarh, were willing to pay WTP amount towards "Environment Fund". This percentage is on lower side in comparison to that reported from developed countries when almost three-quarters (74.90%) of all respondents were willing to pay for an annual entry fee in the urban forest of Allschwill near city of Basel, Switzerland (Kleiber, 2001). Other such kind of surveys related to the urban forest entry fee reported more than 70% (Elsasser, 1996: 72.90%; Winter et al., 1999: 71.60%).

A close scrutiny of stated WTP of the residents reveals that "Professional" category is the group, which includes mainly self-earning lawyers, chartered accountants, engineers, doctors, architects, executives in the fields of marketing, finance and production etc., that has quoted maximum WTP, i.e., Rs. 211 per year per family, whereas persons from "government- service" category have quoted lowest amount of Rs. 103. The percentage of such employees wanting to contribute is also lowest at 51%. Main findings of the CVM survey among residents of the city are presented in Table 17.1.

Table 17.1 Results of CVM survey among people of the city

S. No.	*Govt. Service*	*Private Service*	*Business-men*	*Profes-sional*	*Housewife*	*Overall*
1. Respondents who furnished information as per CV-format	725	725	378	287	243	2358
2. Respondents who agreed to contribute towards "Environment Fund" (EF)	369	521	268	218	136	1512
3. Respondents who declined to contribute towards EF due to different reasons	356	204	110	69	107	846
4. Percentage of respondents, willing to contribute in EF.	51	72	71	76	56	64
5. Mean age (years)	43	30	36	34	38	36
6. Mean net monthly household income (Rs.)	12231	8190	9709	10441	10771	10216
7. Mean WTP per family per year (Rs.)	103	185	169	211	115	153

The main reason, as found during the survey, for this low WTP amount was that they were already paying income tax annually and it was the duty of government to invest in the urban forestry of the city from its resources. In the CV study of BNP, Mumbai, 25 percent of respondents said that government should pay for environmental preservation, whereas 9 per cent did not respond to the offered bids under double-dichotomous bid method and did not disclose why (Hadker et al., 1997). In the present case, however, a few more government employees agreed to contribute, provided central government gave some income-tax relief on such amount. A few respondents of other category, including housewives, complained about corruption in the system due to which they declined to contribute towards such fund. In spite of all these hick-ups, nearly two-thirds of the people were willing to pay the amount with the condition, as mentioned by majority of younger ones, that the government must utilize the amount properly for the purpose it was being collected.

The annual recreational value of city's urban forestry (on 2002-03 prices) comes out to be:

Annual recreational use value = (Mean WTP) (number of families)

= Rs. 153 × 1,80,000.

= Rs. 2,75,40,000.

= Rs. 27.50 million (say).

The above value has been arrived at by taking a conservative estimate of 1.80 lakh families in the city, assuming on an average, four persons per family, totaling to 7.20 lakh and excluding slum population of the city. This annual recreational use value of the city's urban greenery has been estimated by CVM (OE) and conducting questionnaire survey among average city residents, including people of low as well as high income groups, people actually using as well as not regularly using these spots for walking/jogging and family recreation purpose. Therefore, the above estimated economic use value seems to be the lower limit of the annual recreational use value, whereas for the higher limit of this value, Individual Travel Cost Method (ITCM) has to be utilised by conducting survey among people habitual of using these sites regularly in different parts of the city. It is likely that ITCM would yield higher value estimates than CVM because people actually enjoying these spots are coming to these different parks/gardens from different corners of the city as well as from two satellite towns of Panchkula (Haryana state) and Mohali (Punjab state). Their revealed behavior through ITCM would most probably give higher consumer surplus per visit than CVM. Only then the exact range of the annual recreational use value of City's urban greens, from the point of view of residents, can be known.

MULTIPLE REGRESSION ANALYSIS

Both ordinary least squares (OLS) and weighted least squares (WLS) methods have been used to develop relationship between the dependent variable (WTP function for the residents) and other predictive independent variables. As mentioned above, few questions related to environmental resources like, visit to Sukhna Wildlife Sanctuary, position of Chandigarh city in urban forestry with respect to other cities, improvement or deterioration in green cover of the city in terms of tree avenues, parks, landscapes, etc., during last 15-20 years etc., were asked from the respondents. In the present analysis, these variables have been considered along with other socio-economic variables like age, education, income and household size. In total, nine independent variables were considered, the F-statistic was found significant ($F = 40.182; p < 0.00$) which evinced overall significance of the model. The results are summarised in Table 17.2. The fit statistics was not very satisfactory. R^2 value was only 14.5%, while standard error was a bit high, i.e., 222.63. This is indicative of the fact that there was a fairly large variation in the socio-economic variables and the response given by the residents.

Table 17.2 Analysis of variance results of multiple regressions using OLS technique

	DF	*Sum of squares*	*Mean square*	*R^2*	*SEE*
Regression	9	17924950.6	1991661.18	0.145	222.634
Residual	2140	106071197.0	49565.98		
F= 40.182 (p < 0.000)					
Variables	*Value*	*SE*	*t*		*Sig. t*
Constant	168.054	36.028	4.665		0.001
Age	– 4.360	0.506	–8.611		0.001

Income	0.0103	0.001	9.735	0.001
Education	20.981	6.619	3.170	0.002
Green cover Yes=1, No=0	47.677	10.248	4.652	0.001
wls Yes=1, No=0	34.145	9.774	3.493	0.001
wta Yes=1, No=0	76.504	15.900	4.811	0.001
Urban forestry Chandigarh Yes=1, No=0	11.209	11.259	0.996	0.320
Environmental quality Yes=1, No=0	−17.163	10.113	−1.697	0.090
hhs	−22.947	3.690	−6.219	0.001

wta = willingness to accept compensation, wls = wild life sanctuary, hhs = household size

The *t*-statistics for the different regressor (independent) variables showed that the parameter estimates of two variables, i.e., urban forestry Chandigarh and environmental quality were not statistically significant. Therefore, these two variables can be discarded from the model in forecasting (Mantymaa, 1999).

The use of WLS method has slightly improved the performance of the model (Table 17.3). Per capita income was used as source variable for weighting, and weight assigned was its power to 0.5. Standard error was drastically reduced, which indicated much greater precision in the prediction by the model obtained through WLS. Here also, the two variables, i.e., urban forestry and environmental quality did not significantly affect the prediction of the model as can be seen from their *t*-statistics.

Table 17.3 Analysis of variance results using weighted least squares technique

	DF	*Sum of squares*	*Mean square*	R^2	*SEE*
Regression	9	364597.90	40510.88	0.151	30.838
Residual	2140	2035181.40	951.02		
$F = 42.597$ ($p < 0.000$)					

Variables	*Value*	*SE*	*t*	*Sig. t*
Constant	188.839	32.827	5.752	0.001
Age	−4.022	0.467	−8.618	0.001
Income	0.0102	0.0009	10.633	0.001
Education	15.286	5.936	2.575	0.010
Green cover Yes=1, No=0	46.592	9.501	4.904	0.001
wls Yes=1, No=0	32.576	9.093	3.582	0.001
wta Yes=1, No=0	73.510	15.158	4.850	0.001

Contd...

Contd...

Urban forestry Chandigarh Yes=1, No=0	10.315	10.727	0.962	0.336
Environmental quality Yes=1, No=0	−17.561	9.253	−1.898	0.058
hhs	−24.312	3.468	−7.009	0.001

wta = willingness to accept compensation, wls = wild life sanctuary, hhs = household size
Log-likelihood function = −14577.0308

Assessing Reliability and Validity

Mitchell and Carson (1989) proposed that an R^2 value around 0.15 or more is taken as an evidence of reliability. But low *R*-squared values are not necessarily evidence of theoretical inconsistency; they could merely be indicative of less tightly clustered observations around the regression line and stochastic variation in the data, which overshadows the systematic influence of variables. Such low R^2 values and statistically insignificant coefficients on predictor variables are often associated with environmental goods, which attract very small bids (Garrod and Willis, 1999). In some studies, regression models only managed to explain 3% or 4% variation in WTP responses (e.g., Bateman et al., 1995); whilst in some other cases 60% or more of the variance is explained (Garrod and Willis, 1997). In the present study, the R^2 value in case of the residents of the city is 15%. In case of the studies in developing countries, such low values are not uncommon. Murty and Menkhaus (1994) found R^2 values of 0.12 and 0.17 for their study on recreational value of Keoladeo National Park, Bharatpur on domestic and foreign tourists respectively. Other published studies in this field have had similarly low R^2 values (Maille and Mendelsohn, 1993).

CONCLUSION

The mean willingness to pay (WTP) for the betterment of existing green landscape features and for creating new parks/gardens on the part of each family residing in the Chandigarh city has been found to be Rs. 153/- per year for a period of five years, which converts to an annual recreational use value of city's urban forestry assets to Rs. 27.50 million at 2002-03 prices. Socio-economic variables of the city's residents like age, income, education and profession affect WTP for urban greenery.

Acknowledgements

This paper forms part of the authors' doctoral thesis work entitled "Valuing recreational benefits of urban forestry- A case study of Chandigarh city" undertaken during 2002-04. The doctoral degree was awarded by FRI (Deemed University) Dehradun during 2006-07.

REFERENCES

1. Action Plan (2002-03). Greening Chandigarh. Forest Department, Chandigarh Administration, Chandigarh.

2. Arrow, K.J. Solow, R. Leamer, E. Portney, P. Radner, R. and Schuman, H. (1993). Report of the NOAA panel on CV. *Resources for the Future*, Washington DC.
3. Banerjee, A. (2003). Rich like us. *The Tribune* (Chandigarh edition). Aug. 2, 2003.
4. Bateman, I.J. Langford, I.H. Willis, K.G. Turner, R.K. Garrod, G.D. (1995). Elicitation and Truncation effects in C.V studies. *Ecol. Econ.*, 12: 161-179.
5. Carson, R.T. (1991). Constructed markets. In: Braden, J. and Kolstad, C. (Eds.) Measuring the demand for environmental commodities, Amsterdam, Netherlands.
6. Cole, D. (1996). Wilderness recreation use trends, 1965 through 1994. Research paper INT-RP- 488. U.S. Department of Agriculture, Forest Service, Rocky Mountain Research Station, Ogden.
7. Dwyer, John. (1994). Customer diversity and the future demand for outdoor recreation. Gen. Tech. Report. RM-252. U.S Department of Agriculture, Forest Science, and North Central Forest Experiment Station, St. Paul, MN.
8. Elsasser, P. (1996). Der Erholungswert des Waldes: Monetare Bewertung der Erholungsleistung ausgewahlter walder in Deutschland. *Schriften zu Forstokonomie*, Band 11. Sauerlander, Frankfurt.
9. Fisher, A.C. (1996). The conceptual underpinnings of the CVM. In: Bjornstad, D.J. and Kahn, J.R. (Eds.) *The Contingent Valuation of Environmental Resources: Methodological Issues and Research Needs*. Edward Elgar, Cheltanham.
10. Garrod, G.D. and Willis, K.G. (1997). The recreational value of tropical forests in Malaysia. *J. World Forest Resour. Managem.*, 8: 183-201.
11. Garrod, G.D and Willis, K.G. (1999). Economic Valuation of the Environment. Edward Elger Publishing Ltd., Northampton, Massachusetts, USA.
12. Hadker, N. Sharma, S. David, A. and Muralledharan, T.R. (1997). Willingness-to-pay for Borivilli National Park: evidence from a contingent valuation. *Ecol. Econ.*, 21: 105-122.
13. Hanemann, W.M. (1996). Theory versus data in the contingent valuation debate. In: Bjornstad D.J. and Kahn J.R. (Eds.) The Contingent Valuation of Environmental Resources. Edward Elgar, Cheltenham.
14. Harrison, G.W. (1993). General Reactions to the NOAA Report. Reports for the CVM network. Stockholm School of Economics, Stockholm, March 12, 1993, pp. 19-25.
15. Hausman, J.A. (Ed.) (1993). *Contingent valuation: A critical assessment*. Elsevier Science Publishers BV, The Netherlands.
16. Kleiber, O. (2001). Valuation of recreational benefits and visitor conflicts in an urban forest. In: Ecological economic management and planning in regional and urban systems. *Proceedings of 5th International Conference of the International Society for Ecological- Economics (ISEE)*, Sep. 26-29, 2001. Institute of Central Sciences, Russian Academy of Sciences, Moscow, Russia.
17. Maille, P. and Mendelsohn, R. (1993). Valuing Ecotourism in Madagascar. *J. Environ. Managem.*, 38: 213-218.

18. Mäntymaa, E. (1999). Pennies from Heaven? A test of the social budget constraint in a willing to accept compensation elicitation format. *J. For. Econ.*, 5: 169-192.
19. Mitchell, R.C. and Carson, R.T. (1989). Using surveys to value public goods: The CVM. Resources for the Future, Washington D.C.
20. Mohd. Shahwahid H.O., A.G. Awang, M. Rusli, M. Shukri, H. Faridah and Z.H. Mohd (1998). Valuing the recreational benefits of Air Hitam Forest Reserve, Selangor. *Malays. For.*, 61(1): 102-108.
21. More, T.A. Stevens, T. and Allen, P.G. (1998). Valuation of Urban Parks. *Lands. Urban Plan.*, 15(1-2): 139-152.
22. Murty, H.N. and Menkhaus, S.M. (1994). Economic aspects of wildlife protection in developing countries: A case study of Keoladeo National Park, Bharatpur, India. In: *Valuing India's Natural Resources. Society for Promotion of Wastelands Development*, New Delhi, pp. 93-150.
23. Schulze, W.D. McClelland, G. Waldman, D. and Lazo, J. (1996). *Sources of bias in contingent valuation.* In: Bjornstad D.J. and Kahn J.R. (Eds.) The Contingent Valuation of Environmental Resources. Edward Elgar, Cheltenham.
24. Sharma, R. (2003). Chandigarhians have evolved a culture of their own. *The Tribune* (Chandigarh edition). Mar. 29, 2003.
25. Tribune news service (2002). Per capita income highest in Chandigarh. *The Tribune* (Chandigarh edition), Aug. 1, 2002.
26. Willis, K.G. and Benson, J.F. (1989). Recreational value of forests. *Forestry*, 62(2): 93-110.
27. Willis, K.G. (1991). The Recreational value of the forestry commission estate in Great Britain: A Clawson-Knetsh travel cost analysis. *Scott. J. Polit. Econ.*, 38(1): 58-75.
28. Willis, K.G. (1995). Contingent valuation in a policy context: the National Oceanic and Atmospheric Report and its implications for the use of CVMs in policy analysis in britain. In: Willis K.G. and Corkindale J.T. (Eds.) *Environmental Valuation: New Perspectives.* CAB International, Wallingford, Oxfordshire.
29. Winter, P. Laura, P. and Rachel, B. (1999). Anticipated responses to a fee programme: The key is Trust. *J. Leis. Res.*, 31(3): 207-227.

18

Management of Energy Resources in India

A.K. Kandya
Department of Botany, Dr. H.S. Gour University, Sagar, M.P.
E-mail: anand.kandya@gmail.com

ABSTRACT

A status report of the renewable and non-renewable energy resources in India has been given in the present paper. Use of non-conventional resources of energy, which are non-biological in origin, is increasing very fast as the exhaustive resources are becoming lesser and lesser day by day. Development of alternative resources of energy is being done worldwide, whether by the developed nations or by the developing and poor countries. The main reason for this is the exploding enhancement in the per capita requirement of energy which was only 20 K cal per head per day in the medieval period but increased by 100 times and became 2000 K cal per head per day after the development of civilization. This energy is utilized in domestic works, agriculture, industries, transport, government offices, educational and research institutions, etc.

Fuelwood has been the oldest and traditional source of energy for all domestic purposes, viz., cooking food, making jiggery and funerals, etc. But, nowadays, solar energy is being used through solar panels to provide street light, solar cars, heating water in large hotels and as solar cookers. Atomic energy from uranium, thorium and silicon, etc., is being harvested for huge industries, coal, diesel and electricity are being used for agriculture, motor transport and railways, and wind energy is being used for several purposes. Details of all power generation plants and their annual production capacities, extraction of fossil fuels and production of bio-fuels, etc., are given in detail.

At present, there are about 130 lakh biogas plants, 4.75 lakh solar cookers, 2.5 lakh alcohol driven and 40 thousand battery driven vehicles in our country. Solar water heating systems cover a collection area of about 3 crore sq. m., solar photovoltaic system produces about 20 MW of electricity per sq. km. per year. Biomass energy produces about 17000 MW and wind energy produces about 20000 MW of electricity.

For a sustainable use of our energy resources, we need pollution free, cheapest resources of regenerative energy and every person must consume the minimum possible amount of energy.

INTRODUCTION

It is said nowadays that up to the end of this century, due to the application of genetic engineering and biotechnology in agriculture, there will be sufficient food items available for the mankind but sufficient amount of fuel to cook them may not be available. Therefore, the scientists and governments of all developed and developing nations are actively engaged in the development of methods to utilize various non-conventional sources of energy.

According to the principle of physics, every work is done with the help of some energy. We also know that energy is required even at the cell level in an organism. During cell division, chromosomes require energy to move from the equator towards the poles. It is well known that energy has several forms, i.e., electrical energy, light energy, heat energy, magnetic energy and sound energy, etc. But, some other forms are also known such as chemical energy, biological energy, mechanical and kinetic energy, etc. All these forms are inter-convertible. There are several conventional sources of energy which are biological in origin, i.e., biomass or dendro-thermal energy (dry organic matter, dead and live organisms and their products) and fossil fuel (mineral oil, coal and natural gas), etc. They are being used by the man since time immemorial.

In the modern and civilized world today, the most widely used form of energy is electricity as it is a source of light for all domestic and office works, viz., computers, laptops, printers, mixers, grinders, electric press, hair driers, vacuum cleaners, micro-wave ovens, washing machines, induction chulhas and even in funerals etc. Electricity is also used to run trains; petrol and diesel used in transport vehicles like buses, trucks, cars, scooters, motorcycles and other two wheelers and three wheelers also generate electric current to run them. Electric power is also used in agriculture (irrigation pumps, threshers, tractors, harvesters, etc.) and cooling/heating water/air conditioners, room coolers, room heaters, water coolers, refrigerators, etc.

Due to the huge requirement of energy during the development of civilization, several non-conventional sources of energy have been developed which are abiotic in origin, viz., solar energy (radiant energy from the sun), tidal energy (waves of the ocean), wind energy (force of the wind harnessed through the windmills), geothermal energy (heat from the earth), hydroelectrical energy (kinetic energy of the moving water), nuclear energy (atomic energy of uranium etc.) and hydrogen energy (burning of H_2 cakes). At the global level, it has been estimated that about 20% of the world's population residing in the developed nations uses 2/3 (about 66%) part of total energy resources. It is also a fact that about 90% of world's energy supply is by mineral oil, coal, natural gas and uranium, whereas about 10% is by hydroelectricity and biomass. Only less than 1% is by solar and wind energy (Verma, 2004). At present, there are about 130 lakh biogas plants, 4.75 lakh solar cookers, 2.5 lakh alcohol driven and 40 thousand battery driven vehicles in our country. Solar water heating systems are covering a collection area of about 3 crore sq. m. Solar photo-voltaic system produces about 20 MW of electricity per sq. km. per year. Biomass energy produces around 17 thousand MW and wind energy produces about 20 thousand MW of electricity.

Fuelwood Energy: It has been the oldest traditional source of energy for cooking food worldwide. Requirement of fuelwood in our country was 150 million tonnes in 1985, but it doubled (300 million tonnes) in a period of only 20 years (in 2005). Though several methods of non-conventional energy sources were developed, our forests became less productive due to huge destruction for developmental purposes. At present, total fuel-wood production in our country from the natural forests is only 50 million tonnes. Destruction of forests is resulting in increased atmospheric temperature, low rainfall, drought and floods every year in some or the other part of our country.

Hydroelectric Power: About 71% of the earth's surface is covered by water. The flowing water is the best source of energy which is obtained in the easiest way and without any pollution. Water is stored in huge dams and allowed to fall down. The force of flowing water moves the turbines through which electricity is generated. The main hydel projects in our country are: Bhakhra Nangal in Punjab, Hirakud in Orissa, Damodar Valley Project in West Bengal, Kosi Project in Bihar, Nagarjun Sagar Project in Andhra Pradesh and Chambal Project in Rajasthan and Madhya Pradesh. And the total power production every year from all these power plants is about 30936 MW (Table 18.3).

Biogas Plants (Gobar Gas): Energy is obtained from biological wastes gathered from agriculture, industries, animals and domestic sources. Every year, about 30 million tonnes of solid waste and 4400 million cubic metres liquid waste is generated in our country. It reaches the underground water and creates severe water pollution. It also produces green-house gases (GHGs) like methane, carbon dioxide, and carbon monoxide and creates air pollution. All this waste is burnt in landfills to obtain biogas, which is environment-friendly.

Bio-energy (Biomass and Fossil fuel): It has been estimated that about 90% of world's energy supply is by mineral oil, coal, natural gas and uranium, about 10% is by hydro-electricity and biomass and only less than 1% is by solar and wind energy. Annual extraction of coal in the world is 2730 million tonnes. Before 1900, 75% energy demand was fulfilled by coal, but nowadays, only 30% is by coal as other sources are also being utilized with their fullest capacities. Oldest coal deposits in the world are 30 crore-year-old, whereas the most recent deposits are 2 crore-year-old. According to an assessment, the amount of energy used by man in one year, is synthesized by the nature in one million years in the form of fossil fuels, i.e., coal, mineral oil and natural gas. Coal deposit in the world is 27350 billion tonnes which will be available maximum up to 800 years. Almost 50% of the deposits are in USSR, 28% in USA and Canada, and the rest in the remaining parts of the world. For the people of poor nations, petroleum products are very costly. For them, cooking gas and electricity are still dreams. The only form of energy available to them is biomass. As a result, about 43% requirement of energy in developing countries is fulfilled by only fuelwood. Oil and natural gas areas in our country are in Andhra Pradesh, Arunachal Pradesh, Assam, Gujarat, Rajasthan, Tamil Nadu, Daman and Bombay High (off shore). Mineral oil has several by-products like ethylene, propylene and benzene, which are used in the production of plastic, polyvinyl chloride (PVC), synthetic cloth, synthetic rubber and rayon, etc.

Environmental damage is a serious problem in energy production. It is caused not only by fossil fuel but also by nuclear energy and biomass. Maximum air pollution is caused by coal burning because it results in liberation of large amounts of CO_2, SO_2 and smoke. However, comparatively lesser amount of air pollution is generated by mineral oil because it produces lesser quantities of CO_2, SO_2 but some amounts of NO_X and particles are also mixed in the air. But, natural gas is least polluting as compared to coal and mineral oil.

An assessment in our country indicated that out of the total conventional sources of energy (biomass and fossil fuel), maximum (36.4%) is used in industries, 33.8% in transport, 15.7% in domestic purposes, 9.8% in agricultural and the rest 4.3% in other works (Table 18.1).

Table 18.1 Use of conventional energy for different purposes in India

Industries	36.4%
Transport	33.8%
Domestic works	15.7%
Agriculture	9.8%
Others	4.2%

When the percentage of various kinds of energy (oil, electricity and coal) used for different purposes in our country was computed, it was found that even now, coal (78.9%) is used in maximum amount in industries, whereas, in agriculture, its use is zero. Consumption of electrical energy is almost equal in domestic works (29%) and industries (30%). Data in Table 18.2 indicate that oil is used in maximum quantity (31.4%) in transport, whether personal or government vehicles (Sharma, 2009).

Table 18.2 Percent use of various kinds of energy for different purposes in India

Purposes	*Oil*	*Electricity*	*Coal*
Domestic works	18.2	29.0	11.3
Industries	5.0	30.0	18.9
Agriculture	10.0	16.0	0.0
Transport	31.4	16.0	2.6
Others	35.0	9.0	6.6

Power Generation: Power generation capacity in India also increased by tenfolds in the last six decades (Table 18.3). It was only 1400 MW in 1947 but attained the mark of 1,24763 MW in 2007. Out of this, the maximum contribution (80902 MW) is of thermal energy and 30936 MW is of hydroelectrical energy. Windmills (3811 MW) and nuclear energy (2770 MW) have very little contribution.

Table 18.3 Power generation capacity in India

1947	1400 MW
1999	92864 MW
2005	118410 MW
2007	124763 MW

Thermal Power 80902 MW, Hydroelectric Power 30936 MW, Windmills 3811 MW, Nuclear Energy 2770 MW, Other sources 6344 MW

Atomic Energy: Department of atomic energy was established in our country on 3rd August 1954. Since its inception, it is engaged in the development of nuclear power technology, and it has many applications in industry, agriculture, medicine and basic research. It was found by the Department that fissionable uranium (92 U 235) gives 30 lakh times more energy than coal. Helium is even a more rich source of energy and gives 2.4 crore times more energy than coal. But this energy is costly due to the establishment of atomic reactors and processing.

Similarly, disposal of radioactive wastes is a big problem, as the rules for protection from radioactivity are extremely difficult in our country.

Nuclear Energy: Production of nuclear energy started in our country in the third five year plan (1969) at Tarapore with 100 MW capacity. Up to 1985, five more power plants were established in Kalpakkam (TN), Kota (Rajasthan), Karwar (Karnataka), Kakrapur (Guj.) and Narora (UP) with a production capacity of 2270 MW. In the year 2000, energy production increased to 9150 MW. It was also assessed that thorium available in plenty in the sand of Rajasthan may fulfil our need of atomic energy for a long time.

Another rich source of energy is hydrogen cake, which gives 142 K joules / g of energy on burning, and H_2O is obtained as, by-product.

$$H_2 \xrightarrow[O_2]{\hspace{3cm}} H_2O + 284 \text{ K joules per Mol.}$$

There is no pollution during the production of this energy, but storage of hydrogen is very difficult as it is extremely light, otherwise it is available in plenty in water (Asthana and Asthana, 2007).

Solar Energy: It is received on the earth as light and heat. Its average amount is 75 × 1014 kW per day. However, only 0.1% of this is sufficient to fulfil the demand of the whole world. In fact, approximately 33% of the solar radiation is absorbed by the clouds and dust particles, 10% by water vapour, oxygen and ozone in the atmosphere, and 9% is reflected from the earth surface. Thus, only 48% of the solar radiation reaches the earth.

In our country, sun light is available in most parts throughout the year and a single solar panel fixed in a house can fulfill the energy requirement of the whole family. Solar energy can be converted to electrical, mechanical and chemical energy. Panels of solar cells made by silicon are costly, but once installed, they can fulfil the energy demand without any extra expenses.

To popularize the use of solar energy, the government is giving large subsidies on the purchase of solar cookers and solar lanterns for last several years. Similarly, in several remote villages where electrification has not been possible, solar street lights have been fixed up.

Solar energy is the only non-polluting source of energy with unlimited utilization potential. Similarly, hydroelectricity is a cheap source of energy which also does not create pollution.

Wind Energy: The kinetic energy of wind is converted into mechanical energy and then into electrical energy by windmills. Approximately, 1 hp energy is obtained from the wind blowing at a speed of 10 km per hour. But, 8 horsepower energy is obtained from the wind blowing at a speed of 20 km/hour, and 27 hp energy is procured from the wind blowing at a speed of 30 km per hour.

Wind-wave Energy: In our country, the coastal line is about 6000 km long and the potential of wind energy is estimated to be about 60000 MW. This energy can be utilized for pumping water in the rural areas of coastal states. The Department of Non-Conventional Energy Sources (DNES), GOI has established small plants to harvest this energy along with 924 wind pumps as given in Table 18.4.

Table 18.4 Capacity of various wind wave power stations in India

	Place	*Capacity of wind wave power station*
1.	Mandvi (Mah)	3300 KW
2.	Kutch (Guj)	1500 KW
3.	Okha (Guj.)	550 KW
4.	Deogarh	550 KW
5.	Tutikorin (TN)	550 KW
6.	Puri (Odisha)	550 KW

OCEAN (Tidal) Energy: In 1966, France constructed the first tidal electricity plant. Tidal power potential in our country is about 9000 MW as given in Table 18.5.

Table 18.5 Capacity of various Tidal Power Plants in India

	Place	*Capacity of Tidal Power Plant*
1.	Gulf of Kutch (Guj)	800 – 1000 MW
2.	Gulf of Cambay	7000 – 8000 MW
3.	Sunderban (WB)	700 – 800 MW
4.	Lakshadweep Islands	Under construction
5.	Andaman-Nicobar Islands	Under construction

Ocean Thermal Energy Conversion (OTEC): The temperature difference between the surface water (28-30°C) and deep water (5-7°C at 800 to 1000 m deep) in the tropical parts of the oceans is utilized to get continuous, renewable and pollution-free electricity from the floating OTEC power plant. OTEC and wind-wave energy (WWE) can be excellent alternative sources of energy in remote oceanic islands and coastal areas for meeting the requirement of energy as it has been estimated that their total potential can be 50000 MW.

Geothermal Energy: The heat in the interior of the earth can be utilized for power generation, particularly in the volcanic regions and in the regions of hot springs and geysers. A cold storage and 5 MW power plant have been established in Manikaran (Kullu, H.P.). Geological explorations for prospecting locations by drilling into deeper levels are going on in Puga (Ladakh), Tattapani (Sarguja, C.G.), Cambay Basin (Alaknanda Valley, Uttarakhand), and Parvati Valley (H.P.). In all, approximately 350 geothermal springs have been located. Total potential in the developing countries has been estimated to be about 2300 MW.

Dendro-thermal Energy: It is obtained by burning bushes grown on wastelands and waste material of sugarcane obtained after extracting sugar. Wastelands are nowadays being used for growing energy plantations, i.e., fast growing tree species and shrubs having high calorific values. Through gasification system, such plantations are generating electricity (8000 ha plantation produce 1.5 MW electric power per day). At present, there is no power plant for dendrothermal energy in our country because we do not produce that much of biomass which is required to run a project regularly. The biomass is always required at local level, otherwise the transportation makes the production of energy very costly. Only the protection of forests and huge plantation programmes of all kinds, viz., afforestation, reforestation and rehabilitation of degraded forests can give us sufficient amount of biomass to run the dendrothermal power plants regularly and economically efficiently, which can become the cheapest and easiest replenishable natural source of energy. Plantations can be developed in almost all kinds of culturable wastelands and in all edapho-climatic conditions by specialized efforts and techniques.

Biofuel: Due to dwindling fossil fuel, microbial production gained importance. By converting the starch of biomass fuel, alcohol (ethanol) is produced. Waste residues of food processing industries serve as suitable raw material for alcohol production. Similarly, starch-rich materials like potato, corn, sago, wood and agricultural residues are also major source of this energy. Several plant families such as Asclepiadaceae, Apocynaceae, Euphorbiaceae and Convolvulaceae, etc., have been identified as excellent source of biofuel.

Biodiesel: It is a kind of biofuel obtained from the oil derived from vegetable sources. Ratanjot, (*Jatropha curcas*), Karanj (*Pongamia pinnata*) have become the new money plants. Hiking prices of petroleum products has pressed us to develop such environment-friendly sources of energy. India has to import about 70% of its required petroleum every year for about 1 lakh crore rupees. This is a cheaper source and the farmers may get a better earning by cultivating biodiesel plants in the agricultural fields. To look after the policies and commercial production, the National Oilseed and Vegetable Oils Development Board (NOVOD) was established in our country in 2007.

Recommendations: For conserving energy and for a better future, we must change our life style. We must walk more for our short distanced places or must use bicycles as used in China, Japan, USA and several European nations. For longer distances, we must use shared vehicles or public transport system like city buses, metros and trams, etc. Similarly, in our kitchens, we must use solar cookers and energy efficient chulhas.

REFERENCES

1. Asthana, D.K. and Asthana, M. (2007). Energy resources. In – A textbook of environmental studies. S. Chand & Co. Ltd. New Delhi, pp. 121-138.
2. Sharma, P.D. (2009). Energy crisis and non-conventional sources. In – Ecology and Environment. Rastogi Publications, Meerut, pp. 266-283.
3. Verma, D. (Ed.) (2004). Urja sansadhan. In – Paryavaran Chetna, M.P. Hindi Granth Academy, Bhopal, pp. 69-88.

19

Plant Diversity in Sacred Groves of Madhya Pradesh

S.K. Masih

State Forest Research Institute, Polipathar, Jabalpur - 482008 (M.P.)

Email: Shashi_masih@rediffmail.com

ABSTRACT

Sacred groves are the forest patches conserved by the local people linked with their socio-cultural and religious practices. Sacred groves are distributed over a wide ecosystem and help in conservation of rare and endemic species in our country. Various indigenous communities all over the world live in harmony with nature and thus conserve biodiversity. It is believed that such sacred groves have been surviving for several thousand years. Sacred groves act as reservoirs of rare flora and fauna, amid rural and even urban settings. The degree of sanctity of these sacred groves varies in reference to their cultural and religious significance. Threats to the groves include urbanization, overexploitation of resources and environmental destruction from religious practices. There is strong need to initiate people's participation, training for promoting the indigenous traditional knowledge and conserving the biodiversity through this traditional knowledge. Well-preserved sacred groves are storehouses of valuable medicinal and other plants having high economic value, and serve as a refuge to threatened species. The current paper attempts to map out the availability of sacred groves in the state of Madhya Pradesh. It provides the inventory of various sacred groves, their total area, species composition and other details, which are important for the future conservation of these sacred groves.

INTRODUCTION

In ancient times, sacred groves were places of sanctuary and worship for the druids. Like a temple or chapel set within the natural world, they were places of spiritual refuge places to calm the mind, refresh the spirit and give comfort in times of distress. Druids today continue this tradition of seeking tranquility in woods and forests, in which they meditate and hold ceremonies. Many contemporary druids are creating new sacred groves in their gardens, on their farms or on public land.

A sacred grove is usually dedicated to a deity or a 'mother goddess' who is supposed to protect and preside over the grove. It is believed that such sacred groves have been surviving for several thousand years. The degree of sanctity of these sacred forests varies. In some forests the deadwood may be picked up, but never the live trees or their branches and even the animals and birds are not disturbed. The Garo and Khasi tribes of north-eastern India completely prohibit any human interference in their sacred groves. The gonds of Central India prohibit cutting of a tree but allow fallen branches to be used.

In India, sacred groves are found all over the country in abundance. Although, there has been no comprehensive study on the sacred groves of the entire country, experts estimate the total number of sacred groves in India could be in the range of 1,00,000 – 1,50,000 (Malhotra, et al., 1998). Sacred groves act as reservoirs of rare flora and fauna, amid rural and even urban settings. Threats to the groves include urbanization, over-exploitation of resources and environmental destruction from Hindu religious practices, while many of the groves are looked upon as abode of Hindu gods and goddess. But in remote forest areas and tribal dominated localities various sacred groves were in the name of their ancient diety such as *van deo, gram deo, grah deo, kul deo, vraksha deo*, etc.

The survey work of sacred groves was initiated in undivided Madhya Pradesh by Prof. Madhav Gadgil in Sarguja district. No further exploration work was done in this field till 2001. From 2001 onwards, State Forest Research Institute, Jabalpur has initiated the catalouging of exisiting sacred groves, inventory of plant species and their biodiversity assessment work in some districts of Madhya Pradesh. Whereas some leading work of India has been done by Alfrod, J. R. B., 1996; Amirthalingam, M., 2000, 2002 & 1998; Bennet, et al., 1992; Bhakat and Pandit, 2003; Bhasin, 1999; Bhowmik, 1963; Boraiah, et al., 2002; Chandrashekhara and Sankar, 1998; Gadgil and Gokhale, 2000 and Gadgil and Vartak, 1973 & 1975.

STUDY SITES

The study sites of the present task are Mandla, Dindori, Chhindwara and Hoshangabad districts of Madhya Pradesh. These districts are dominated by the various tribal communities. The differences in these communities spread over in various parts of the districts. It is clearly seen on the basis of their heredity, lifestyle and cultural traditions, their social, economic structure, religious beliefs and their language and speech. Due to the different linguistic, cultural and geographical environment, its peculiar complications, the diverse tribal groups of these districts largely cut off from the mainstream of development. The main tribal groups are Gond, Bhil, Baiga, Korku, Bhadia, Halba, Kaul, Mariya, Sahariya, etc.

MATERIAL AND METHODS

The important tribal localities, pilgrim places and biodiversity rich areas of Chhindwara, Hoshangabad, Mandla and Dindori districts of Madhya Pradesh were selected with the help of reconnaissance survey. Status survey of existing biodiversity and identification of sacred groves were done for the first time in these districts. The information related to location, climatic condition, physiographic features and importance of the area was collected. Socio-religious importance was ascertained with the help of local tribals. During important festivals, melas and other religious gatherings, the areas were surveyed to find out the relationship of tribal people and the sacred groves. Cultural relationship was also studied to know the importance of sacred groves.

RESULTS AND DISCUSSIONS

Total 175 sacred groves representing 38 forest ranges were identified from Chhindwara, Hoshangabad, Mandla and Dindori districts of Madhya Pradesh. Out of which, districts Chhindwara, Hoshangabad, Mandla and Dindori represent 49, 78, 30 and 18 sacred groves from 12, 12, 9 and 5 forest ranges (Fig. 19.1).

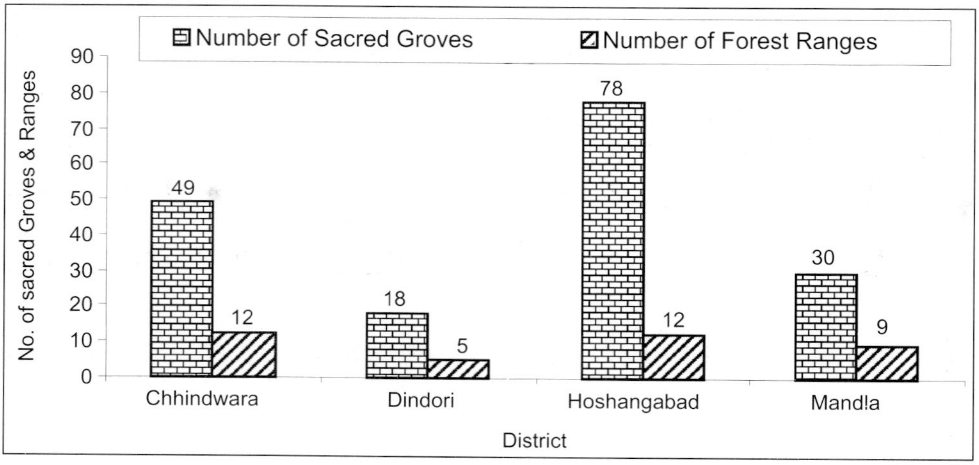

Fig. 19.1 Number of sacred groves & forest ranges

The details of identified sacred groves is given in Table 19.1 with names of sacred grove, district and forest ranges, forest compartment numbers/name of forest beat in which the grove is situated, name of god and goddess worshiped, name of other associated god and goddess worshiped along with leading god, approximate area covered by the grove (in m^2), year of existence of grove (in number of years) and location of grove from neighbouring village.

The findings also revealed that 127 sacred groves are worshiped in the name of local gods and 46 for local goddess. Whereas, sacred groves, namely, Aala Okhali (Seoni Malwa, Hoshangabad) and Churna Gundi (Park Pachmarhi, Hoshangabad) are water bodies worshiped for regular supply of water during hot summer in the name of unknown god. (Fig. 19.2)

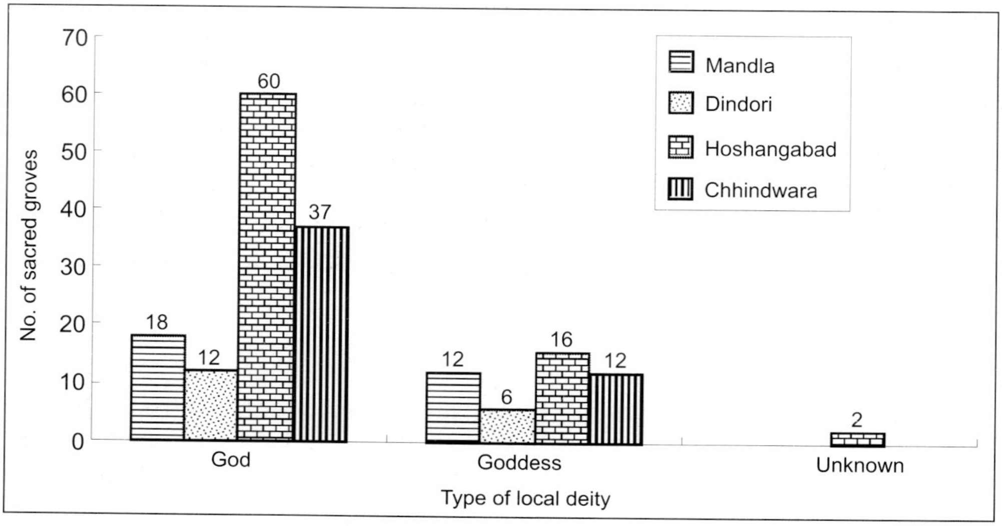

Fig. 19.2 Number of god and goddess of identified sacred groves in M.P.

Table 19.1 List of sacred groves identified from four forest divisions in Madhya Pradesh

S. No.	*Name of Sacred Grove*	*Name of District*	*Name of Forest Range*	*Comptt. Number/ Name of Beat*	*Name Of Deity Worshiped*	*Name of other Deity*	*Approx. area in M^2*	*No. of years of existence*	*Distance from the near village in km.*	*Important plant species conserved in the grove*
1.	Tilaksindur (Rudradham)	Hoshangabad	Itarsi	PF-180	Sankar	Bherowji, Parvati, Ganesh	2000	88	6	*Aegle marmelos, Mangifera indica, Buchanania lanzan, Emblica officinalis, Diospyros melanoxylon, Terminalia tomentosa, Artocarpus hetrophyllus, Helecteres isora, Chlorophytum arundinaceum, Woodfordia floribunda*
2.	Sarasdeo Mandir	Hoshangabad	Itarsi	PF-134	Sankarji	Sarasdeo	2500	100	3	*Curculigo orchioides, Ziziphus oenopelia, Cyperus iria, Borhaavia diffusa, Mitragyna parvifolia, Nyctanthes arbortristis*
3.	Chourashi Deo	Hoshangabad	Itarsi	RF-112	Shankar	NA	400	100	2	*Buchanania lanzan, Emblica officinalis, Terminalia chebula, Lagerstroemia parviflora, Chloroxylon swietenia, Anogeissus latifolia, Cassia fistula, Helecteres isora, Chlorophytum arundinaceum, Curculigo orchioides*
4.	Chitlimai	Hoshangabad	Sukhtawa	PF-46	Chitlimai	NA	100	40	2	*Hemidesmus indicus, Solanum nigrum, Bauhinia malabarica, Boerhavia diffusa, Calotropis gigantea*
5.	Badbalebaba (Gwalibaba)	Hoshangabad	Sukhtawa	PF-73	Gwalibaba	NA	100	40	3	*Ficus bengalensis, Terminalia tomentosa, Diospyros melanoxylon, Lagerstroemia parviflora*

Contd...

S. No.	Name of Sacred Grove	Name of District	Name of Forest Range	Comptt. Number/ Name of Beat	Name Of Deity Worshiped	Name of other Deity	Approx. area in M^2	No. of years of existence	Distance from the near village in km.	Important plant species conserved in the grove
6.	Jilehari Baba	Hoshangabad	Sukhtawa	RF-91	Jilehari Baba	NA	2500	150	4	*Anogeisus latifolia, Zizyphus jujube, Diospyros malanoxylon, Buchanania lanzan, Wrightia tinctoria*
7.	Hanumanji (Bajrangbali)	Hoshangabad	Sukhtawa	PF-93	Hanumanji	NA	400	100	3	*Mangifera indica, Delonix regia, Sterculia urens, Woodfordia floribunda, Curculigo orchioides, Lagerstroemia parviflora, Schleichera oleosa*
8.	Jamuniya Baba	Hoshangabad	Sukhtawa	RF-86	Jamuniya Baba	NA	100	50	1	*Madhuca indica, Terminalia arjuna, Anogeissus latifolia, Ficus racemosa, Ziziphus oenopelia, Borhaavia diffusa, Mitragyna parvifolia, Smilax zeylanica*
9.	Basaniya Baba	Hoshangabad	Sukhtawa	RF-89	Basaniya Baba	NA	400	50	4	*Syzygium cumini, Butea monosperma, Madhuca indica*
10.	Majhi Baba	Hoshangabad	Sukhtawa	Revenue land	Shankar	NA	100	50	1	*Bombax ceiba, Datura strumarium, Mangifera indica, Acalypha indica, Adhatoda vasica, Amaranthus spinosus*
11.	Hanumanji	Hoshangabad	Sukhtawa	RF-66	Hanumanji	NA	400	5	1.5	*Terminalia tomentosa, Diospyros melanoxylon, Ficus religiosa, Lagerstroemia parviflora, Madhuca indica, Buchanania lanzan, Aegle marmelos, Anogeissus latifolia, Holarrhena antydysenterica*

12.	Kaliji	Hoshangabad	Sukhtawa	RF-42	Kaliji	NA	400	35	2	*Schleichera oleosa, Anthocephalus cadamba, Anogeissus latifoliada, Holoptelia integrifolia, Diospyros melanoxylon, Butea monosperma*
13.	Khedapati Devi	Hoshangabad	Sukhtawa	RF-42	Durgaji	NA	400	35	3	*Tamarindus indica, Diospyros melanoxylon, Dioscorea bulbifera, Gymnema sylvestre*
14.	Sadhu Baba	Hoshangabad	Sukhtawa	RF-64	Sadhu Baba	NA	100	50	4	*Schleichera oleosa, Diospyros melanoxylon, Butea monosperma*
15.	Khedapati Devi	Hoshangabad	Sukhtawa	RF-64	Khedapati Devi	NA	100	50	2	*Mangifera indica, Tamarindus indica, Boerhavia diffusa, Carissa opaca, Catharanthus roseus*
16.	Mama Bhanja	Hoshangabad	Sukhtawa	RF-70 & RF-67	Mama Bhanja	NA	2500	40	2	*Anogeissus latifolia, Schleichera oleosa, Diospyros melanoxylon, Lagerstroemia parviflora*
17.	Daiyat Baba	Hoshangabad	Sukhtawa	RF-47	Daiyat Baba	NA	400	45	2	*Terminalia tomentosa, Madhuca indica, Diospyros melanoxylon, Anogeissus latifolia*
18.	Mithiya Baba	Hoshangabad	Sukhtawa	RF-48	Mithiya Baba	NA	100	50	3	*Catharanthus roseus, Cymbopogon martini, Holoptelia integrifolia, Diospyros melanoxylon, Butea monosperma*
19.	Khedapati Mai	Hoshangabad	Sukhtawa	Revenue land	Khedapati Mai	NA	400	100	5	*Tectona grandis, Schleichera oleosa, Anogeissus latifolia, Diospyros melanoxylon, Butea monosperma*
20.	Bhootkahu	Hoshangabad	Seoni Malwa	Revenue land	Bhootkahu	NA	400	100	4	*Terminalia tomentosa, Tectona grandis, Schleichera oleosa, Anthocephalus cadamba, Alstonia scholaris, Ailanthus excelsa*

Contd...

S. No.	Name of Sacred Grove	Name of District	Name of Forest Range	Comptt. Number/ Name of Beat	Name Of Deity Worshiped	Name of other Deity	Approx. area in M^2	No. of years of existence	Distance from the near village in km.	Important plant species conserved in the grove
21.	Peer Baba	Hoshangabad	Seoni Malwa	186	Peer Baba	NA	100	30	1	*Terminalia tomentosa, Tectona grandis, Schleichera oleosa, Anogeissus latifolia, Holoptelia integrifolia, Diospyros melanoxylon, Butea monosperma*
22.	Jango Bai	Hoshangabad	Seoni Malwa	184	Jango Bai	NA	2500	100	1	*Terminalia tomentosa, Tectona grandis, Schleichera oleosa, Holoptelia integrifolia, Butea monosperma*
23.	Aala Okhli (Water reservoir)	Hoshangabad	Seoni Malwa	190	Unknown God	NA	400	150	4	*Terminalia tomentosa, Tectona grandis, Schleichera oleosa, Butea monosperma, Ziziphus oenopelia, Ziziphus mauritina, Borhaavia diffusa, Mitragyna parvifolia, Nyctanthes arbortristis, Smilax zeylanica*
24.	Gadbad Baba	Hoshangabad	Banapura	RF-14	Gadbad Baba	NA	400	100	3	*Terminalia tomentosa, Tectona grandis, Schleichera oleosa, Holoptelia integrifolia, Diospyros melanoxylon, Butea monosperma*
25.	Panal Shiv Mandir	Hoshangabad	Banapura	264	Shiv Bhagwan	NA	400	50	3	*Anogeissus latifolia, Holoptelia integrifolia, Diospyros melanoxylon, Butea monosperma, Ailanthus excelsa, Cuscuta reflexa*

26.	Babdi	Hoshangabad	Banapura	RF-4	Rori Devi	NA	2500	100	2	*Terminalia tomentosa, Tectona grandis, Schleichera oleosa, Diospyros melanoxylon, Butea monosperma*
27.	Chicha Baba (Imli Wale Baba)	Hoshangabad	Banapura	RF-14	Chicha Baba	NA	100	100	2	*Terminalia tomentosa, Tectona grandis, Schleichera oleosa, Anogeissus latifolia, Holoptelia integrifolia, Butea monosperma*
28.	Garden Baba	Hoshangabad	Banapura	264	Garden Baba	NA	100	100	4	*Terminalia tomentosa, Tectona grandis, Schleichera oleosa, Diospyros melanoxylon, Butea monosperma, Phoenix acualis*
29.	Sayyad Baba ki Mazar	Hoshangabad	Bankhedi	P-340	Sayyad Baba	Shaill Saheb Baba	2500	150	2	*Tectona grandis, Chlorophytum arundinaceum, Buchanania lanzan, Madhuca indica, Lagerstroemia parviflora, Annona squamosa, Chloroxylon swieteria, Butea monosperma*
30.	Siddha Baba (Dolariya)	Hoshangabad	Babaii	Revenue land	Siddha Baba	Hanuman, Nandi, Shani, Shankar	2500	60	1.5	*Tectona grandis, Terminalia tomentosa, Azadirachta indica, Curculigo orchioides, Buchanania lanzan, Emblica officinalis, Madhuca indica, Helicteres isora, Soymida febrifuga, Lagerstroemia parviflora, Woodfordia floribunda, Angogeissus latifolia, Butea monosperma*

Contd...

S. No.	Name of Sacred Grove	Name of District	Name of Forest Range	Comptt. Number/ Name of Beat	Name Of Deity Worshiped	Name of other Deity	Approx. area in M^2	No. of years of existence	Distance from the near village in km.	Important plant species conserved in the grove
31.	Banjari Mai	Hoshangabad	Bagda	231	Banjari Mata	NA	400	50	1	*Tectona grandis, Helicteres isora, Cassia tora, Annona squamosa, Basella rubra, Annona squamosa, Alstonia scholaris*
32.	Imliwale Baba	Hoshangabad	Bagda	RF-220	Imliwale Baba	NA	400	50	2	*Tectona grandis, Helicteres isora, Annona squamosa, Desmodium velutinum, Celosia argentea, Cardiospermum halicacabum, Basella rubra*
33.	Dugariya Wale Baba	Hoshangabad	Bagda	RF-224	Dugariya Wale Baba	NA	100	50	4	*Tamarindus indica, Tectona grandis, Butea monosperma, Syzygium cumini, Lantana camara, Aristida funiculate, Azadirachta indica, Bauhinia purpurea, Butea superba*
34.	Khokhla Mata	Hoshangabad	Bagda	RF-224	Khokhla Mata	NA	100	100	3	*Tamarindus indica, Tectona grandis, Butea monosperma, Madhuca indica, Ampelocissus tomentosa, Caesalpinia bonducella, Cassia fistula, Chenopodium album*
35.	Siddha Baba	Hoshangabad	Bagda	RF-224	Siddha Baba	NA	400	50	3	*Terminalia tomentosa, Angogeissus latifoliada, Butea monosperma, Coccinia grandis, Crotalaria juncea, Cyperus scariosus, Desmodium velutinum*

36.	Banjari Mata	Hoshangabad	Bagda	RF-229	Banjari Mata	NA	400	60	4	*Tectona grandis, Terminalia tomentosa, Adina cordifolia, Mangifera indica, Anogeissus latifolia, Aegle marmelos, Terminalia bellirica, Emblica officinalis, Buchanania lanzan, Diospyros melanoxylon, Madhuca indica, Soymida febrifuga, Lagerstroemia parviflora, Sterospermum personatum, Sterculia urens, Woodfordia fructicosa, Terminalia tomentosa, Butea monosperma, Chlorophytum arundinaceum*
37.	Jamunjhiria	Hoshangabad	Sohagpur	RF-258	Siddha Baba	NA	400	100	2	*Tectona grandis, Terminalia tomentosa, Adina cordifolia, Mangifera indica, Anogeissus latifolia, Aegle marmelos, Terminalla bellirica, Emblica officinalis, Buchanania lanzan, Diospyros melanoxylon, Madhuca indica, Soymida febrifuga, Lagerstroemia parviflora, Gardenia latifolia, Sterculia urens, Woodfordia fruiticosa, Terminalia tomentosa, Butea monosperma, Chlorophytum arundinaceum*

Contd...

S. No.	Name of Sacred Grove	Name of District	Name of Forest Range	Comptt. Number/ Name of Beat	Name Of Deity Worshiped	Name of other Deity	Approx. area in M^2	No. of years of existence	Distance from the near village in km.	Important plant species conserved in the grove
38.	Jamani Deo	Hoshangabad	Sohagpur	RF-258	Jamani Deo	Banjari Mata	2500	100	3	*Tectona grandis, Terminalia tomentosa, Adina cordifolia, Aegle marmelos, Emblica officinalis, Diospyros melanoxylon, Madhuca indica, Soymida febrifuga, Lagerstroemia parviflora, Gardenia latifolia, Sterculia urens, Woodfordia fruiticosa, Terminalia tomentosa, Butea monosperma, Chlorophytum arundinaceum, Bombax ceiba, Ampelocissus tomentosa*
39.	Khedapati Laliya	Hoshangabad	Sohagpur	RF-275	Khedapati	NA	400	75	1	*Tectona grandis, Aegle marmelos, Tamarindus indica, Terminalia tomentosa, Acacia leucophloca, Butea monosperma, Feronia limonia, Diospyros melanoxylon, Emblica officinalis, Buchanania lanzan, Lagerstroemia parviflora, Schleichera oleosa*
40.	Mandra Baba	Hoshangabad	Sohagpur	RF-287	Khode Baba (Mandra Baba)	NA	400	50	3	*Mangifera indica, Mitragyna parvifolia, Aegle marmelos, Terminalia tomentosa, Soymida febrifuga, Madhuca indica, Emblica officinalis, Tectona grandis, Miliusa tomentosa, Adina cordifolia*

41.	Telibaba	Hoshangabad	Sohagpur	RF-286	Telibaba	NA	400	60	1.5	*Mitragyna parvifolia, Aegle marmelos, Terminalia tomentosa, Madhuca indica, Lagerstroemia parviflora, Acacia leucophloca, Diospyros melanoxylon*
42.	Siddha Baba	Hoshangabad	Sohagpur	RF-285	Siddha Baba	NA	400	100	3	*Terminalia arjuna, Tectona grandis, Madhuca indica, Anogeissus latifolia, Buchanania lanzan, Emblica officinalis, Aegle marmelos, Curculigo orchioides, Hemidesmus indicus, Woodfordia floribunda, Helicteres isora, Evolvulus alsinoides, Parthenium hysterophorus, Ephorbia hirita, Tephroria pupurea*
43.	Mahadeo	Hoshangabad	STR Pachmarhi	PF-252	Shankarji	Parwati, Kal Bherav, Chandika Devi	2500	200	8	*Mangifera indica, Mitragyna parvifolia, Aegle marmelos, Terminalia tomentosa, Madhuca indica, Careya arborea, Boerhavia diffusa, Barleria prionitis, Asparagus racemosus, Apluda mutica, Ampelocissus tomentosa*
44.	Gupth Mahadeo	Hoshangabad	STR Pachmarhi	PF-252	Shankarji	Parwati, Kal Bherav, Chandika Devi	400	200	8	*Mangifera indica, Mitragyna parvifolia, Aegle marmelos, Terminalia tomentosa, Madhuca indica, Lagerstroemia parviflora, Acacia leucophloca, Diospyros melanoxylon*

Contd...

S. No.	Name of Sacred Grove	Name of District	Name of Forest Range	Comptt. Number/ Name of Beat	Name Of Deity Worshiped	Name of other Deity	Approx. area in M^2	No. of years of existence	Distance from the near village in km.	Important plant species conserved in the grove
45.	Chawraghad	Hoshangabad	STR Pachmarhi	PF-252	Shankarji	Parwati , Ganesha	2500	200	8	*Mangifera indica, Mitragyna parvifolia, Aegle marmelos, Terminalia tomentosa, Tectona grandis, Miliusa tomentosa, Lagerstroemia parviflora*
46.	Rock painting Mahadeo	Hoshangabad	STR Pachmarhi	P-269	Shankarji	Parwati , Ganesha	400	200	8	*Terminalia tomentosa, Buchanania lanzan, Madhuca indica, Lagerstroemia parviflora, Curculigo orchioides, Mangifera indica, Emblica officinalis, Angogeissus latifolia, Shorea rabusta, Pheonix sylvestris*
47.	Khedapati Mai	Hoshangabad	STR Pachmarhi	P-269	Khedapati	NA	100	100	1	*Mangifera indica, Aegle marmelos, Terminalia tomentosa, Soymida febrifuga, Madhuca indica, Calotropis procera*
48.	Banjari Mata	Hoshangabad	STR Pachmarhi	P-254	Banjari Mata	Gadivanbaba, Vandevi	100	100	3	*Dandrocalamus strictus, Mangifera indica, Madhuca indica, Emblica officinalis, Diospyros melanoxylon, Shorea robusta, Terminalia chebula, Cassia fistula, Buchnania lanzan, Phoenix sylvestre*

49.	Shanka Bhagwan ki Pindi	Hoshangabad	STR Pachmarhi	P-240	Shankarji	NA	100	100	3	*Terminalia tomentosa, Gmelina arborea, Mangifera indica, Emblica officinalis, Syzygium cumini, Manilkara hexandra, Buchanania lanzan, Lagerstroemia parviflora, Anogeissus latifolia, Syzygium cumini*
50.	Sidda Baba	Hoshangabad	STR Pachmarhi	P-234	Sidda Baba	NA	100	100	2	*Terminalia tomentosa, Gmelina arborea, Mangifera indica, Emblica officinalis, Syzygium cumini, Manilkara hexandra, Buchanania lanzan, Terminalia bellirica, Lagerstroemia parviflora, Anogeissus latifolia, Syzygium cumini*
51.	Majar	Hoshangabad	STR Pachmarhi	P-232	Peerbaba	NA	400	30	2	*Terminalia tomentosa, Gmelina arborea, Mangifera indica, Emblica officinalis, Syzygium cumini, Manilkara hexandra, Buchanania lanzan, Terminalia bellirica, Lagerstroemia parviflora, Anogeissus latifolia, Syzygium cumini*
52.	Gajandeo	Hoshangabad	STR Pachmarhi	P-240	Gajandeo	NA	400	50	1	*Terminalia tomentosa, Gmelina arborea, Mangifera indica, Emblica officinalis, Syzygium cumini, Manilkara hexandra, Buchanania lanzan, Terminalia bellirica, Lagerstroemia parviflora, Anogeissus latifolia*

Contd...

S. No.	*Name of Sacred Grove*	*Name of District*	*Name of Forest Range*	*Comptt. Number/ Name of Beat*	*Name Of Deity Worshiped*	*Name of other Deity*	*Approx. area in M^2*	*No. of years of existence*	*Distance from the near village in km.*	*Important plant species conserved in the grove*
53.	Nagadwari	Hoshangabad	STR Pachmarhi	P-307	Nagmaharaj	NA	100	100	10	*Terminalia tomentosa, Gmelina arborea, Mangifera indica, Emblica officinalis, Syzygium cumini, Manilkara hexandra, Lagerstroemia parviflora, Anogeissus latifolia*
54.	Majar	Hoshangabad	STR Pachmarhi	P-308	Peerbaba	Baba shab	100	50	10	*Terminalia tomentosa, Gmelina arborea, Mangifera indica, Emblica officinalis, Syzygium cumini, Manilkara hexandra, Buchanania lanzan, Terminalia bellirica, Lagerstroemia parviflora, Anogeissus latifolia, Hygrophila auriculata, Imperata cylindrica, Grewia hirsuta, Euphorbia hirta, Dillenia pentagyna, Datura stramonium, Calotropis procera*
55.	Tirathdham	Hoshangabad	Pachmarhi	Revenue land	Elders And Dead Members	Baba Shab	400	150	10	*Terminalia tomentosa, Gmelina arborea, Mangifera indica, Emblica officinalis, Syzygium cumini, Manilkara hexandra, Terminalia bellirica, Lagerstroemia parviflora, Anogeissus latifolia*

56.	Khermai	Hoshangabad	Park Panchmarhi	252	Gramdevi	Bherav, Kaliji, Ambamai	100	50	1	*Madhuca indica, Ficus glomarata, Bombax ceiba, Euphorbia hirita, Desmodium trifolium, Parthenium hysterophorus, Aegle marmelos*
57.	Nishan Garh	Hoshangabad	Park Panchmarhi	256	Lord Ganesha	NA	2500	500	1	*Shorea rabusta, Anogeissus pendula, Holoptelea integrifolia, Ficus racemosa, Ziziphus jujuba, Ziziphus oenopelia, Ziziphus mauritina, Borhaavia diffusa, Mitragyna parvifolia, Smilax zeylanica*
58.	Siddhababa	Hoshangabad	Park Panchmarhi	253	Siddhababa	Nil	100	20	8	*Syzygium cumini, Mangifera indica, Holoptelea integrifolia, Bauhinia vahlii, Anogeissus pendula, Emblica officinalis, Boswellia serrata, Cassia fistula*
59.	Paschim Dwar	Hoshangabad	Park Panchmarhi	259	Shivji	Sheshnag, Ganeshji, Parwatiji, Nandi dev, Shrngi Bhrngi, Shivgarh	4000	1000	3	*Mangifera indica, Elaeodendron glaucum, Syzygium cumini, Terminalia tomentosa, Riccia* sp., *Ficus racemosa, Anogeissus pendula*
60.	Swarg Dwar	Hoshangabad	Park Panchmarhi	259	Sheshnag	Shivji, Parwatiji	400	1000	4	*Mangifera indica, Manilkara hexandra, Syzygium cumini, Helicteres isora and Thysaonlaena maxima*
61.	Chintaman (Lord Ganesh)	Hoshangabad	Park Panchmarhi	259	Lord Ganesh	Nagin, Paiman, Dhuni Baba, Harushesh.	4000	1000	5	*Mangifera indica, Psidium guajava, Syzygium cumini, Ficus racemosa, Anogeissus pendula, Manilkara hexandra, Milletia auriculata*

Contd...

S. No.	Name of Sacred Grove	Name of District	Name of Forest Range	Comptt. Number/ Name of Beat	Name Of Deity Worshiped	Name of other Deity	Approx. area in M^2	No. of years of existence	Distance from the near village in km.	Important plant species conserved in the grove
62.	Chitrashala Mandir	Hoshangabad	Park Panchmarhi	259	Durga Mata	Sheshnag, Shivaji	1500	1000	6	*Mangifera indica, Psidium guajava, Syzygium cumini, Ficus racemosa, Anogeissus pendula, Manilkara hexandra, Milletia auriculata*
63.	Nag murti/Nag phauri	Hoshangabad	Park Panchmarhi	260	Ganeshji	Sheshnag, Nandi	100	500	7	*Mangifera indica, Ficus rumphii, Phyllanthus emblica emblicaena maxima, Eranthemum purpurascens, Elephantopus scaber*
64.	Amba Mai	Hoshangabad	Park Panchmarhi	Revenue land	Durga Ji	Laxmi, Shankar, Saibaba, Kaliji	2000	200	2	*Mangifera indica, Psidium guajava, Manilkara hexandra, Bombax ceiba, Ficus rumphii, Cassia fistula, Plumeria rubra, Hibicus rosa senensis.*
65.	Shankarji (Jatashankar)	Hoshangabad	Park Panchmarhi	Revenue land	Shakarji	Parwatiji, Ganeshji, Sheshnag	2000	100	1.5	*Mangifera indica, Ficus racemosa, Anogeissus pendula, Datura stramonium, Catheranthes roseus, Psidium guajava, Bauhinia vahlii, Cryptolepis buchanani, Aegle marmelos, Manilkara hexandra, Agave, Shorea robusta rest, Syzygium cumini, Adhatoda vasica, Morus alba*

66.	Pandav Gufa (Budh Kalin)	Hoshangabad	Park Panchmarhi	Revenue land	Budh Kalin Sput	Budhha scraches and other sign.	5000	1000	1.5	*Mangifera indica, Syzygium cumini, Ficus bengalensis, Chloroxylon swietenia, Ficus rumphii, Cassia fistula, Madhuca indica, Hisbiscus rosa sinensis, Emblica officinalis, Bauhinia variegata, Agave sissalana*
67.	Gutti baba	Hoshangabad	Bhora	214	Gutti Baba	Shankar, Parwati and Ganesha	2000	60	1	*Terminalia tomentosa, Leucaena leucocephala, Butea monosperma, Eucalyptus* spp., *Tectona grandis, Lagerstroemia parviflora, Eranthemum purpurascens, Mangifera indica*
68	Saimal Baba	Hoshangabad	Bhora	219	Saimal Baba	No	100	165	8	*Shorea robusta, Dendrocalamus strictus, Butea monosperma, Sida acuta, Sida cordifolia, Anogeissus pendula, Diospyros melanoxylon*
69.	Jhoola Wali Mata	Hoshangabad	Bhora	217	Jhoola Wali Mata	Ban Devi	100	250	5	*Tectona grandis, Tamarindus indica, Butea monospermah, Schleicher oleosa, Dendrocalamus strictus, Phoenix acualis, Lagerstroemia parviflora*
70.	Churna Gundi	Hoshangabad	Park Panchmarhi	216	Unknown God	NA	100	200	3	*Mangifera indica, Syzygium cumini, Manilkara hexandra, Ficus rumphii, Ficus racemosa, Terminalia arjuna, Cynodon dactylon, Schleichera oleosa, Careya arborea, Syzygium heyneanum, Phylanthes nirurii*

Contd...

S. No.	Name of Sacred Grove	Name of District	Name of Forest Range	Comptt. Number/ Name of Beat	Name Of Deity Worshiped	Name of other Deity	Approx. area in M^2	No. of years of existence	Distance from the near village in km.	Important plant species conserved in the grove
71.	Sakhut Baba	Hoshangabad	Bhora	198	Sakhut Baba	NA	100	300	4	*Tamarindus indica, Terminalia tomentosa, Diospyros melanoxylon, Tectona grandis, Buchanania lanzan, Eranthemum purpurascens*
72.	Jhinjhini Mahal	Hoshangabad	Kamti	275	Jhinjhini Baba	NA	300	1000	4	*Terminalia tomentosa, Tectona grandis, Dendrocalamus strictus, Chloroxylon swietenia, Syzygium cumini, Acacia catechu, Mangifera indica, Manilkara hexandra, Diospyros melanoxylon, Lantana camara, Casia tora*
73.	Nagdeo	Hoshangabad	Kamti	293	Nagdev	NA	200	250	15	*Chloroxylon swietenia , Dendrocalamus strictus, Manilkara hexandra, Terminalia tomentosa, Diospyros melanoxylon, Buchanania lanzan, Lagerstroemia parviflora*
74.	Sidha Baba	Hoshangabad	Park Panchmarhi	309	Sidha Baba	NA	100	100	10	*Shorea rabusta, Syzygium cumini, Mangifera indica, Anogeissus pendula, Terminalia chebula, Chloroxylon swietenia*
75.	Gond Baba (Gond Bangla)	Hoshangabad	Park Panchmarhi	Revenue land	Gond Baba	NA	200	500	1	*Litsea glutinosa, Mangifera indica, Syzygium cumini, Hisbiscus rosa sinensis, Morus alba, Momordica dioca, Leea crispa, Justicia betonica, Ipomoea cornea*

76.	Karbala	Hoshangabad	Park Panchmarhi	Revenue land	Shahidana Karwala	NA	200	15	2	*Anogeissus pendula, Syzygium cumini, Phoenix acualis, Careya arborea, Thevitia nerifolia, Psidium guajava, Cassia fistula, Elaeodendron glaucum, Cynodon dactylon*
77.	Siddhababa	Hoshangabad	Park Panchmarhi	254	Siddhababa	NA	20	100	4	*Terminalia tomentosa, Anogeissus pendula, Phoenix sylvestris, Dendrocalamus strictus, Hibicus rosa senensis, Jasminum sambac, Jasminum grandiflorum, Emblica officinalis, Syzygium cumini*
78.	Sitanahani	Hoshangabad	Park Panchmarhi	252	Sankarji	NA	100	5000	5	*Terminalia tomentosa, Leucaena leucocephala, Ziziphus jujuba, Butea monosperma, Eucalyptus* spp., *Tectona grandis, Lagerstroemia parviflora, Mangifera indica*
79.	Nandagiri Shiv Mandir	Chhindwara	Damua	P - 437	Nandagiri Shiv	Sankar, Krishna, Hanuman	100	50	0.5	*Diospyros melanoxylon, Tectona grandis Tamarindus indica , Aegle marmelos, Ficus benghalensis, Bauhinia vahlii, Azadrachta indica, Lagerstroemia perviflora, Anogeissus latifolia, Syzygium cumini, Terminalia arjuna, Butea monosperma, Madhuca indica*

Contd...

S. No.	Name of Sacred Grove	Name of District	Name of Forest Range	Comptt. Number/ Name of Beat	Name Of Deity Worshiped	Name of other Deity	Approx. area in M^2	No. of years of existence	Distance from the near village in km.	Important plant species conserved in the grove
80.	Gorakhnath	Chhindwara	Damua	P-363	Gorakhnath	NA	100	100	6	*Terminalia tomentosa, Diospyros melanoxylon Tectona grandis, Aegle marmelos, Bauhina variegata, Ficus benghalensis, Azadrachta indica, Embilica officinalis, Lagerstroemia perviflora, Anogeissus latifolia, Terminalia arjuna Acacia catachu, Butea monosperma, Madhuca indica*
81.	Gorakhnath and Machhindernath	Chhindwara	Damua	P-363	Gorakhnath And Machhindernath	NA	50	150	5	*Terminalia tomentosa, Diospyros melanoxylon, Tectona grandis, Tamarindus indica, Aegle marmelos, Bauhina vareigata, Ficus benghalensis, Sterculia urens, Azadrachta indica, Embilica officinalis, Lagerstroemia perviflora, Anogeissus latifolia, Syzygium cumini, Terminalia arjuna, Acacia catechu, Butea monosperma, Albizia procera, Madhuca indica*

82.	Sidhnathbaba	Chhindwara	Damua	P-449	Sidhnathbaba	NA	500	75	0.5	*Terminalia tomentosa, Diospyros melanoxylon, Tectona grandis, Tamarindus indica, Aegle marmelos, Bauhina vareigata, Ficus benghalensis, Sterculia urens, Azadrachta indica, Embilica officinalis, Lagerstroemia perviflora, Anogeissus latifolia, Terminalia arjuna, Acacia catachu, Butea monosperma, Albizia procera, Madhuca indica*
83.	Badi Chandi Mata	Chhindwara	Tamia	P-202	Chandimata	NA	50	100	3	*Terminalia tomentosa, Diospyros melanoxylon, Tectona grandis, Bauhina vareigata, Ficus benghalensis, Sterculia urens, Azadrachta indica, Embilica officinalis, Syzygium cumini*
84.	Choti Chandi Mata	Chhindwara	Tamia	P-209	Chandimata	NA	50	50	0.5	*Terminalia tomentosa, Diospyros melanoxylon, Tectona grandis, Ficus benghalensis, Azadrachta indica, Embilica officinalis, Lagerstroemia perviflora, Anogeissus latifolia, Pterocarpus marsupium, Syzygium cumini, Terminalia arjuna, Acacia catachu, Butea monosperma, Madhuca indica, Ficus infectoria, Hamidesmus indicus, Ziyiphus jujuba, Xanthium strumarium*

Contd...

S. No.	*Name of Sacred Grove*	*Name of District*	*Name of Forest Range*	*Comptt. Number/ Name of Beat*	*Name Of Deity Worshiped*	*Name of other Deity*	*Approx. area in M^2*	*No. of years of existence*	*Distance from the near village in km.*	*Important plant species conserved in the grove*
85.	Majhar (Golaiwali Majhar)	Chhindwara	Tamia	P-217	Peerbaba	NA	50	75	0.5	*Ficus benghalensis, Terminalia tomentosa, Diospyros melanoxylon, Tectona grandis Azadrachta indica, Embilica officinalis, Lagerstroemia perviflora, Anogeissus latifolia, Butea monosperma*
86.	Tultula Nainadevi	Chhindwara	Tamia	P-215	Nainadevi	Durga, Hanuman, Shankar	10	75	5	*Diospyros melanoxylon, Tectona grandis, Tamarindus indica, Aegle marmelos, Bauhina vareigata, Azadrachta indica, Embilica officinalis, Lagerstroemia perviflora, Anogeissus latifolia, Butea monosperma, Madhuca indica Vernonia cineria, Lantana camara, Ampelocissus latifolia, Achyranthes aspara, Calotropis gigantea*
87.	Banjarimaii	Chhindwara	Tamia	P-215	Banjarimaii	NA	10	100	3	*Terminalia tomentosa, Diospyros melanoxylon, Tectona grandis, Ficus benghalensis, Lagerstroemia perviflora, Anogeissus latifolia, Terminalia arjuna, Butea monosperma, Vernonia cineria, Lantana camara, Dendrophthoe falcate, Careya arborea, Cassia fistula*

88.	Hinglaj Devi	Chhindwara	Jamaii	RF-471/472	Hinglaj Devi	Hanuman, Shardadevi, Shankar, Krishna	200	100	4	*Terminalia tomentosa, Tectona grandis, Bauhina vareigata, Azadrachta indica, Lagerstroemia perviflora, Anogeissus latifolia, Butea monosperma, Birdelia retusa, Buchanania lanzan, Cuscuta reflexa, Dendrocalamus strictus*
89.	Junnardeo Mandir	Chhindwara	Jamaii	RF-455	Junnardeo Mahadeo	Shankar, Hanuman	200	100	3	*Terminalia tomentosa, Diospyros melanoxylon, Azadrachta indica, Lagerstroemia perviflora, Anogeissus latifolia, Butea monosperma, Dalbergia paniculata, Bauhinia vahlii, Helicteres isoradesma diandrum, Clematis triloba, Gardenia gummifera*
90.	Hanuman Mandir	Chhindwara	Tamia		Hanuman	NA	10	35	3	*Tamarindus indica, Aegle marmelos, Ficus benghalensis, Mangifera indica, Syzygium cumini, Terminalia arjuna, Vernonia cineria, Lantana camara, Zizyphus jujuba, Xanthium strumarium, Vitex negundo, Tribulus terrestris*
91.	Deorani Dai	Chhindwara	Parasia	RF-687	Deorani Dai	Shardadevi	50	40	2	*Terminalia tomentosa, Tectona grandis, Bauhina vareigata, Sterculia urens, Lagerstroemia perviflora, Anogeissus latifolia, Syzygium cumini, Terminalia arjuna, Vernonia cineria, Lantana camara, Bauhinia retusa, Arundo donex, Aristolochia indica, Temerix dioica*

Contd...

S. No.	Name of Sacred Grove	Name of District	Name of Forest Range	Comptt. Number/ Name of Beat	Name Of Deity Worshiped	Name of other Deity	Approx. area in M^2	No. of years of existence	Distance from the near village in km.	Important plant species conserved in the grove
92.	Sanideo Mandir	Chhindwara	Parasia	RF-724	Shanideo	NA	50	40	3	*Terminalia tomentosa, Diospyros melanoxylon, Tectona grandis, Tamarindus indica, Aegle marmelos, Lagerstroemia perviflora, Anogeissus latifolia, Pterocarpus marsupium, Syzygium cumini, Terminalia arjuna, Borhavia diffusa, Cordia dichotoma*
93.	Hanuman Mandir	Chhindwara	Parasia	RF-724	Hanuman	NA	10	25	4	*Ficus religiosa, Lagerstroemia perviflora, Anogeissus latifolia, Syzygium cumini, Terminalia arjuna, Dodonea viscosa, Colebrookia oppositifolia, Smilax zeylanica, Ventilago calyculata, Zizyphus jujube*
94.	Bajrangbali Mandir	Chhindwara	Parasia	RF - 720	Bajrangbali	NA	50	20	8	*Syzygium cumini, Terminalia arjuna, Butea monosperma, Madhuca indica, Ficus religiosa, Zizyphus jujube, Tribulus terrestris, Madhuca indica*
95.	Mahulwale Bajrangbali	Chhindwara	Parasia	RF - 721	Bajrangbali	NA	50	15	3	*Lagerstroemia perviflora, Lantana camara, Buchanania lanzan Saccharum spontaneum Amaranthus spinosus, Agave americana, Murraya koenigii, Nyctanthes arbor tristis*

96.	Pathraii Ram Mandir	Chhindwara	Parasia	RF - 722	Ram	Lakshman, Sita, Hanuman	50	40	3	*Tectona grandis, Lagerstroemia perviflora, Anogeissus latifolia, Syzygium cumini, Terminalia arjuna, Vernonia cineria, Lantana camara, Nyctanthes arbortristis, Themeda guadrivalvis, Ventilago calyculata, Zizyphus jujube, Vitex negundo, Soymida febrifuga*
97.	Durga Mandir	Chhindwara	Parasia	RF - 720	Durga	NA	50	100	1.5	*Terminalia tomentosa Diospyros melanoxylon, Tectona grandis, Lagerstroemia perviflora, Anogeissus latifolia, Syzygium cumini, Terminalia arjuna, Acacia catachu, Butea monosperma, Madhuca indica, Thespesia lampas*
98.	Saiiyadbaba ki Majhar	Chhindwara	Parasia	RF-720	Saiiyadbaba	NA	50	50	2	*Aegle marmelos, Bauhina vareigata, Azadrachta indica, Embilica officinalis, Lagerstroemia perviflora, Anogeissus latifolia, Syzygium cumini, Madhuca indica, Schleichera oleosa, Randia dumetorum, Pogostemon benghalensis*
99.	Shiv Mandir	Chhindwara	Pandurna	P-2051	Shiv	NA	20	100	0.5	*Tectona grandis, Annona squamata, Butea monosperma, Azadrachta indica, Dandrocalamus strictus, Grewia rothii, Milletia auriculata, Mangifera indica, Kydia calycina*

Contd...

S. No.	Name of Sacred Grove	Name of District	Name of Forest Range	Comptt. Number/ Name of Beat	Name Of Deity Worshiped	Name of other Deity	Approx. area in M^2	No. of years of existence	Distance from the near village in km.	Important plant species conserved in the grove
100.	Bhuli Mandir	Chhindwara	Pandurna	P-2017	Shiv	Durga, Dadaji	500	50	2	*Tectona grandis, Lagerstroemia perviflora, Chlorlxylon sweitenia, Butea monosperma Cynodon dactylon, Desmodium trifolium, Casia tora, Lantana camara, Indigofera pulchella, Hymenodictyon excelsum, Holarrhena antidysentrica, Erianthus munja, Dodonea viscosa*
101.	Dadaji Ka Mandir	Chhindwara	Pandurna	1779	Dadaji	NA	500	100	4	*Tectona grandis, Syzygium cumini, Gmelina arborea, Butea monosperma, Buchanania lanzan, Eragrotis tenella, Woodfordia fruiticosa, Xanthium strumarium, Syzygium heyneanum, Dendrocalamus strictus, Randia dumetorum*
102.	Pollerdeo	Chhindwara	Sausar	1882	Pollerdeo	Hanuman	50	50	5	*Tectona grandis, Calotropis procera, Syzygium cumini, Gmelina arborea, Butea monosperma, Buchanania lanzan, Pennisetum hohenackeri, Pogostemon benghalensis, Tamarindus indica, Desmodium pulchllum, Flacourtia indica, Grewia rothii*

103.	Bheemsen Devta	Chhindwara	Sausar	1881	Bheemsen	NA	20	50	2	*Tectona grandis, Azadrachta indica, Syzygium cumini, Butea monosperma, Buchanania lanzan, Ventilago calyculata, Tribulus terrestris, Terminalia tomentosa, Solanum nigrum.*
104.	Lambadeo	Chhindwara	Ambada	1904	Hanuman	NA	50	50	1	*Azadrachta indica, Calotropis procera, Syzygium cumini, Gmelina arborea, Oscimum sanctum, Butea monosperma, Buchanania lanzan, Ougenia oogeinensis, Solanum nigrum*
105.	Bagdeo / Banjari Mata	Chhindwara	Sausar	1922	Bagdeo, Banjari Mata	Bagdeo, Banjari Mata	50	50	2	*Tectona grandis, Butea monosperma, Aegle marmelos, Terminalia tomentosa, Ficus relegiosa, Mitragyna parviflora, Cassia fistula, Azadrachta indica, Derris scandens, Dendrocalamus strictus, Cassia tora*
106.	Badadeo	Chhindwara	Sausar	1921	Badadeo	NA	50	50	1	*Anogeissus latifolia, Tectona grandis, Lagerstroemia parviflora, Butea monosperma, Buchanania lanzan, Embilica officinalis, Butea monosperma, Cuscuta reflexa, Apluda mutica, Bridelia retusa*

Contd...

S. No.	Name of Sacred Grove	Name of District	Name of Forest Range	Comptt. Number/ Name of Beat	Name Of Deity Worshiped	Name of other Deity	Approx. area in M^2	No. of years of existence	Distance from the near village in km.	Important plant species conserved in the grove
107.	Chipadghondi	Chhindwara	Sausar	1952	Chipadghondi	NA	20	50	1	*Anogeissus latifolia, Lagerstroemia perviflora, Madhuca indica, Terminalia tomentosa, Diospyros melanoxylon, Soymida febrifuga, Butea monosperma, Aegle marmelos, Asparagus racemosus, Soymida febrifuga*
108.	Banjarimata	Chhindwara	Sillewani	1714	Banjarimata	NA	100	80	2	*Tectona grandis, Buchanania lanzan, Terminalia bellirica, Aegle marmelos, Butea monosperma, Terminalia tomentosa, Soymida febrifuga, Ziziphus oenopelia, Ziziphus mauritina, Borhaavia diffusa, Mitragyna parvifolia, Smilax zeylanica, Syzium cumini*
109.	Taj Wale Baba	Chhindwara	Sillewani	1716	Taj Wale Baba	NA	50	90	3	*Tectona grandis, Buchanania lanzan, Embilica officinalis, Semecarpus anacardium, Terminalia chebula, Terminalia bellirica, Butea monosperma, Annona squamata, Terminalia tomentosa, Eclipta prostrate, Euphorbia neriifolia, Ixora arborea, Grewia tiliaefolia*

110.	Shankar Van	Chhindwara	Bicchua	1505	Shankar	Durga	500	80	0.5	*Tectona grandis, Terminalia tomentosa, Madhuca indica, Butea monosperma, Aegle marmelos, Lagerstroemia perviflora, Datura alba, Solanum nigrum, Achyranthes aspera, Terminalia arjuna, Adina cordifolia, Mangifera indica, Azadrachta indica, Ficus relegiosa*
111.	Ambamaii	Chhindwara	Bicchua	1501	Ambamaii	NA	50	10	1	*Madhuca indica, Butea monosperma, Aegle marmelos, Lagerstroemia parviflora, Terminalia arjuna, Ficus relegiosa, Gardenia resinifera, Grewia tiliaefolia, Helieteres isora, Heteropogon contortus, Ichnocarpus frutescens, Jatropha gossypifolia curcas, Ricinus communis*
112.	Siddha Baba	Chhindwara	Parasia	682	Siddha Baba	Hanuman	100	100	6	*Tectona grandis, Lagerstroemia perviflora, Anogeissus latifolia, Schleichera oleosa, Azadrachta indica, Madhuca indica, Butea monosperma, Bauhinia vahlii, Semecarpus anacardium, Tamarix dioica, Zizyphus jujube, Thespesia lampas*

Contd...

S. No.	Name of Sacred Grove	Name of District	Name of Forest Range	Comptt. Number/ Name of Beat	Name Of Deity Worshiped	Name of other Deity	Approx. area in M^2	No. of years of existence	Distance from the near village in km.	Important plant species conserved in the grove
113.	Matadai	Chhindwara	Parasia	688	Matamaii	Durga	100	80	2	*Tectona grandis, Butea monosperma, Terminalia tomentosa, Madhuca indica, Soymida febrifuga. Anogeissus latifolia, Legerstromia perviflora, Aegle marmelos, Ventilago calyculata, Tribulus terrestris, Woodfordia fruiticosa, Vitex negundo*
114.	Bausai	Chhindwara	Harraii	1107	Bausaideo	NA	50	50	2	*Tectona grandis, Lagerstroemia perviflora, Anogeissus latifolia, Buchanania lanzan, Solanum nigrum, Ventilago calyculata, Vitex negundo, Xanthium strumarium, Scheichera oleosa*
115.	Matamandir Kokanpat	Chhindwara	Harraii	1118	Matamandir	Dadaji, Durga	500	100	2	*Tectona grandis, Lagerstroemia perviflora, Anogeissus latifolia, Buchanania lanzan, Embilica officinalis, Shorea robustahe, Diospyros melanoxylon, Arundo donax, Madhuca indica, Mangifera indica, Ocimum sanctum, Ricinus communis, Syzium cumini, Vernonia cinerea, Xanthium strumarium*

116.	Maihar Wali Mata	Chhindwara	Amarwada	1204	Maihar Wali Mata	NA	200	50	1	*Butea monosperma, Aegle marmelos, Nyctanthus arbortristis, Ougenia oogeinensis Lagerstroemia perviflora, Ficus relegiosa, Bauhinia malabarica, Borhavia diffusa, Butea parviflora, Annona squamosa, Adhatoda vasica, Cymbopogon martini, Diospyros melanoxylon, Gardenia gummifera, Hymenodictyon excelsum*
117.	Baratgadh	Chhindwara	Amarwada	P - 1137	Sidhababa	NA	50	70	1	*Mangifera indica, Ziziphus jujube, Butea monosperma, Terminalia tomentosa, Tectona grandis, Tamarindus indica, Azadrachta indica, Annona squamata, Mitragyna parviflora, Diospyros melanoxylon, Buchanania lanzan, Xanthium strumerium, Calotropis procera, Solanum nigrum, Schleichera oleosa, Madhuca indica, Litsea gluinosa, Grewia hirsuta, Eragrostis interrupta, Clematis triloba*
118.	Dulha Deo	Chhindwara	Amarwada	1204	Dulha Deo	NA	100	25	1	*Mangifera indica, Ficus benghalensis, Heteropogon contortus, Azadrachta indica, Michelia champaca, Tamarindus indica, Ziziphus Jujuba poniculata, Dalbergia paniculata Calotropis procera, Dendrocalamus strictus, Xanthium strumarium, Ficus religiosa*

Contd...

S. No.	Name of Sacred Grove	Name of District	Name of Forest Range	Comptt. Number/ Name of Beat	Name Of Deity Worshiped	Name of other Deity	Approx. area in M^2	No. of years of existence	Distance from the near village in km.	Important plant species conserved in the grove
119.	Kuwari Bhimsen Baba	Chhindwara	Chaurai	1375	Bhimsen Baba	NA	50	50	2	*Butea monosperma, Madhuca indica, Aegle marmelos, Tectona grandis, Diospyros melanoxylon, Annona squamata, Cynodon dactylon, Celastrus paniculata, Careya arborea, Dioscorea demona, Xanthium strumarium, Zizyphus jujube, Vitex negundo, Terminalia chebula, Solanum nigrum*
120.	Chauki ghat	Chhindwara	Chaurai	1375	Sharda	NA	50	22	0.5	*Terminalia arjuna, Tectona grandis, Aegle marmelos, Butea monosperma, Tamarindus indica, Schleichera oleosa, Xanthium strumarium, Celastrus paniculata, Careya arborea, Dioscorea daanona, Zizyphus jujube, Vitex negundo, Terminalia chebula, Solanum nigrum*
121.	Jhikkudhana Dada	Chhindwara	Pandruna	P-1774	Kabir	NA	300	50	1	*Ficus relegiosa, Syzygium cumini, Psidium guajaba, Dandrocalamus strictus, Mangifera indica, Euphorbia neriifolia, Ficus glomerata, Eragrotis tenella, Carissa opaca, Calotropis gigantea, Themeda guadrivalvis, Eclipta prostrate, Dodonea viscose*

122.	Sharda Saliwada	Chhindwara	Amarwada	1204	Sharda Saliwada	Shiva, Hanumanji, Kalimata	500	50	3	*Butea monosperma, Lagerstroemia perviflora, Tectona grandis, Annona squamata, Gmelina arborea, Tamarindus indica, Azadrachta indica, Ficus benghalensis, Mangifera indica, Dandrocalamus strictus, Diospyros melanoxylon, Semecarpus anacardium, Terminalia tomentosa*
123.	Banjari Mata	Chhindwara	Amarwada	1224	Banjari Mata	NA	200	50	1	*Terminalia tomentosa, Anogeissus latifolia, Tectona grandis, Legerstromia parviflora, Ougenia oojeinensis, Heteropogon contortus, Indigofera pulchella, Cynodon dactylon, Flacourtia indica, Imperata cylindrica, Xanthium strumarium*
124.	Majar	Chhindwara	Amarwada	1224	Peer Baba	NA	500	10	2	*Terminalia tomentosa, Angogeissus latifolia, Tectona grandis, Legerstromia parviflora, Acacia catachu, Ougenia oojeinensis, Litsea gluinosa, Shorea rabusta, Terminalia arjuna, Betea monosperma, Abrus precatorius, Gardenia latifolia, Heteropogon contortus, Indigofera pulchella, Cymbopogon martini, Bauhinia purpurea, Aristida setaceae, Adina cardifolia*

Contd...

S. No.	Name of Sacred Grove	Name of District	Name of Forest Range	Comptt. Number/ Name of Beat	Name Of Deity Worshiped	Name of other Deity	Approx. area in M^2	No. of years of existence	Distance from the near village in km.	Important plant species conserved in the grove
125.	Dhuniwale Baba	Chhindwara	Pandruna	-	Dhuniwale	NA	50	50	0.5	*Mangifera indica, Azadrachta indica, Terminalia chebula, Ficus benghalensis, Carissa opaca, Madhuca indica, Dandrocalamus strictus, Ocimum sanctum, Daedalacanthus puroraseens, Aristida setaceae, Annona squamosa, Aegle marmelos*
126.	Sidha baba	Chhindwara	Amarwada	1155	Sidha Baba	NA	400	50	1	*Mangifera indica, Butea monosperma, Schleichera oleosa, Mitragyna parviflora, Madhuca indica, Azadrachta indica, Diospyros melanoxylon, Semecarpus anacardium, Ougenia oogeinensis, Ficus relegiosa, Annona squamosa, Aristolochia indica, Arundo donax, Briedelia retusa, Cymbopogon martini*
127.	Dongerdeo	Chhindwara	Chaurai	RF - 1375	Dongerdeo	NA	200	50	1	*Diospyros melanoxylon, Tectona grandis, Tamarindus indica, Aegle marmelos, Bauhina vareigata, Ficus benghalensis, Azadrachta indica, Embilica officinalis, Lagerstroemia perviflora, Anogeissus latifolia, Syzygium cumini, Terminalia arjuna, Butea monosperma, Albizzia procera, Madhuca indica*

128.	Amajhiria	Dindori	Karangia	Jagatpur	Bholenath	Budhi Mata	100	50	4	*Terminalia tomentosa, Boswelia serrata, Ficus benghalensis, Ficus infectoria, Schleichera oleosa*
129.	Amoleshwar	Dindori	Shapur	Danitola	Bholenath	Budhi Mata	2000	60	4	*Schleichera oleosa, Ficus religiosa, Shorea robusta, Syzygium cumini, Aegle marmelos*
130.	Babadongri	Dindori	Bajag	Sani Baghraji	Siddha Baba	Nil	100	70	2	*Shorea robusta, Aegle marmelos, Butea monosperma, Terminalia chebula, Bauhinia racemosa*
131.	Babagupha	Dindori	Bajag	Raksintola	Budi Mata	Nil	100	100	3	*Shorea robusta, Terminalia tomentosa, Schleichera oleosa, Randia dumatorum, Achyranthes aspara, Zizyphus xylopyrus, Bauhinia vahlii, Nyctanthus arbortris tis, Cassia fistula, Terminalia bellirica, Madhuca latifolia, Mangifera indica*
132.	Banjari	Dindori	Bajag	Chada	Banjari Mata	Budhi Mata	50	30	3	*Shorea robusta, Terminalia tomentosa, Schleichera oleosa, Anogeissus latifolia, Terminalia tomentosa, Randia dumatorum, Zizyphus xylopyrus, Cassia fistula*

Contd...

S. No.	Name of Sacred Grove	Name of District	Name of Forest Range	Comptt. Number/ Name of Beat	Name Of Deity Worshiped	Name of other Deity	Approx. area in M^2	No. of years of existence	Distance from the near village in km.	Important plant species conserved in the grove
133.	Kishangadhi	Dindori	Samnapur	Samnapur	Budhimata	Kishanbaba	50	100	1.5	*Shorea robusta, Terminalia tomentosa, Schleichera oleosa, Anogeissus latifolia, Dioscorea daemoni*
134.	Dagona	Dindori	Samnapur	Dagoni	Budi Maiya	Nil	100	50	1.5	*Aegle marmelos, zizyphus numularis, Syzygium cumini, Madhuca latifolia, Shorea robusta, Randia dumatorum, Achyranthes aspara, Cassia fistula*
135.	Devidand	Dindori	Bajag	Pandripani	Ganesh, Hanuman	Nil	100	80	3	*Shorea robusta, Terminalia tomentosa, Bombax ceiba, Terminalia chebula, Madhuca latifolia, Randia dumatorum, Zizyphus xylopyrus, Nyctanthus arbortris tis*
136.	Madiya	Dindori	Bajag	Thadpathara	Khermata	Thakurdeo	50	30	2	*Buchanania lanzan, Rhyncosia minima, Clitoria ternatea, Centella asiatica, Cassia tora, Canscora diffusa, Boerhavia diffusa, Basella rubra, Argemone maxicana, Amaranthus spinosus*
137.	Maharishi	Dindori	Bajag	Phitari	Siddha Baba	Budhi Mata	5000	60	5	*Bombax ceiba, Euphorbia hirta, Dodonaea viscosa, Desmodium pulchellum, Datura metel, Curculigo orchioides, Cleistanthus collinus*

138.	Padaria dongri	Dindori	Bajag	Neemtola	Siddha Baba	Nil	100	100	2	*Terminalia bellirica, Terminalia chebula, Shorea robusta, Madhuca latifolia, Mangifera indica, Randia dumatorum, Achyranthes aspara, Zizyphus xylopyrus, Bauhinia racemosa, Cassia fistula*
139.	Saranggarh	Dindori	Bajag	Sarangpur	Mahavir	Shiva	50	30	3	*Butea monosperma, Terminalia chebula, Syzygium cumini, Ficus glomarata, Mangifera indica, Achyranthes aspara, Zizyphus xylopyrus, Cassia fistula*
140.	Sharda Mandir	Dindori	Shapur	Shahpura	Sharda Mata	Mansha Devi, Shivji, Mahaveer	100	40	1.5	*Ficus religiosa, Azadirachta indica, Ficus banghalansis, Emblica officinalis, Butea monosperma, Mangifera indica, Randia dumatorum, Achyranthes aspara, Cassia fistula*
141.	Siddha Baba	Dindori	Karangia	Jagatpur	Siddha Baba	Nil	50	60	2	*Mangifera indica, Syzygium cumini, Shorea robusta, Ficus banghalansis, Mitragyna parvifolia, Schleichera oleosa, Euphorbia hirta*
142.	Siddha Baba	Dindori	Karangia	Bondar	Siddha Baba	Nil	100	50	3	*Schleichera oleosa, Ficus glomarata, Aegle marmelos, Syzygium cumini, Bombax ceiba, Olax scandens, Musa paradisiaca, Morinda tinctoria, Miliusa tomentosa*

Contd...

S. No.	Name of Sacred Grove	Name of District	Name of Forest Range	Comptt. Number/ Name of Beat	Name Of Deity Worshiped	Name of other Deity	Approx. area in M^2	No. of years of existence	Distance from the near village in km.	Important plant species conserved in the grove
143.	Siddha baba	Dindori	Karangia	Karangia	Siddha Baba	Budhi Mata	100	70	1	*Syzygium cumini, Bombax ceiba, Shorea robusta, Schleichera oleosa, Terminalia bellirica, Randia dumatorum, Achyranthes aspara, Zizyphus xylopyrus, Cassia fistula*
144.	Banjarimata	Dindori	Chada	Chada	Banjari Mata	Nil	50	80	2	*Jatropha carcus, Miliusa tomentosa, Kydia calycina, Indigofera pulchella, Gymnema sylvestre, Grewia hirsuta, Euphorbia hirta*
145.	Thakurdeo	Dindori	Bajag	Tatar	Thakurdeo	Khermata	100	80	3	*Ficus religiosa, Azadirachta indica, Ficus banghalansis, Emblica officinalis, Mangifera indica, Randia dumatorum, Achyranthes aspara, Zizyphus xylopyrus, Bauhinia vahlii, Nyctanthus arbortris tis, Cassia fistula*
146.	Babapat	Mandla	Anjaniya	Baghrodi	Siddha Baba	Nil	10	90	3	*Anogeissus latifolia, Terminalia tomentosa, Diospyros melanoxylon, Zizyphus jujuba, Buchanania lanzan, Anogeissus latifolia, Holarrhaena antydesenterica,*

										Celastrus paniculata, Madhuca latifolia, Zizyphus jujuba, Tectona grandis, Semecarpus anacardium, Lagerstroemia parviflora, Butea monosperma, Achyranthes aspara
147.	Baiga Baba	Mandla	Nainpur	Paili	Bhole Baba	Nil	50	80	2	*Ixora indica, Phyllanthus emblica, Buchanania lanzan, Diospyros melanoxylon, Dandrocalamus strictus; Anogeissus latifolia, Holarrhaena antydesenterica, Celastrus paniculata, Madhuca latifolia, Zizyphus jujuba, Lagerstroemia parviflora, Butea monosperma*
148.	Bakramundi	Mandla	Bhichhiya	Mawai	Banjari Mata	Nil	50	More than 500	3	*Shorea robusta, Boswelia serrata, Ficus religiosa, Zizyphus jujuba, Syzygium cuminii, Bombax ceiba, Curcuma caisia, Actinopteris radiata, Helicteres isora*
149.	Sidha Baba	Mandla	Bhamni	Bhatiatola	Sidha Baba	Nil	50	220	1.5	*Acacia catechue, Azadiracta indica, Thevetia nerifolia, Psidium guajaba, Saraca indica, Terminalia tomentosa, Madhuca latifolia, Ficus religiosa, Mangifera indica, Randia dumatorum, Achyranthes aspara, Zizyphus xylopyrus, Bauhinia vahlii, Nyctanthus arbortris tis, Cassia fistula*

Contd...

S. No.	Name of Sacred Grove	Name of District	Name of Forest Range	Comptt. Number/ Name of Beat	Name Of Deity Worshiped	Name of other Deity	Approx. area in M^2	No. of years of existence	Distance from the near village in km.	Important plant species conserved in the grove
150.	Ratnahipat	Mandla	Bhamni	Bhamni	Sidha Baba	Nil	50	90	3	*Anogeissus latifolia, Pterocarpus marsupium Lagerstroemia parviflora, Ficus religiosa, Cleistanthus colinus, Tectona grandis, Diospyros melanoxylon, Azadiracta indica, Buchanania lanzan, Anona squamosa*
151.	Banjari Mata	Mandla	Mawai	Mawai	Banjari Mata	Nil	50	230	3	*Shorea robusta, Diospyros melanoxylon, Terminalia tomentosa, Boswelia serrata, Psidium guajaba, Zizyphus jujuba, Ficus infectoria, Schleichera oleosa, Mallotus phillipensis, Ventilago denticulata, Chlorophytum tuberosum*
152.	Sidha Baba	Mandla	Bhamni	Dungaria	Sidha Baba	Nil	100	170	3	*Terminalia tomentosa, Pterocarpus marsupium, Phyllanthus emblica, Lagerstroemia parviflora, Dandrocalamus strictus, Oroxylum indicum, Erythrina* sp., *Chloroxylon switeniana*
153.	Banjari Mata	Mandla	Aherwada	Babeha	Banjari Mata	Sidha Baba	50	70	2	*Schleichera oleosa, Phyllanthus emblica, Ficus religiosa, Zizyphus jujuba, Syzygium cuminii, Ougenia oojenensis, Madhuca latifolia, Diospyros melanoxylon, Terminalia tomentosa, Bombax ceiba, Azadiracta indica*

154.	Banjari Mata	Mandla	Mandla	Mohania patpara	Banjari Mata	Baba Shab	50	230	2	*Dandrocalamus strictus, Tectona grandis, Terminalia tomentosa, Lagerstroemia parviflora, Madhuca latifolia, Semecarpus anacardium, Butea monosperma, Mangifera indica, Randia dumatorum, Achyranthes aspara, Zizyphus xylopyrus, Bauhinia vahlii, Schleichera oleosa, Ficus benghalensis*
155.	Barhapat	Mandla	Nainpur	Dhanora	Barhadeo	Nil	100	300	2	*Ficus religiosa, Zizyphus jujuba, Syzygium cuminii, Citrus lemon, Ougenia oojenensis, Madhuca latifolia, Diospyros melanoxylon, Terminalia tomentosa, Bombax ceiba, Buchanania lanzana, Ficus religiosa, Ailanthus exelsa, Azadiracta indica*
156.	Behar	Mandla	Bhichhiya	Medatal	Baba Deo	Nil	50	70	2	*Azadiracta indica, Schleichera oleosa, Anogeissus latifolia, Ixora indica, Phyllanthus emblica, Buchanania lanzan, Diospyros melanoxylon, Dandrocalamus strictus, Holarrhaena antydesenterica, Celastrus paniculata, Madhuca latifolia, Zizyphus jujuba, Tectona grandis, Semecarpus anacardium, Lagerstroemia parviflora, Butea monosperma*

Contd...

S. No.	Name of Sacred Grove	Name of District	Name of Forest Range	Comptt. Number/ Name of Beat	Name Of Deity Worshiped	Name of other Deity	Approx. area in M^2	No. of years of existence	Distance from the near village in km.	Important plant species conserved in the grove
157.	Belpat	Mandla	Anjaniya	Anjaniya	Bholenath, Devidai, Hanumanji	Bajarimata	50	70	2	*Tectona grandis, Mitragyna parviflora, Anogeissus latifolia, Terminalia tomentosa, Lagerstroemia parviflora, Diospyros melanoxylon, Dandrocalamus strictus, Vitex nigundo, Madhuca latifolia, Desmodium trifolium, Evolvulus alsinoides*
158.	Chitrahi Pat	Mandla	Nainpur	Chougan	Bada Maharaj	Nil	500	40	2	*Aegle marmelos, Phyllanthus emblica, Madhuca latifolia, Gmelina arborea, Buchanania lanzan, Albizzia labback, Terminalia tomentosa, Zizyphus jujuba, Cissempelos pareira, Holarrhaena antydesenterica*
159.	Devi Madiya	Mandla	Bichhiya	Bargi	Banjari Mata, Dhuni Baba	Nil	10	190	3	*Tectona grandis, Mitragyna parviflora, Buchanania lanzani, Anogeissus latifolia, Terminalia tomentosa, Lagerstroemia parviflora, Diospyros melanoxylon, Phyllanthus emblica, Carreya arborea, Madhuca latifolia, Pongamia pinnata*

160.	Sidha Baba	Mandla	Bichhiya	Shijhora	Sidh Baba	Nil	100	50	2	*Tinospora cordifolia, Diospyros melanoxylon, Anogeissus latifolia, Lagerstroemia parviflora, Terminalia bellirica, Terminalia tomentosa, Schleichera oleosa, Phyllanthus emblica, Curculigo orchioides, Amorphlus virdis, Tegetus eracta, Elephantophus scaber, Dioscorea bulbifer, Leea microphylla*
161.	Hanuman tekri	Mandla	Bichhiya	Devgaon	Mahavir	Nil	100	80	2	*Terminalia tomentosa, Boswelia serrata, Psidium guajaba, Zizyphus jujuba, Ficus infectoria, Schleichera oleosa, Curcuma aungustifolia, Mallotus phillipensis, Ventilago denticulata, Curculigo orchioides*
162.	Jagdambani ashram	Mandla	Bichhiya	Raja Karanjiya	Jagdambamai	Bholenath, Hanuman, Parvati	50	100	3	*Syzygium cuminii, Ougenia oojenensis, Madhuca latifolia, Diospyros melanoxylon, Terminalia tomentosa, Bombax ceiba, Baccopa monerii, Tephrosia purpurea, Trema orientalis*
163.	Jhijham ashram	Mandla	Bichhiya	Nainpur	Jhijham Devi	Bholenath, Hanuman, Parvati	100	60	4	*Ficus religiosa, Zizyphus jujuba, Syzygium cuminii, Citrus lemon, Ougenia oojenensis, Madhuca latifolia, Diospyros melanoxylon, Terminalia tomentosa, Bombax ceiba, Curcuma caisia, Actinopteris radiata, Ailanthus exelsa, Helicteres isora*

Contd...

S. No.	Name of Sacred Grove	Name of District	Name of Forest Range	Comptt. Number/ Name of Beat	Name Of Deity Worshiped	Name of other Deity	Approx. area in M^2	No. of years of existence	Distance from the near village in km.	Important plant species conserved in the grove
164.	Jungwani Baba	Mandla	Nainpur	Devgaon	Shivji	Naagdevta	50	60	3	*Syzygium cuminii, Shorea robusta, Terminalia tomentosa, Ficus benghalensis, Mitragyna parviflora, Madhuca latifolia, Schleichera oleosa, Cassia fistula, Diospyros melanoxylon, Ichnocarpus frutescens., Acorus calamus, Ventilago denticulata, Elenthopapus scaber, Dioscorea bulbifer, Dioscorea pentaphylla, Triumfetta rhomboidae, Curculigo orchioides*
165.	Kamdhenu	Mandla	Mohgaon	Devgaon	Kamdhenua Devi	Bajrangbali	100	35	2	*Tectona grandis, Mitragyna parviflora, Buchanania lanzani, Anogeissus latifolia, Terminalia tomentosa, Lagerstroemia parviflora, Diospyros melanoxylon, Phyllanthus emblica, Carreya arborea, Dandrocalamus strictus*

166.	Khairati	Mandla	Bichhiya	Lohta	Bada Deo	Nil	500	320	2	*Aegle marmelos, Anogeissus latifolia, Grewia hirsuta, Terminalia tomentosa, Boswelia serrata, Psidium guajaba, Zizyphus jujuba, Ficus infectoria, Schleichera oleosa, Curcuma aungustifolia, Mallotus phillipensis Ventilago denticulata, Curculigo orchioides, Elephantophus scaber*
167.	Kurlupat	Mandla	Nainpur	Imlitoa	Vandevi	Patbaba	10	280	3	*Grewia hirsuta, Schleichera oleosa, Dandrocalamus strictus, Lannea grandis, Mimosa pudica, Mucuna pruriens, Ocimum canum, Ougeinia oogeinsis, Syzygium cumini, Tamarindus indica*
168.	Mudiya pat	Mandla	Bichhiya	Amwar	Banjari Mata	Nil	50	270	3	*Shorea robusta, Tectona grandis, Mitragyna parviflora, Buchanania lanzani, Anogeissus latifolia, Terminalia tomentosa, Lagerstroemia parviflora, Diospyros melanoxylon, Phyllanthus emblica, Carreya arborea, Gardenia gummifera, Dandrocalamus strictus, Vitex nigundo, Madhuca latifolia, Aegle marmelos*

Contd...

S. No.	Name of Sacred Grove	Name of District	Name of Forest Range	Comptt. Number/ Name of Beat	Name Of Deity Worshiped	Name of other Deity	Approx. area in M^2	No. of years of existence	Distance from the near village in km.	Important plant species conserved in the grove
169.	Navnadar	Mandla	Bhamni	Jarga	Siddha Baba	Nil	100	500	3	*Ficus religiosa, Zizyphus jujuba, Syzygium cuminii, Ougenia oojenensis, Madhuca latifolia, Diospyros melanoxylon, Terminalia tomentosa, Bombax ceiba, Achyranthes aspara, Curcuma caisia, Actinopteris radiata, Baccopa monerii, Azadiracta indica, Helicteres isora*
170.	Chakkaiyapat	Mandla	Bhamni	Indri	Shankarji	Nil	50	250	1.5	*Anogeissus latifolia, Butea monosperma, Lagerstroemia parviflora, Diospyros melanoxylon, Buchanania lanzan, Marsdenia hamiltonii, Pithecellobium dulce*
171.	Shiv Mandir	Mandla	Mohgaon	Devgaon	Shivji	Nil	50	50	3	*Aegle marmelos, Ficus banghalensis, Psidium guajaba, Azadiracta indica, Mangifera indica, Eucalyptus* sp.
172.	Sidha baba	Mandla	Mandla	Podilinga	Narmada Maiya	Babadev, Shardadevi Narmada	100	500	2	*Anogeissus latifolia, Lagerstroemia parviflora, Tectona grandis, Diospyros melanoxylon, Azadiracta indica, Anona squamosa, Aegle marmelos, Vitex nigundo, Mangifera indica*

173.	Sitaraptan	Mandla	Mandla	Sitarpattan	Badebaba	Nil	100	150	3	*Azadiracta indica, Ficus benghalensis, Anogeissus latifolia, Lagerstroemia parviflora, Ficus religiosa, Tectona grandis, Diospyros melanoxylon, Gymnema sylvestre, Holoptelea integrifolia, Lannea grandis*
174.	Surajkund	Mandla	Mawai	Brijgaon	Siddhapat,	Chousat yogini	100	470	3	*Ficus religiosa, Zizyphus numularis, Syzygium cuminii, Ougenia oojenensis, Terminalia tomentosa, Buchanania lanzana, Curcuma caisia, Actinopteris radiata, Baccopa monerii, Solanum surretence, Helicteres isora*
175.	Upka	Mandla	Chabi	Chabi	Budi Mata	Nil	100	340	1.5	*Butea monosperma, Lagerstroemia parviflora, Tectona grandis, Diospyros melanoxylon, Azadiracta indica, Buchanania lanzan, Anona squamosa*

Area occupied by sacred groves varies considerably. Some have occupied only 10 m^2 area whereas others are much longer and larger in shape also. The identified grove ranges from 10 m^2 to 5000 m^2. Table 19.2 shows the maximum 107 (61%) sacred groves up to 100 m^2 area and 47 (27%) sacred groves are up to 500 m^2 area. Whereas other sacred groves up to 2000 m^2 and 5000 m^2 area represents 6 (3%) and 15 (9%) respectively.

Table 19.2 Number of sacred groves occupying different area classes (in m^2) under different district

Approximate area (m^2)	*Number of sacred groves in different districts*				*Total*
	Mandla	*Dindori*	*Hoshangabad*	*Chhindwara*	
A – 10	3	–	–	4	7
B – 20	–	–	1	3	4
C – 50	14	6	–	22	42
D – 100	11	10	27	6	54
E – 200	–	–	3	5	8
F – 300	–	–	1	1	2
G – 400	–	–	27	1	28
H – 500	2	–	–	7	9
I – 1500	–	–	1	–	1
J – 2000	–	1	4	–	5
K – 2500	–	–	11	–	11
L – 3000	–	–	–	–	–
M – 4000	–	–	2	–	2
N – 5000	–	1	1	–	2
Total	**30**	18	78	49	175

The analysis of the collected data with reference to age of grove is also found quite variable as shown in Table 19.3. It can be seen that out of total 175 groves maximum 67 groves are under age class A and B (1-50 and 51-100 years old). Whereas, 6, 9, 6, 4 and 16 groves are represented to C, D, E, F and G age classes respectively.

Table 19.3 Number of sacred groves under different years of existence in various districts

Year of existence class of sacred groves	*Number of sacred groves*				*Total*
	Mandla	*Dindori*	*Hoshangabad*	*Chhindwara*	
A – 1-50	4	7	26	30	67
B – 51-100	10	11	28	18	67
C – 101-150	1	–	4	1	6

D – 151-200	2	–	7	–	9
E – 201-250	4	–	2	–	6
F – 251-300	3	–	1	–	4
G – More than 500	6	–	10	–	16
Total	**30**	**18**	**78**	49	175

The distances of the groves from the nearby adjoining villages were also analyzed and found that, out of total 175 groves 85.7% groves are located within a distance of 4 km and rest of the groves (14.3%) are located at above 4 km distance Table 19.4. It can be seen from this finding that villagers prefered to establish sacred groves near village, so they can easily walk down this distance within couple of time.

Table 19.4 Number of sacred groves under different distance classes (in km)

Distance class from village (in km)	*Number of Sacred Groves*				*Total*
	Mandla	*Dindcri*	*Hoshangabad*	*Chhindwara*	
A – 0.5	–	–	–	8	8
B – 1	–	1	12	11	24
C – 1.5	3	3	5	1	12
D – 2	12	5	15	11	43
E – 3	14	6	16	9	45
F – 4	1	2	12	3	18
G – 5	–	1	4	3	8
H – 6	–	–	2	2	4
I – 7	–	–	1	–	1
J – 8	–	–	6	1	7
K – 9	–	–	–	–	0
L – 10	–	–	4	–	4
M – More than 10	–	–	1	–	1
Total	**30**	**18**	**78**	**49**	**175**

An inventory of existing plant species available in the groves resulted total 250 plant species. Highest 231 plant species belonging to 176 genera and 80 families were recorded from 78 sacred groves of Hoshangabad district. Whereas 141 species, 114 genera and 47 families (from 49 sacred groves of Chhindwara); 140 species, 109 genera and 55 families (from 30 sacred groves of Mandla) and 126 species, 90 genera and 52 families from 18 sacred groves of Dindori district. The good regeneration of plant species mentioned in Table 19.1 indicated that the protection and conservative practices were followed by the local inhabitants in the name of their deity. The plant category-wise details of species are highlighted in Table 19.5.

Table 19.5 Number of plant recorded under different plant categories from sacred groves

Plant category	*Name of district*			
	Mandla	*Dindori*	*Hoshangabad*	*Chhindwara*
Trees	65	46	84	43
Shrubs	11	14	43	53
Herbs	47	43	63	30
Climbers	11	13	25	18
Grasses	3	6	14	16
Epiphytes / Parasites	3	1	3	2
Total	**140**	**126**	**231**	**141**

India State of Forest Report – 2009 (MoEF, 2009) indicated that the current forest cover with the previous assessment decreased up to 39 km^2, which pointed out that the dwindling of natural forest area is increasing rapidly. The natural biodiversity is also affected due to various biotic, abiotic and biological factors. It is an urgent need to conserve the biodiversity in the natural forest areas in general and species in particular. The studies conducted from 2002 to 2009 in four districts are reccommended that, these 175 sacred groves are playing an important role in the local biodiversity conservation and it is proposed to conserve such areas as:

- Tourist spot / hotspot.
- Development of sacred groves as medicinal and aromatic plants, non-timer forest produce and rare and endangered species Gene bank.
- In situ conservation habitats for native floral and faunal species particularly avi fauna.

Acknowledgements

I take this opportunity to express my gratefulness and deep sense of gratitude to Dr. Ram Prakash, APCCF, Director, State Forest Research Institute, Jabalpur (M.P.) for providing technical and moral support to accomplish this task. I am also thankful to the Member Secretary, Madhya Pradesh State Biodiversity Board, Bhopal (M.P.) for providing financial assistance. I express my special thanks to DFO of Chhindwara, Hoshangabad, Mandla and Dindori forest divisions along with their SDOs, Range Officers and local field staff members for their kind cooperation and support provided during the course of investigation.

REFERENCES

1. Alfrod, J.R.B. (1996). "Sacred Groves of Meghalaya". Calcutta.
2. Amrithalingam, M. (1998). Sacred Trees *of Tamilnadu*, C.P.R. Environmental Education Centre, Chennai.

3. Amirthalingam, M. (2000). Folklore of Sacred Groves. *Indian Folklife* Vol. 1(3): 8-9.
4. Amirthalingam, M. (2002). Sacred groves in the environmental protection. *Kisan World* Vol. 29(3): 37.
5. Bhowmik, P.K. (1963). "*Lodhas of West Bengal*" Punthi Pustak, Calcutta.
6. Bennet S.S.R., Gupta, P.C. and Rao, R.V. (1992). "*Origin of Tree Worship*", Venerated Plants, Indian Council of Forestry Research and Education, Dehra Dun, India. 1-35,
7. Bhasin, V. (1999). "Religious and cultural perspective of a sacred site - Sitabari in Rajasthan". *J. Hum. Ecol.* 10: 329-340.
8. Boraiah, K.T., Vasudeva, R. Bhagwat, A.S. and Kushalappa, C.G. (2002). "Regeneration of threatened flora among the sacred groves of Kodagu, Karnataka, South India", *My Forest*, 38 (2): 123-128.
9. Bhakat, R.K. and Pandit, P.K. (2003). "Role of a sacred grove in conservation of medicinal plants", *Indian Forester*, Vol. 129 (2): 224-232.
10. Chandrashekhara, U.M. and Sankar, S. (1998). "Structure and functions of sacred groves: Case Studies in Kerala", In: Ramakrishnan P.S., Saxena K.G., and Chandrashekhara, U.M. (Eds.), *Conserving the Sacred for Biodiversity Management*, Oxford and IBH Publishing Co. Pvt. Ltd., 323-335.
11. Gadgil, M. and Vartak, V.D. (1973). "*Groves Dedicated to the Gods*", The Illustrated Weekly of India 21.
12. Gadgil, M. and Vartak, V.D. (1975). "Sacred groves of India: A plea for continued conservation". *Journal of Bombay Natural History Society* 72: 314-320.
13. Gadgil, M. and Gokhale, Y. (2000). "*Sacred Elements of Nature in India*", setting biodiversity conservation priorities for India. pp. 691-707.
14. Malhotra, K.C. Gokhale, Y. and Chatterjee, S. (1998). *Cultural and ecological dimensions of sacred groves in India.* INSA, New Delhi.

Index